Tree Growth in Relation to Climate Change

Tree Growth in Relation to Climate Change

Guest Editors

Yassine Messaoud
Jan Světlík
Giorgio Alberti

MDPI

Basel • Beijing • Wuhan • Barcelona • Belgrade • Novi Sad • Cluj • Manchester

Guest Editors

Yassine Messaoud
Laval University
Quebec, QC
Canada

Jan Světlík
Mendel University in Brno
Brno
Czech Republic

Giorgio Alberti
Università degli Studi di Udine
Udine
Italy

Editorial Office
MDPI AG
Grosspeteranlage 5
4052 Basel, Switzerland

This is a reprint of the Special Issue, published open access by the journal *Forests* (ISSN 1999-4907), freely accessible at: https://www.mdpi.com/journal/forests/special_issues/3K1X8XC2SL.

For citation purposes, cite each article independently as indicated on the article page online and as indicated below:

Lastname, A.A.; Lastname, B.B. Article Title. *Journal Name* **Year**, *Volume Number*, Page Range.

ISBN 978-3-7258-2785-5 (Hbk)
ISBN 978-3-7258-2786-2 (PDF)
https://doi.org/10.3390/books978-3-7258-2786-2

Cover image courtesy of Oleg Gapeenko

Contents

Preface

This SI of *Forests* represents a collection of relevant articles from different scientific researchers, focusing on tree height growth associated with climate change around the world, thus including different species from temperate and boreal biomes. It explores the impact of climate change on tree species growth, embracing different aspects of response to drought conditions such as mortality and anatomical and physiological adaptations. Case studies including genetic and dendrochronological features are also discussed. All these represent paramount insights, improving our comprehension of how tree species will respond to climate change worldwide.

Yassine Messaoud, Jan Světlík, and Giorgio Alberti
Guest Editors

forests

MDPI

Editorial

Tree Growth in Relation to Climate Change: Understanding the Impact on Species Worldwide

Yassine Messaoud

Faculty of Natural Resources Management, Lakehead University, 955 Oliver Road, Thunder Bay, ON P7B 5E1, Canada; ymessaou@lakeheadu.ca

Climate change is one of the most important environmental issues of our time, which has profound effects on ecosystems all over the world [1]. Given its vital role in forest dynamics and carbon sequestration, the effect of climate change on tree growth has drawn the most attention of all its effects [2]. Understanding the various aspects linked to climate change and their impact on tree growth is crucial for forecasting future forest composition and dynamic and efficiently managing forest resources, explaining its significant and increasing interest not only at the species scale but also for biodiversity and ecosystem services [3,4].

Climate is known to influence tree growth since it controls photosynthesis, cell division, and tree metabolism [5,6]. Notwithstanding, climate acts at different spatial and temporal scales [2,7,8], adding to the soil condition and microfauna dynamic [9,10]. Climate variables, including temperature, precipitation, and moisture availability, all affect tree growth [11]. These variables can alter over time as a result of climate change and are not constant in different locations [2]. As a result, different tree species are responding differently throughout the world in terms of growth [2,12,13]. Thus, the structure and function of forests may be significantly impacted by this difference in growth patterns [12,14,15].

Tree growth response is also species-specific, linked to growth habit [2] and autecology [14,16,17]. In addition, this response depends on the stand structure and composition [18–20]. Nevertheless, there is a general decline trend for trees facing drought conditions [21,22] or in more heat-limiting (high latitude and altitude, or in cold-air pooling) environments coupled with cloudiness [23] or snowpack [24]. Thus, these factors weaken the tree and cause mortality [25–27]. Tree growth is a key for competitive advantage and thus is considered as a proxy for predicting future species and vegetation migration [28].

Tree growth tends to decrease as we move towards higher latitudes, attributed to colder temperatures, shorter growing seasons, and reduced light availability in these regions [29]. As we ascend in elevation, tree growth typically slows down due to lower temperatures and reduced nutrient availability in the soil [6,29]. However, tree species adapted to high-altitude environments may exhibit unique growth patterns and adaptability to changing climate conditions [30–32]. In addition, drought events are becoming increasingly common under climate change scenarios [33]. Drought limits the availability of water, leading to physiological stress and decreased tree growth [34–36]. In extreme cases, prolonged drought can even lead to tree mortality [37–39].

Research on how climate change can affect tree growth in Europe, Asia, and America has revealed that significant timber species could lose adequate habitat in the next decades and suffer from a combination of rising trends and decreased precipitation throughout the growing season [40,41]. Some species, on the other hand, might actually widen their range and even exhibit better growth rates in some areas of their current range [42,43]. They might also demonstrate promise for commercial species when they become established in new areas. Furthermore, the intricate relationships between tree growth and temperature, water, and nutrients increase the complexity of tree growth [2,44]. Forest management might be adjusted to the new growing conditions that are gradually but surely emerging by

Citation: Messaoud, Y. Tree Growth in Relation to Climate Change: Understanding the Impact on Species Worldwide. *Forests* **2024**, *15*, 1601. https://doi.org/10.3390/f15091601

Received: 19 August 2024
Revised: 8 September 2024
Accepted: 9 September 2024
Published: 11 September 2024

utilizing mixed forests' potential resilience to changes in growth conditions [45]. In order to further our understanding of how climate change may affect tree growth, this Special Issue was initiated.

The impact of climate change on tree growth has implications at multiple scales—local regional, and large-scale. At the local and regional level, changes in tree growth can affect the distribution and composition of tree species [46]. Some species may become more dominant, while others may decline, potentially leading to shifts in entire ecosystems and disrupting species interactions [47]. At a larger scale, climate change can influence the location of ecotones—transitional areas between different ecosystems. As tree species respond differently to changing climate conditions, ecotones can shift, potentially affecting species migration patterns and altering biodiversity patterns [48,49].

While climate change poses significant challenges to tree growth, some species are showing remarkable resilience and adaptability [50–52]. Tree species with broader ecological niches and genetic diversity are often better equipped to cope with changing conditions [53]. Understanding the adaptive capacity of different species and promoting diverse forests can enhance the overall resilience of forests to climate change [54]. All these explain the complexity of tree growth associated with climate change-induced and the contradictory findings from different studies [55–57].

As the climate continues to change, it is essential to monitor and understand the impact on tree growth worldwide [58,59]. By studying the relationships between climate variables and growth patterns, we can better predict future forest dynamics and develop sustainable management strategies. It is crucial for policymakers, scientists, and land managers to collaborate and implement adaptive measures to preserve the resilience and functionality of forests in the face of climate change.

Numerous studies were mostly conducted in more limiting environments, i.e., higher latitudes [60–62] or elevations [63–65] and with more common and widespread species, such as black spruce, trembling aspen in North America [66–68], Norway spruce, Scots pine, and common beech in Europe [69,70], and larches in Siberian Asia [71,72]. Although these previous studies are very relevant to assess how trees respond to climate change, knowledge gaps remain because there is uncertainty about the response of less common species to climate change-induced events in other parts of the world, such as in the southern hemisphere [73]. In addition, the question still arises about how trees react to climate change in physiological and phenological terms [74].

All in all, this Special Issue offers readers a variety of research investigating the relationships between climate change and tree growth in various biomes (boreal, temperate cool subalpine), on various continents (North and South America, Europe, and Asia), at various scales (from local populations to latitudinal and altitudinal gradient studies), and with various species (evergreen and deciduous conifer and broadleaf). As much as the editors enjoyed putting this Special Issue together, we hope readers will too. As well, this Special Issue gives a good insight that deals with the above-mentioned aspects and to increase our knowledge and to have ample information on how tree species responded to climate change. Likewise, it will promote discussion, exchange, and debate between scientists and also make forest decision-makers aware for future management.

Funding: This research received no external funding.

Conflicts of Interest: The author declares no conflicts of interest.

References

1. Malhi, Y.; Franklin, J.; Seddon, N.; Solan, M.; Turner, M.G.; Field, C.B.; Knowlton, N. Climate change and ecosystems: Threats opportunities and solutions. *Philos. Trans. R. Soc. B Biol. Sci.* **2020**, *375*, 20190104. [CrossRef] [PubMed]
2. Messaoud, Y.; Reid, A.; Tchebakova, N.M.; Goldman, J.A.; Hofgaard, A. The Historical Complexity of Tree Height Growth Dynamic Associated with Climate Change in Western North America. *Forests* **2022**, *13*, 738. [CrossRef]
3. Ding, J.; Eldridge, D. Intensifying aridity induces tradeoffs among biodiversity and ecosystem services supported by trees. *Glob Ecol. Biogeogr.* **2024**, e13894. [CrossRef]

4. Zheng, J.; Yang, J.; Jia, H.; Lyu, L.; Langzhen, J.; Zhang, Q.-B. Changes of growth-climate relationships of Smith fir forests along an altitudinal gradient. *J. For. Res.* **2024**, *35*, 76. [CrossRef]
5. Schmiege, S.C.; Griffin, K.L.; Boelman, N.T.; Vierling, L.A.; Bruner, S.G.; Min, E.; Maguire, A.J.: Jensen, J.; Eitel, J.U.H. Vertical gradients in photosynthetic physiology diverge at the latitudinal range extremes of white spruce. *Plant Cell Environ.* **2023**, *46*, 45–63. [CrossRef]
6. Pallardy, S.G. *Physiology of woody plants*; Elsevier: London, UK, 2008.
7. Sang, Z.; Sebastian-Azcona, J.; Hamann, A.; Menzel, A.; Hacke, U. Adaptive limitations of white spruce populations to drought imply vulnerability to climate change in its western range. *Evol. Appl.* **2019**, *12*, 1850–1860. [CrossRef]
8. Izmir, Ş.; Jevsenak, J.; Krajnc, L.; Hafner, P.; Köse, N. Distinct spatial patterns in climate-growth relationships, vegetation and resilience indices of Black pine (Pinus nigra J.F. Arnold) from its northern and southern distribution range. *Dendrochronologia* **2024**, *88*, 126236. [CrossRef]
9. Zhang, W.; Gou, X.; Zhang, F.; Liu, W.; Zhang, Y.; Gao, L. Divergent responses of Qinghai spruce (Picea crassifolia) to recent warming along elevational gradients in the central Qilian Mountains, Northwest China. *J. Geogr. Sci.* **2023**, *33*, 151–168. [CrossRef]
10. Sachsenmaier, L.; Schnabel, F.; Dietrich, P.; Eisenhauer, N.; Ferlian, O.; Quosh, J.; Richter, R.; Wirth, C. Forest growth resistance and resilience to the 2018–2020 drought depend on tree diversity and mycorrhizal type. *J. Ecol.* **2024**, *112*, 1787–1803. [CrossRef]
11. Messaoud, Y.; Chen, H.Y. The influence of recent climate change on tree height growth differs with species and spatial environment. *PLoS ONE* **2011**, *6*, e14691. [CrossRef]
12. Jiang, Y.; Marchand, W.; Rydval, M.; Matula, R.; Janda, P.; Begović, K.; Thom, D.; Fruleux, A.; Buechling, A.; Pavlin, J.; et al. Drought resistance of major tree species in the Czech Republic. *Agric. For. Meteorol.* **2024**, *348*, 109933. [CrossRef]
13. Zhang, H.; Wang, K.; Zeng, Z.; Du, H.; Zou, Z.; Xu, Y.; Zeng, F. Large-scale patterns in forest growth rates are mainly driven by climatic variables and stand characteristics. *For. Ecol. Manag.* **2019**, *435*, 120–127. [CrossRef]
14. Basu, S.; Stojanović, M.; Jevšenak, J.; Buras, A.; Kulhavý, J.; Hornová, H.; Světlík, J. Pedunculate oak is more resistant to drought and extreme events than narrow-leaved ash in Central European floodplain forests. *For. Ecol. Manag.* **2024**, *561*, 121907. [CrossRef]
15. Dupont-Leduc, L.; Power, H.; Fortin, M.; Schneider, R. Climate interacts with the trait structure of tree communities to influence forest productivity. *J. Ecol.* **2024**, *112*, 1758–1773. [CrossRef]
16. Jevsenak, J.; Saražin, J. Pinus halepensis is more drought tolerant and more resistant to extreme events than Pinus nigra at a sub-Mediterranean flysch site. *Trees* **2023**, *37*, 1281–1286. [CrossRef]
17. Kunert, N.; Hajek, P.; Hietz, P.; Morris, H.; Rosner, S.; Tholen, D. Summer temperatures reach the thermal tolerance threshold of photosynthetic decline in temperate conifers. *Plant Biol.* **2022**, *24*, 1254–1261. [CrossRef]
18. Soubeyrand, M.; Marchand, P.; Duchesne, L.; Bergeron, Y.; Gennaretti, F. Interactions between climate, soil and competition drive tree growth in Quebec forests. *For. Ecol. Manag.* **2024**, *555*, 121731. [CrossRef]
19. Vospernik, S.; Heym, M.; Pretzsch, H.; Pach, M.; Steckel, M.; Aldea, J.; Brazaitis, G.; Bravo-Oviedo, A.; Del Rio, M.; Löf, M.; et al. Tree species growth response to climate in mixtures of Quercus robur/Quercus petraea and Pinus sylvestris across Europe—A dynamic, sensitive equilibrium. *For. Ecol. Manag.* **2023**, *530*, 120753. [CrossRef]
20. Vospernik, S.; Vigren, C.; Morin, X.; Toigo, M.; Bielak, K.; Brazaitis, G.; Bravo, F.; Heym, M.; Del Río, M.; Jansons, A.; et al. Can mixing Quercus robur and Quercus petraea with Pinus sylvestris compensate for productivity losses due to climate change? *Sci. Total Environ.* **2024**, *942*, 173342. [CrossRef]
21. Barber, V.A.; Juday, G.P.; Finney, B.P. Reduced growth of Alaskan white spruce in the twentieth century from temperature-induced drought stress. *Nature* **2000**, *405*, 668–673. [CrossRef]
22. Linares, J.C.; Tíscar, P.A.; Camarero, J.J.; Taïqui, L.; Viñegla, B.; Seco, J.I.; Merino, J.; Carreira, J.A. *Tree Growth Decline on Relict Western-Mediterranean Mountain Forests: Cause and Impacts*; Nova Science Publishers Inc.: Hauppauge, NY, USA, 2011; pp. 91–110.
23. Yang, J.; Zhao, B.; Zheng, J.; Zhang, Q.; Li, Y.; Ma, F.; Fang, O. Linkage between spruce forest decline and cloud cover increase in the Qilian Mountains of the northeastern Tibetan Plateau. *Trees* **2023**, *37*, 1097–1106. [CrossRef]
24. Ettinger, A.K.; Ford, K.R.; HilleRisLambers, J. Climate determines upper, but not lower, altitudinal range limits of Pacific Northwest conifers. *Ecology* **2011**, *92*, 1323–1331. [CrossRef] [PubMed]
25. Bigler, C.; Gavin, D.G.; Gunning, C.; Veblen, T.T. Drought induces lagged tree mortality in a subalpine forest in the Rocky Mountains. *Oikos* **2007**, *116*, 1983–1994. [CrossRef]
26. Petrov, I.; Shushpanov, A.S.; Im, S.; Golyukov, A.S. Climatic aspect of fir (*Abies sibirica* Ledeb.) mortality in the Eastern Sayan Mountains. In Proceedings of the VII International Scientific Conference, Tomsk, Russia, 28–30 September 2020; pp. 94–96.
27. Maher, C.T.; Nelson, C.R.; Larson, A.J.; Leys, B. Winter damage is more important than summer temperature for maintaining the krummholz growth form above alpine treeline. *J. Ecol.* **2019**, *108*, 1074–1087. [CrossRef]
28. Tang, Y.; Du, E.; Guo, H.; Wang, Y.; Peñuelas, J.; Reich, P.B. Rapid migration of Mongolian oak into the southern Asian boreal forest. *Global Chang. Biol.* **2024**, *30*, e17002. [CrossRef]
29. Adams, J.M. *Vegetation-Climate Interaction: How Vegetation Makes the Global Environment*; Springer: Berlin/Heidelberg, Germany, 2007.
30. Li, M.-Y.; Leng, Q.-N.; Hao, G.-Y. Contrasting patterns of radial growth rate between Larix principis-rupprechtii and *Pinus sylvestris* var. mongolica along an elevational gradient are mediated by differences in xylem hydraulics. *For. Ecol. Manag.* **2021**, *497*, 119524. [CrossRef]
31. Taneda, H.; Funayama-Noguchi, S.; Mayr, S.; Goto, S. Elevational adaptation of morphological and anatomical traits by Sakhalin fir (*Abies sachalinensis*). *Trees* **2019**, *34*, 507–520. [CrossRef]

32. Zhou, Q.; Shi, H.; He, R.; Liu, H.; Zhu, W.; Yu, D.; Zhang, Q.; Dang, H. Plastic and adaptive response of carbon allocation to temperature change in alpine treeline trees. *Environ Exp Bot.* **2023**, *208*, 105271. [CrossRef]
33. Vicente-Serrano, S.M.; Beguería, S.; López-Moreno, J.I. A Multiscalar Drought Index Sensitive to Global Warming: The Standardized Precipitation Evapotranspiration Index. *J. Clim.* **2010**, *23*, 1696–1718. [CrossRef]
34. Balducci, L.; Deslauriers, A.; Giovannelli, A.; Beaulieu, M.; Delzon, S.; Rossi, S.; Rathgeber, C.B. How do drought and warming influence survival and wood traits of Picea mariana saplings? *J. Exp. Bot.* **2015**, *66*, 377–389. [CrossRef]
35. Kelsey, K.C.; Redmond, M.D.; Barger, N.N.; Neff, J.C. Species, Climate and Landscape Physiography Drive Variable Growth Trends in Subalpine Forests. *Ecosystems* **2017**, *21*, 125–140. [CrossRef]
36. Wang, X.; Pederson, N.; Chen, Z.; Lawton, K.; Zhu, C.; Han, S. Recent rising temperatures drive younger and southern Korean pine growth decline. *Sci. Total Environ.* **2019**, *649*, 1105–1116. [CrossRef] [PubMed]
37. Hogg, E.H.; Brandt, J.P.; Michaelian, M. Impacts of a regional drought on the productivity, dieback, and biomass of western Canadian aspen forests. *Can. J. For. Res.* **2008**, *38*, 1373–1384. [CrossRef]
38. Worrall, J.J.; Egeland, L.; Eager, T.; Mask, R.A.; Johnson, E.W.; Kemp, P.A.; Shepperd, W.D. Rapid mortality of Populus tremuloides in southwestern Colorado, USA. *For. Ecol. Manag.* **2008**, *255*, 686–696. [CrossRef]
39. Galiano, L.; Martínez-Vilalta, J.; Lloret, F. Drought-Induced Multifactor Decline of Scots Pine in the Pyrenees and Potential Vegetation Change by the Expansion of Co-occurring Oak Species. *Ecosystems* **2010**, *13*, 978–991. [CrossRef]
40. Vieira, J.; Nabais, C.; Campelo, F. Dry and hot years drive growth decline of *Pinus halepensis* at its southern range limit in the Moroccan High Atlas Mountains. *Trees* **2022**, *36*, 1585–1595. [CrossRef]
41. Wilson, E.C.; Cousins, S.; Etter, D.R.; Humphreys, J.M.; Roloff, G.J.; Carter, N.H. Habitat and climatic associations of climate-sensitive species along a southern range boundary. *Ecol. Evol.* **2023**, *13*, e10083. [CrossRef]
42. Puchałka, R.; Paź-Dyderska, S.; Jagodziński, A.M.; Sádlo, J.; Vítková, M.; Klisz, M.; Koniakin, S.; Prokopuk, Y.; Netsvetov, M.; Nicolescu, V.-N.; et al. Predicted range shifts of alien tree species in Europe. *Agric. For. Meteorol.* **2023**, *341*, 109650. [CrossRef]
43. Gazda, A.; Kościelniak, P.; Hardy, M.; Muter, E.; Kędra, K.; Bodziarczyk, J.; Frączek, M.; Chwistek, K.; Różański, W.; Szwagrzyk, J. Upward expansion of distribution ranges of tree species: Contrasting results from two national parks in Western Carpathians. *Sci. Total. Environ.* **2019**, *653*, 920–929. [CrossRef]
44. Messaoud, Y.; Asselin, H.; Bergeron, Y.; Grondin, P. Competitive Advantage of Black Spruce Over Balsam Fir in Coniferous Boreal Forests of Eastern North America Revealed by Site Index. *For. Sci.* **2014**, *60*, 57–62. [CrossRef]
45. Hörl, J.; Keller, K.; Yousefpour, R. Reviewing the performance of adaptive forest management strategies with robustness analysis. *For. Policy Econ.* **2020**, *119*, 102289. [CrossRef]
46. van Tiel, N.; Fopp, F.; Brun, P.; van den Hoogen, J.; Karger, D.N.; Casadei, C.M.; Lyu, L.; Tuia, D.; Zimmermann, N.E.; Crowther, T.W.; et al. Regional uniqueness of tree species composition and response to forest loss and climate change. *Nat. Commun.* **2024**, *15*, 4375. [CrossRef] [PubMed]
47. Boonman, C.C.F.; Serra-Diaz, J.M.; Hoeks, S.; Guo, W.-Y.; Enquist, B.J.; Maitner, B.; Malhi, Y.; Merow, C.; Buitenwerf, R.; Svenning, J.-C. More than 17,000 tree species are at risk from rapid global change. *Nat. Commun.* **2024**, *15*, 166. [CrossRef] [PubMed]
48. Beckage, B.; Osborne, B.; Gavin, D.G.; Pucko, C.; Siccama, T.; Perkins, T. A rapid upward shift of a forest ecotone during 40 years of warming in the Green Mountains of Vermont. *Proc. Natl. Acad. Sci. USA* **2008**, *105*, 4197–4202. [CrossRef]
49. Kark, S. Effects of Ecotones on Biodiversity. In *Reference Module in Life Sciences*; Elsevier: Amsterdam, The Netherlands, 2017.
50. Sarmoum, M.; Camarero, J.J.; Abdoun, F. Aridification increases growth resistance of Atlas cedar forests in NW Algeria. *For. Ecol. Manag.* **2024**, *556*, 121730. [CrossRef]
51. Chapin, F.S.; McGuire, A.D.; Ruess, R.W.; Hollingsworth, T.N.; Mack, M.C.; Johnstone, J.F.; Kasischke, E.S.; Euskirchen, E.S.; Jones, J.B.; Jorgenson, M.T.; et al. Resilience of Alaska's boreal forest to climatic change. *Can. J. For. Res.* **2010**, *40*, 1360–1370.
52. Vitasse, Y.; Bottero, A.; Cailleret, M.; Bigler, C.; Fonti, P.; Gessler, A.; Lévesque, M.; Rohner, B.; Weber, P.; Rigling, A.; et al. Contrasting resistance and resilience to extreme drought and late spring frost in five major European tree species. *Glob. Chang. Biol.* **2019**, *25*, 3781–3792. [CrossRef]
53. Aitken, S.N.; Yeaman, S.; Holliday, J.A.; Wang, T.; Curtis-McLane, S. Adaptation, migration or extirpation: Climate change outcomes for tree populations. *Evol. Appl.* **2008**, *1*, 95–111. [CrossRef]
54. Vacek, Z.; Vacek, S.; Cukor, J. European forests under global climate change: Review of tree growth processes, crises and management strategies. *J. Environ. Manag.* **2023**, *332*, 117353. [CrossRef]
55. Kašpar, J.; Tumajer, J.; Altman, J.; Altmanová, N.; Čada, V.; Čihák, T.; Doležal, J.; Fibich, P.; Janda, P.; Kaczka, R.; et al. Major tree species of Central European forests differ in their proportion of positive, negative, and nonstationary growth trends. *Glob. Chang. Biol.* **2024**, *30*, e17146. [CrossRef]
56. Lloyd, A.H.; Fastie, C.L. Spatial and Temporal Variability in the Growth and Climate Response of Treeline Trees in Alaska. *Clim. Chang.* **2002**, *52*, 481–509. [CrossRef]
57. Bosela, M.; Lukac, M.; Castagneri, D.; Sedmák, R.; Biber, P.; Carrer, M.; Konôpka, B.; Nola, P.; Nagel, T.A.; Popa, I.; et al. Contrasting effects of environmental change on the radial growth of co-occurring beech and fir trees across Europe. *Sci. Total. Environ.* **2017**, *615*, 1460–1469. [CrossRef]
58. Allen, C.D.; Breshears, D.D.; McDowell, N.G. On underestimation of global vulnerability to tree mortality and forest die-off from hotter drought in the Anthropocene. *Ecosphere* **2015**, *6*, 1–55. [CrossRef]

59. Anderegg, W.R.L.; Schwalm, C.R.; Biondi, F.; Camarero, J.J.; Koch, G.W.; Litvak, M.; Ogle, K.; Shaw, J.D.; Shevliakova, E.; Williams, A.P.; et al. Pervasive drought legacies in forest ecosystems and their implications for carbon cycle models. *Science* **2015**, *349*, 528–532. [CrossRef] [PubMed]
60. Karlsson, F. Climate Change and the Norway Spruce Treeline in the Central Scandinavian Mountains. Treeline Evolution, Tree Growth and a Blue Intensity Summer Temperature Reconstruction. Master Thesis, University of Gothenburg, Gothenburg, Sweden, 2022.
61. Goldblum, D.; Rigg, L.S. The Deciduous Forest—Boreal Forest Ecotone. *Geogr. Compass* **2010**, *4*, 701–717. [CrossRef]
62. Jensen, J. The Environmental Drivers of White Spruce Growth and Regeneration at Arctic Treeline in a Changing Climate. Ph.D. Thesis, Columbia University, New York, NY, USA, 2023.
63. Smith, W.K.; Germino, M.J.; Johnson, D.M.; Reinhardt, K. The Altitude of Alpine Treeline: A Bellwether of Climate Change Effects. *Bot. Rev.* **2009**, *75*, 163–190. [CrossRef]
64. Jochner, M.; Bugmann, H.; Nötzli, M.; Bigler, C. Tree growth responses to changing temperatures across space and time: A fine-scale analysis at the treeline in the Swiss Alps. *Trees* **2017**, *32*, 645–660. [CrossRef]
65. Enderle, L.; Gribbe, S.; Muffler, L.; Weigel, R.; Hertel, D.; Leuschner, C. A warmer climate impairs the growth performance of Central Europe's major timber species in lowland regions. *Sci. Total Environ.* **2024**, *941*, 173665. [CrossRef]
66. Drobyshev, I.; Gewehr, S.; Berninger, F.; Bergeron, Y. Species specific growth responses of black spruce and trembling aspen may enhance resilience of boreal forest to climate change. *J. Ecol.* **2012**, *101*, 231–242. [CrossRef]
67. Chagnon, C.; Wotherspoon, A.R.; Achim, A. Deciphering the black spruce response to climate variation across eastern Canada using a meta-analysis approach. *For. Ecol. Manag.* **2022**, *520*, 120375. [CrossRef]
68. Morin-Bernard, A.; Achim, A.; Coops, N.C.; White, J.C. Integration of tree-ring data, Landsat time series, and ALS-derived topographic variables to quantify growth declines in black spruce. *For. Ecol. Manag.* **2024**, *557*, 121765. [CrossRef]
69. Briceno-Elizondo, E.; Garcia-Gonzalo, J.; Peltola, H.; Matala, J.; Kellomäki, S. Sensitivity of growth of Scots pine, Norway spruce and silver birch to climate change and forest management in boreal conditions. *For. Ecol. Manag.* **2006**, *232*, 152–167. [CrossRef]
70. Pretzsch, H.; Schütze, G.; Biber, P. Drought can favour the growth of small in relation to tall trees in mature stands of Norway spruce and European beech. *For. Ecosyst.* **2018**, *5*, 20. [CrossRef]
71. Dulamsuren, C.; Wommelsdorf, T.; Zhao, F.; Xue, Y.; Zhumadilov, B.Z.; Leuschner, C.; Hauck, M. Increased Summer Temperatures Reduce the Growth and Regeneration of Larix sibirica in Southern Boreal Forests of Eastern Kazakhstan. *Ecosystems* **2013**, *16*, 1536–1549. [CrossRef]
72. Kharuk, V.I.; Petrov, I.A.; Golyukov, A.S.; Dvinskaya, M.L.; Im, S.T.; Shushpanov, A.S. Larch growth across thermal and moisture gradients in the Siberian Mountains. *J. Mt. Sci.* **2023**, *20*, 101–114. [CrossRef]
73. Hansson, A.; Shulmeister, J.; Dargusch, P.; Hill, G. A review of factors controlling Southern Hemisphere treelines and the implications of climate change on future treeline dynamics. *Agric. For. Meteorol.* **2023**, *332*, 109375. [CrossRef]
74. Qi, X.; Treydte, K.; Saurer, M.; Fang, K.; An, W.; Lehmann, M.; Liu, K.; Wu, Z.; He, H.S.; Du, H.; et al. Contrasting water-use strategies to climate warming in white birch and larch in a boreal permafrost region. *Tree Physiol.* **2024**, *44*, tpae053. [CrossRef]

forests

MDPI

Article

The Environmental and Genetic Controls of Increment Suggest a Limited Adaptability of Native Populations of Norway Spruce to Weather Extremes

Roberts Matisons *, Juris Katrevičs, Pauls Zeltiņš, Diāna Jansone and Āris Jansons

Latvian State Forest Research Institute 'Silava', 111 Rigas Str., LV-2169 Salaspils, Latvia; diana.jansone@silava.lv (D.J.); aris.jansons@silava.lv (Ā.J.)
* Correspondence: robism@inbox.lv; Tel.: +371-29789581

Abstract: In the Baltics, warming is expected to burden the growth of Norway spruce *Picea abies*, with weather anomalies/extremes having strong triggering effects, which can be mitigated by tree breeding. Within the region, breeding programmes have been aiming for productivity, yet being conservative, their sustainability depends on the adaptability of native genotypes, which is unclear. The adaptability of genotypes can be assessed through local adaptations and phenotypic plasticity, with the sensitivity of increment depicting the conformity of genotypes and environments. To assess the adaptability of native populations to anticipated climates, local genetic adaptation and phenotypic plasticity of the weather sensitivity of the radial increment were assessed by the methods of time series analysis and quantitative genetics based on three clonal trials (low-density single-tree plot plantations of grafted clones of native plus trees) representing the local climatic gradient in Latvia. The growth of trees was sensitive to the moisture availability in summer and the thermal regime in winter, yet coinciding anomalies in both were associated with abrupt changes in tree ring width. These environmental effects differed among the clones, indicating genetic controls over the sensitivity of increment, which, however, decreased under a warmer climate, suggesting a limited adaptability of local populations to warming. Still, the weather-growth relationships showed moderate phenotypic plasticity, suggesting some mid-term adaptability. Accordingly, supplementation of breeding populations via assisted gene transfer with the genotypes that are adapted to warmer and drier climates appears crucial.

Keywords: *Picea abies*; heritability; weather-growth relationships; meteorological anomalies; local adaptation

Citation: Matisons, R.; Katrevičs, J.; Zeltiņš, P.; Jansone, D.; Jansons, Ā. The Environmental and Genetic Controls of Increment Suggest a Limited Adaptability of Native Populations of Norway Spruce to Weather Extremes. *Forests* **2024**, *15*, 15. https://doi.org/10.3390/f15010015

Academic Editors: Yassine Messaoud, Jan Světlík and Giorgio Alberti

Received: 14 November 2023
Revised: 15 December 2023
Accepted: 18 December 2023
Published: 20 December 2023

1. Introduction

In the eastern Baltic region, the projected changes in tree species abundance are giving rise to concerns about the sustainability of Norway spruce *Picea abies* Karst. [1,2], which is an economically important species [3]. The decline in Norway spruce, though, has been predicted by the bioclimatic models, which can be biased due to uneven adaptability of populations [4,5]. On the contrary, an increase in forest productivity has been projected as the vegetation season extends [6], thus adding to the uncertainties. Accordingly, estimates of the genetic adaptation and phenotypic plasticity of local populations are crucial for their adaptability assessment [7,8], as well as being indicative of the marginality of the growing conditions [2,9]. Such estimates can be made by the methods of quantitative genetics [10], with provenance/progeny trials providing empirical material for broad-sense estimates [11,12]. In this regard, weather anomalies/extremes resulting from a shift/extension of the local climatic gradient [13,14] are considered the triggers for genetic adaptations [9], with the follow-up responses revealing the resilience of growth [15,16]. Such information is crucial for adaptive management [17] and for tree breeding in particular [10,11,18].

In the eastern Baltic region, forest tree breeding programmes are conservative, i.e., they rely on native genotypes and their genetic diversity [19–21], as well as genetic controls for the sets of the most relevant traits [22,23]. Hence, the presence of small-scale genetic adaptations is crucial [7,8,24]. Currently, the breeding indices emphasize productivity and stem quality traits [21–23], which are a cumulative representation of the conformity of genotypes with the past climates [25,26]. Under the accelerating environmental changes, such information might be outdated [27,28], and hence, traits highlighting compatibility with the climate are crucial [8,9]. Increment, which is an integral estimate of the conformity of genotypes with the environment [29,30], can therefore be considered an informative adaptability proxy [31,32].

Radial increment is a widely used proxy of forest growth and productivity, as it can be conveniently reconstructed from measurements of tree ring widths (TRW) [29,33]. Furthermore, TRW is a complementary source of information on the genetic and environmental control of growth [30,31]. However, radial increment is a product of site, genetics, age, weather, disturbances, etc., and hence, elaborate mathematical methods (e.g., time series deconstruction) have been developed for the separation of variance [15,34]. Considering the discrete nature of weather anomalies, specific standardization methods have been advised for the quantification of responsiveness and resilience of growth, ensuring reliable retrospections [16,34,35].

Norway spruce tolerates a wide range of edaphic conditions, but favors mesotrophic and eutrophic sites [36]. The species is sensitive to water deficit [7,24,37]; still, under a temperate climate, its growth is limited by both warmth and moisture availability [38,39]. Furthermore under temperate climate, the weather controls of growth of spruce are complex, with temperature in the dormancy period having carry-over effects [2,39], which might alter its susceptibility to water shortage in summer [36,40]. In Northern Europe, warming has been projected to improve the growth of Norway spruce [6,25,36]; however, the complex effects of the intensifying water shortages are becoming devastating [3,37,41]. This is particularly so in the case of anomalously warm and dry summers [37], highlighting the ability of genotypes to deal with water deficit as being paramount for their sustainability [8,27,42].

For Norway spruce populations, some local genetic adaptation to drought has been shown under close-to-marginal (tailing) conditions, which, however, rapidly decreases at the distribution limit [2,7]. In the Baltics, local provenances (considered nonmarginal) with trees differing in dimensions and stem/crown properties have been distinguished [43,44]. However, nonmarginal tree populations can have a common sensitivity to the principal weather drivers of growth to maximize growth [45]. Accordingly, estimates of the adaptability of such populations are crucial for the assessment of the sustainability of conservative breeding programmes and management [10,18,28]. Considering the specialization of populations [7], local information is highly anticipated.

This study aimed to estimate the local genetic adaptation and phenotypic plasticity of the native populations (local provenances) of Norway spruce in Latvia in terms of the sensitivity of increment to weather anomalies and drought conditions in particular. We hypothesized that local genotypes would show only moderate differences in sensitivity to weather anomalies, indicating a limited adaptability of the native population to the anticipated climatic changes.

2. Material and Methods

2.1. Trials and Sampling

Three clonal trials (initially intended as seed orchards, but left as clonal trials due to lack of tree topping) of local genotypes (plus trees) located in the mid-part of Latvia near Tukums (TUK), Biksti (BKT), and Kalsnava (KLN), were studied (Figure 1). The trials were established in 1964 (TUK and KLN) and 1967 (BKT) by planting two-year grafts of local plus trees originating within 35 km from the trials (in TUK and KLN trials), as well as from the western and eastern local provenance regions of Latvia (in BKT trial). Accordingly, these were the oldest clonal trials of the species within the region that were available, with their age being sufficient for time series analysis of increment. In all cases, rootstocks grown in local nurseries were used. Each trial had a unique set of clones (without overlap; the number of clones differed). The trials were established on mesotrophic former agricultural land (meadows) with a flat topography and deep freely draining sandy/silty soil. The initial spacing was 5 × 5 m (trees ha^{-1}), and the design was randomized single-tree plots. The trials contained from 32 to 84 clones, initially represented by seven to 30 ramets (trees).

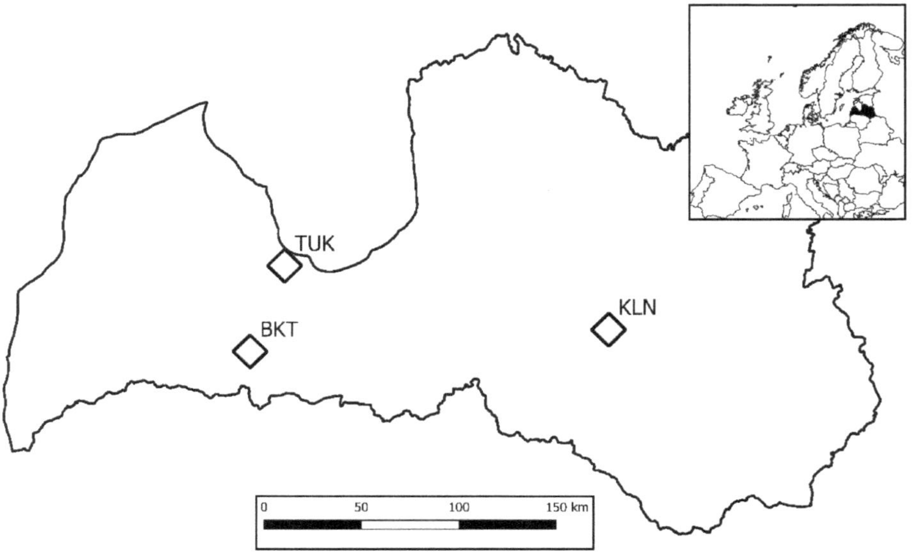

Figure 1. Location of the studied clonal trials of local plus trees of Norway spruce in Latvia.

The trials represented the local continentality gradient from maritime (TUK and BKT trials) to inland (KLN trial) climate; the mean annual temperature (±st. dev.) was 7.4 ± 0.7 and 6.4 ± 0.7 °C, respectively (Supplementary Material, Figure S1). January and July were the coldest and warmest months, with temperatures ranging from −4.3 ± 2.8 to −2.2 ± 2.6 and from 17.9 ± 1.7 to 18.0 ± 1.6 °C in KLN and TUK/BKT trials, respectively. The mean annual precipitation was comparable among the trials, ranging from 681 ± 82 to 700 ± 80 mm/year in TUK/BKT and KLN, respectively. The annual distribution of precipitation was also similar among the trials, as approximately half of the annual precipitation fell during the summer months (June–September, 75 ± 16 mm/month). The climatic changes were expressed as the warming during the dormancy period, and hence as the extension of the vegetation period (by ca. two weeks during the 20th century), and particularly as the increasing heterogeneity of a summer precipitation regime with the extension of hot precipitation-free periods [13,42].

Based on the data from a trial inventory conducted in 2010–2014, 19, 20, and 77 clones, which represented the distribution of tree dimensions of the plantations and had more than four ramets, were selected for sampling in KLN, TUK, and BKT trials, respectively. Between 5 and 22 ramets, representing the stem diameter distribution, were selected for each clone. From each of the selected trees, two increment cores from randomly oriented opposite sides of the stem were collected at breast height using a 5 mm increment borer. It was ensured that at least one core per tree contained the pith (with a maximum offset of three tree-rings). Only visually healthy nonleaking trees were sampled, though. The sampling was carried out in 2015 in the KLN and TUK trials and in 2018 in the BKT trial. In the laboratory, increment cores were mounted, and their surface was progressively grinded to ensure recognition of tree ring borders. The measurements of TRW were done manually using the LINTAB 6 measuring table (RinnTech, Heidelberg, Germany). All measurements were done by the same person; TRW was measured with 0.01 mm accuracy.

2.2. Data Analysis

To ensure the quality of measurements and exact dating of increment crucial for assessment of linkages with meteorological data, time series of TRW were cross-dated graphically and statistically. The metrics of environmental strength (e.g., expressed population signal, signal-to-noise ratio, interseries correlation) and variability of TRW (mean sensitivity, gini coefficient, first-order autocorrelation, etc.) were calculated for the datasets, and their subsets were calculated based on a detrended (by a flexible cubic spline) time series. To relate changes (their abruptness) in increment (TRW) with weather anomalies, as well as to assess the effects of local adaptation in it, the complex approach suggested by Schwarz et al. (2020) [16] and Jetschke et al. (2019) [34] was used for the identification of weather-driven growth changes. To estimate the abruptness of changes in increment, pointer year (PY) analysis based on the relative growth change was used [46]. The analysis was conducted for each clone. The time series of TRW were converted into the time series of growth changes relative to the preceding five years. The relative changes in TRW of ≥40% were considered an indicator of an event year.

A PY for a clone was considered "significant" if ≥50% of trees showed coherent (positive or negative) event year. Such mild threshold values were set due to favorable growing conditions and productive growth. Additionally, time series of mean relative growth deviations were calculated for each clone. Rolling z-scores for 30-year windows were calculated for the meteorological variables to identify anomalies (z-score > 2.0). A bootstrapped (semiparametric bootstrap, 1000 iterations) Pearson correlations analysis between the mean relative deviation in TRWs for clones and local meteorological variables was conducted for the screening of meteorological conditions relating to changes in growth.

The gridded (0.5° latitude/longitude) meteorological data, mean monthly temperature, and precipitation sums were acquired from the online repository of the Climatic Research Unit of the University of East Anglia [47]. Local data (for the grid points located closer than 0.25° from the trials) were acquired. The standardized precipitation evapotranspiration index (SPEI) was calculated with respect to the three preceding months to describe the occurrence of drought conditions [48]. Considering the carry-over effects of weather conditions [2,39], the meteorological variables were arranged into the climatic window from June in the year preceding the formation of increment to September of the year of growth. Additionally, composite variables representing conditions of climatic years (October–September), winter (December–February), spring (March and April), summer (June–September), and growing period (June and July) were calculated.

To evaluate the resilience of TRW in response to the identified PYs and their local differences, resilience ("RRR") analysis was performed [15,16]. The resilience indices (resistance, recovery, resilience, and relative resilience) were calculated for each tree with respect to four years before/after a PY. The local provenance effect of the resilience indices in the common negative PYs was estimated using mixed effects models, which were fitted using the restricted maximum likelihood approach [49]. Affiliation of the clones to local provenance region, the western and eastern part of Latvia, which differ in dimensions and stem quality of trees [43,44], was used as the fixed effect. Stem diameter at breast height was included as a covariate (fixed) to control for tree size. Trial, clone (nested), and year (crossed) were included in the models as the random effects to account for the spatiotemporal dependencies. The models were based on the pooled data from the studied trials (incomplete design). Wald's type II χ^2 test was used to estimate the significance of the differences between the two local provenance regions.

To evaluate the strength of genetic control over responses and resilience of increment in relation to weather anomalies, which differed by nature, as well as "normal" variation, heritability indices [50,51] were calculated on an annual basis. For the separation of variance components, a simple random effect was used, where relative changes in TRW as well as the calculated resilience indices of trees were used as the response, and clone was used as the random effect. The models in general form were as follows:

$$r_{ijk} = \mu + (c_k) + \varepsilon_k, \quad (1$$

where (c_k) is the random effect of clone in the i-th year. The models were fitted using th maximum likelihood approach. Broad-sense heritability (H^2) was estimated as the ratio o the variance of clone and the total variance [51]. To evaluate genetic variability in growt responses, the clone/provenance coefficient of variation (CCV) was calculated as the rati of the square root of variance of a clone to the absolute mean value of the growth respons Considering that the weather-related drivers of changes in increment differed by the trial the analysis was conducted separately for each of the trials.

To generalize genetic effects on relative changes and resilience of growth across th local climatic gradient, variance separation was carried out by mixed models based on th data from all the trials combined. The models in general form were as follows:

$$r_{ijk} = \mu + t_i + y_j + t_i : y_j + \left(c_{k(i)}\right) + \left(c_{k(i)} : y_j\right) + \left(c_{k(i)} : y_j : t_i\right) + \varepsilon_{ijk}, \quad (2$$

where t_i is the fixed effect of trial, y_j is the fixed effect of year, t_i:y_j is their interaction $(c_{k(i)})$ is the random effect of clone nested in trial, $c_{k(i)}$:y_j is the random effect of interactior between clone and year, and $c_{k(i)}$:y_j:t_i is the random effect of the interaction among clone year, and trial. Random effects represent random intercepts. The relative growth change and tolerance indices of trees were used as the responses. The heritability estimates wer calculated as stated above. Data analysis was conducted in R v. 4.2.2 [52], using package "dplR" [53], "pointRes" [46], and "lme4" [49].

3. Results

The measurements of TRWs were of a good quality, and the time series representin 1175 of 1260 initially sampled trees (93.2%) were successfully cross-dated; accordingl the trials were represented by data from 211 to 629 trees (Table 1). Due to the unever number of trees per clone, the metrics of the environmental signals that were capture by the time series of TRW ranged widely, particularly under the harsher climate in th KLN trial. Nevertheless, the datasets of the trials and clones mostly were representativ of environmental variation, as the EPS values exceeded the arbitrary threshold of 0.85 The lower values of EPS for some clones were likely due to the lower number of trees nevertheless, when averaged for the trial, the values were sufficient. The strength o environmental forcing of the TRW indicated by the SNR of the datasets also ranged widel indicating the uneven sensitivity of the clones; however, it tended to be higher in th TUK trial. The gini coeffects were low, implying a low interannual variation of TRW, a the growth of trees has been quite rapid. However, the intermediate values of the mear sensitivity of the time series, as well as the low autocorrelation, indicated a nonlagge responsiveness of increment to environmental forcing.

During the periods covered by the time series of TRW, a few PYs were identifie indicating some abrupt changes in radial growth (Figure 2). However, the incidence o PYs largely differed among the clones, and such differences, as indicated by the commor "significant" PYs, were stronger under the cooler climate of the KLN trial. In contras common PYs were more frequent under the warmer climates of the BKT and TUK trial where up to 100 and 65% of clones showed them, respectively. Nevertheless, commor tendencies in the occurrence of PYs were observed in all trials, as the positive PY occurre when trees were young, and the negative PYs occurred as the trees were ageing. Th spatial synchrony (among the trials) of the positive PYs was limited, indicating the effect of local environmental conditions. Most of the negative PYs, identified during the secon part of the reference intervals, were common among the trials, indicating large-scal environmental effects.

Table 1. Statistics of the time series of tree ring width of Norway spruce clones in three low-density grafted clonal trials in Latvia. The agreement statistics are calculated for time series detrended by flexible cubic spline. The mean value and the range of estimates (among clones; in brackets) are shown.

Trial	Kalsnava (KLN)	Tukums (TUK)	Biksti (BKT)
Number of clones	19	20	77
Timespan of series	1969–2015	1971–2015	1981–2018
Total number of cross-dated trees	211	335	629
Number of ramets (trees) per clone	11.1 (6–18)	16.7 (7–24)	8.3 (5–22)
Mean tree ring width, mm	4.00 (3.58–4.91)	4.62 (3.74–5.44)	5.68 (4.22–7.02)
Standard deviation in tree ring width, mm	1.61 (1.28–2.16)	1.49 (1.21–1.76)	2.00 (1.33–2.99)
Mean sensitivity of time series	0.22 (0.19–0.28)	0.25 (0.21–0.29)	0.23 (0.17–0.29)
Gini coefficient of time series	0.13 (0.10–0.16)	0.14 (0.12–0.15)	0.12 (0.09–0.17)
First-order autocorrelation of time series	0.13 (0.10–0.16)	0.14 (0.12–0.15)	0.12 (0.09–0.17)
Mean interseries correlation (r-bar)	0.40 (0.22–0.57)	0.48 (0.31–0.62)	0.50 (0.20–0.68)
Expressed Population Signal (EPS)	0.87 (0.75–0.94)	0.93 (0.81–0.97)	0.85 (0.82–0.97)
Signal-to-noise ratio (SNR)	8.02 (3.00–17.25)	16.21 (4.07–32.45)	9.02 (3.73–38.06)

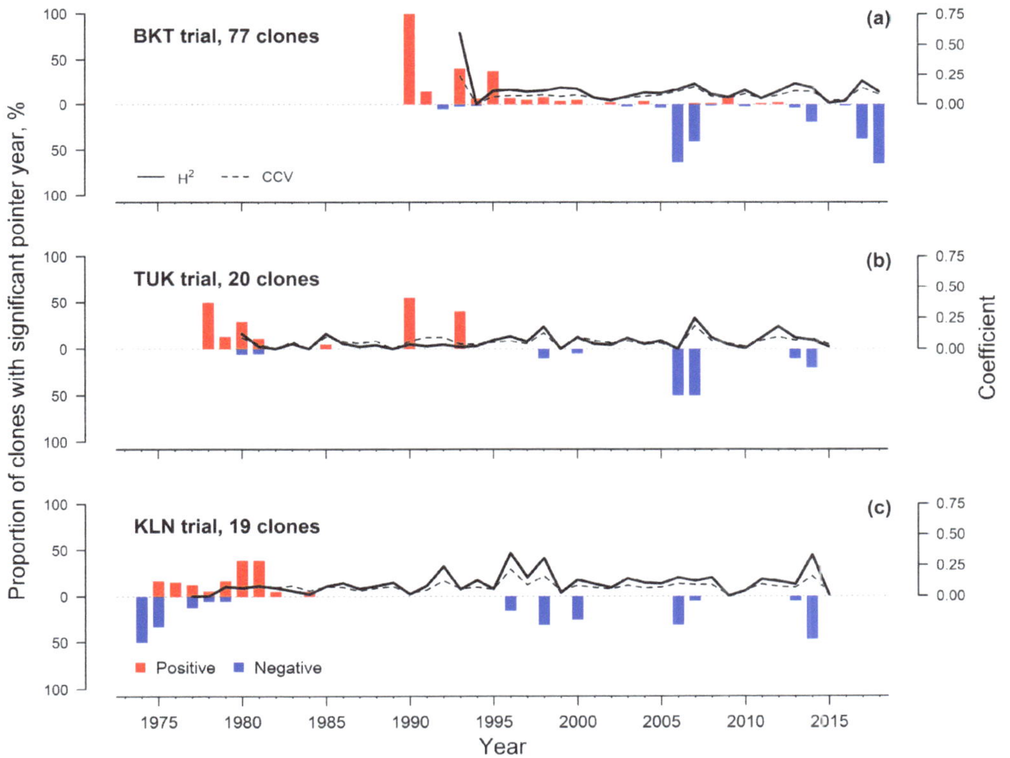

Figure 2. The proportion of clones showing "significant" pointer year in TRW in the studied clonal trials (**a–c**) of local plus trees of Norway spruce in Latvia (bars). Lines indicate broad-sense clonal heritability (H^2; solid line) and coefficient of genetic (clonal) variation (CCV) of relative changes in TRW calculated on an annual basis. Heritability coefficients are calculated for the periods represented by ≥5 clones and ≥50 observations.

The identified PYs occurred in years with a few coinciding weather anomalies (Table 2), indicating complex cumulative effects of meteorological conditions, as well as the robustness of radial growth regarding single moderate climatic disturbances. Generally, the positive PYs (growth improvements) occurred when winters were cold and summers were cool or moist. The negative PYs were associated with hot and dry summers coupled with an increased temperature during the dormancy period, which, however, differed by clone. On an annual basis, the relative growth changes (mean growth deviations for the clones) showed correlations with the meteorological variables, reflecting periods during and before the formation of increment, thus indicating direct and carry-over effects of weather conditions (Figure 3). However, the strength of weather limitations on the TRW differed by trials as indicated by the proportion of clones with significant correlations.

Table 2. Meteorological anomalies according to the national almanack and gridded climatic data (monthly meteorological variables; CRU TS4) coinciding with the estimated "significant" pointer years in tree ring width of local clones of Norway spruce in clonal trials in Latvia during the period of 1971–2018. For the monthly weather anomalies, rolling 30-year z-scores are shown in brackets. Only the anomalies with z-score > 2.00 are shown. MAT—mean annual temperature, SPEI—standardized precipitation evapotranspiration index.

Year	Pointer Year	Almanack	Gridded Data, Monthly Variables
1974	neg., KLN	Cold and heat records in winter and summer	Prec. spring (−2.34)
1975	neg., KLN	Warmth records in winter, contrasting spring temp., dry summer	MAT (2.49), SPEI Mar. (−2.41)
1978	pos., TUK	Low temp. records in spring and summer, moist summer	Prec. Aug. (2.14)
1980	pos., TUK, KLN	Low. temp. records in winter and spring, cool spring, moist summer	Prec. prev. Jul. (2.19), SPEI Aug. (2.46)
1981	pos., KLN	Cold records in spring, heat records in summer, warm and moist summer	Prec. Mar. (2.17), prec. Jun. (2.25)
1990	pos., BKT, TUK	Warmth records in winter, warm winter, spring	Temp. Feb. (2.14), temp. Mar. (2.10), temp. winter (2.68)
1993	pos., BKT, TUK	Low temp. records in winter and summer, cold year	Temp. prev. Oct. (−2.31), temp. May (2.34), SPEI Mar. (2.04)
1995	pos., BKT	Warmth records in spring and summer, contrasting summer temp.	Temp. prev. Jul. (2.32), prec. prev. Jul. (−2.11)
1998	neg., KLN	Warmth records in winter, moist summer	Temp. prev. Aug. (2.46), prec. Jun. (2.1), prec. veg. seas. (2.07)
2000	neg., KLN	Warmth records in winter, spring, and summer	Temp. prev. Jun. (2.29), temp. Apr. (2.37), SPEI. Jun. (2.06)
2006	neg.	Cold records in winter, heat records in July	-
2007	neg., BKT, TUK	Temp. contrasts in winter and spring	Temp. prev. Dec. (2.26), temp Mar. (2.43), prec. Jan. (2.33)
2014	neg.	Warmth record in spring, contrasting temp. May, heat records in summer, warm and dry summer	SPEI. prev. Sep. (2.06), SPEI Jan. (2.30), SPEI. veg. seas (−2.55)
2017	neg., BKT	Warmth records in winter and spring, low temp. records in summer, warm winter, cool summer.	Prec. Sep. (2.26)
2018	neg., BKT	Heat records in spring and summer, dry and warm summer	Temp. May (2.51), temp. summer (2.34)

Under the harsher climate of the KLN trial, clonal differences regarding the meteorological forcing of the TRW were the strongest. Accordingly, the negative correlation with precipitation in May was the most common, being significant for 63% of the clones (Figure 3). Between 32 and 42% of the clones showed correlations with moisture availability (SPEI) in the preceding summer (July to September). The correlation with summer precipitation, indicating a direct effect of moisture availability on increment, was significant for one-third of the clones. A few clones showed correlations with conditions during winter. Under the warmer coastal climate of the TUK trial, correlations with precipitation in February and June (positive) and temperature during the growing season and summer (negative) were the most common, being significant for up to 52% of clones. Such correlations, as well as similar effects of precipitation, indicated growth limitations by summer water availability.

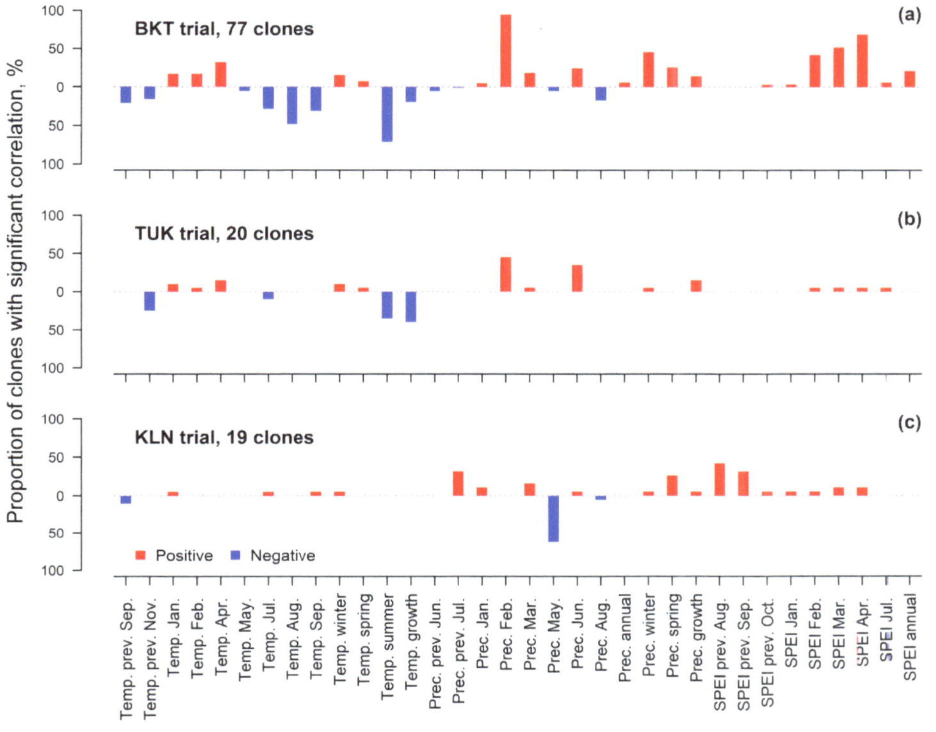

Figure 3. The proportion of clones showing significant correlations between the mean deviation in tree ring width (relative growth changes) and meteorological variables in three clonal trials of local plus trees of Norway spruce in Latvia (**a**–**c**) during the period of 1985–2015. Only the variables estimated with significant correlations are shown. SPEI—standardized precipitation evapotranspiration index.

Considerably stronger weather limitations on the TRW and, hence, the weakest clonal differences were suggested by the weather-growth correlations under warmer inland climate in the BKT trial (Figure 3). This was particularly so considering the higher number of clones tested, which represented genotypes from the local and neighboring provenance region. In this trial, the number of meteorological variables showing significant weather–growth correlations was also higher, likely due to a wider representation of the genotypes (Table 1). In the trial, 95% of the clones showed a positive correlation with February precipitation, and the majority of them also showed a positive correlation with SPEI in spring months (Figure 3). The temperature in summer showed negative correlations for 72% of the clones. Additionally, positive correlations with temperature in winter/spring and precipitation in summer months were significant for 20–30% of the genotypes. Nevertheless, the signs of correlations were consistent for the clones.

The resilience of growth regarding the negative PY was similar among the studied populations (local provenances), as only slight differences in resilience components were estimated (Figure 4, Table 3). Overall, the mean values of the resilience components indicated the ability of the genotypes to restore growth to pre-disturbance levels, even though the negative PY were considered. As trees were young and TRW had an explicit age trend, the resilience of the TRW was slightly below 1.0, and hence, relative resilience was low. However, resilience was specific for the genotypes and local conditions, as well as the nature of the water anomaly underlying PY, as indicated by the structure of random variances. The recovery was comparable to the inverse values of the resistance, confirming that the pre-disturbance levels of TRW were rapidly achieved. The resistance was the only component, which differed among the local populations and indicated higher sensitivity of the genotypes from the eastern population, likely to the intensifying water shortages (Table 2).

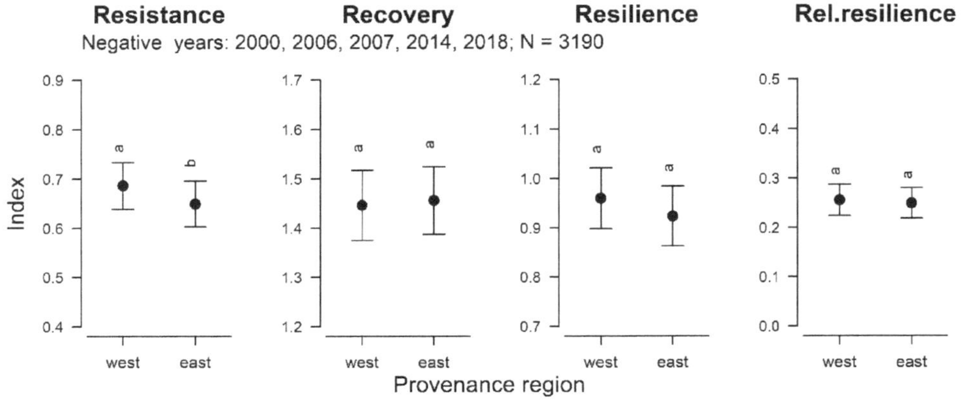

Figure 4. Estimated marginal means of the resilience components of tree ring width of native genotypes of plus trees of Norway spruce, generalized across the studied three clonal trials in Latvia in relation to the negative pointer years during the period of 1985–2018. N—number of observations. Dissimilar letters indicate the significance of the differences (p-value < 0.05).

Table 3. The fixed effect of local provenances (provenance region) and stem diameter at breast height on the resilience components of tree ring width of native genotypes of plus trees of Norway spruce generalized across the studied three clonal trials in Latvia in relation to negative pointer years during the period of 1985–2018, as well as the random variance related to trial design and year of increment.

	Resistance		Recovery		Resilience		Relative Resilience	
Fixed effects								
	χ^2	p-value	χ^2	p-value	χ^2	p-value	χ^2	p-value
Local provenance	4.7	<0.05	0.1	0.82	2.8	0.09	0.1	0.74
Stem diameter at breast height	38.7	<0.05	56.4	<0.05	0.1	0.81	32.2	<0.05
Random effects, variance								
Tree	0.0010		0.0001		0.0355		0.0076	
Clone	0.0022		0.0105		0.0013		0.0023	
Provenance	0.0001		0.0001		0.0001		0.0001	
Year	0.0029		0.0082		0.0023		0.0007	
Trial	0.0045		0.0035		0.0083		0.0014	
Residual	0.0353		0.2224		0.0156		0.0419	
Model statistic, R^2								
Marginal	0.018		0.025		0.005		0.017	
Conditional	0.245		0.114		0.753		0.238	

The clonal differences in the occurrence of PY (Figure 2) and sensitivity of increment (Figure 3) indicated genetic controls over the growth responses to weather anomalies; however, the annual heritability estimates (both H^2 and CCV) were low to intermediate (Figure 2). The CCV was generally lower than H^2, indicating limited genetic plasticity. The heritability estimates mostly peaked during the negative PY, yet the peaks and the overall values tended to be lower under the warmer climate, indicating diminishing genetic effects. Accordingly, the highest annual heritability estimates for the relative changes in TRW were higher under the coolest climate in the KLN trial while being the lowest under warmer conditions in BKT. Still, in the BKT trial, the highest heritability was estimated in response to the positive PY of 1993, which was a cold year (Table 2). The occurrence of peaks in heritability estimates showed some specifics among the PY, indicating complex meteorological triggers of genetic effects. In the BKT and TUK trials, heritability peaked in 2007, although a stronger PY occurred a year before, while in the KLN trial, the peak in 1992 was not associated with the identified PYs. Similar trends occurred in the BKT and TUK trials in 2012. The heritability of the resilience components of TRW showed a pattern that is highly similar to the relative changes in growth, although the heritability estimates for resilience were low (Supplementary Material, Figure S2).

The components of clone/genetic variance and the broad-sense heritability coefficients estimated for the entire reference periods were low, showing weak genetic effects on the responsiveness of TRW to weather anomalies (Table 4). The variance components, however, showed that such low estimates were likely due to a high random variance, as the models did not include fixed effects of environmental variability. Irrespective of the random variance, the clonal variance was low compared to the other components, implying a lack of systematic responses. On the other hand, the variance components related to the interactions involving clones, which indicate the genotype-by-environment interactions, were higher, yet their structure differed for the resilience components of TRW. Clone-by-year interaction was the highest for relative growth changes, resistance, and relative resistance of TRW, indicating specific responses to meteorological anomalies. For recovery, the interaction between clone, year, and trials was the strongest, implying local specifics in the effects of weather anomalies.

Table 4. Components of genetic/clonal variance for relative changes and resilience components in tree ring width of native genotypes of plus trees of Norway spruce, generalized across the studied three clonal trials in Latvia in relation to negative pointer years calculated on an annual basis for the entire period of 1985–2018. H^2—broad sense heritability, CCV—clonal coefficient of variation.

	Relative Changes	Resistance	Recovery	Resilience	Relative Resilience
Variance components					
Clone-trial-year interaction	0.0007	0.0178	0.0468	0.0001	0.0026
Clone-year interaction	0.0035	0.0164	0.0001	0.0221	0.0261
Clone	0.0002	0.0001	0.0001	0.0016	0.0001
Residual	0.0642	0.0680	0.1339	0.1260	0.0857
Heritability estimates					
H^2	0.0035	0.0001	0.0001	0.0104	0.0001
CCV	0.0160	0.0001	0.0001	0.0401	0.0001

4. Discussion

4.1. Representativity of the Datasets

The increment of the studied genotypes of Norway spruce was sensitive to weather fluctuations, implying climatic controls over the growth, thus allowing assessment of sustainability under anticipated climatic changes [29,31,33]. The high percentage of the cross-dated time series of TRW, as well as the statistics on its interannual variation (Table 1), highlighted the presence of explicit weather forcing of radial increment, as well as the informativity of the datasets [53,54]. The intermediate values of the mean sensitivity of the time series confirmed their sufficiency for weather-growth analysis [54]. Although tree growth under temperate conditions depends on nutrient reserves, which results in the autocorrelation of growth [54], the estimate was low implying immediate and plastic responses to environmental changes. The clonal differences in the time series statistics (Table 1) also indicated a diversity of responses to environmental fluctuations, hinting at the adaptability of the local populations [7,8].

The intermediate strength of the PYs (Figure 2), considering limited interannual variability (Table 1), implied a general rigidness of increment, particularly as the trees were young and grew under optimal conditions (trial). The incidence of the PY with several weather anomalies (Table 2) also suggested the robustness of TRW to single climatic disturbances [34,35]. The relatively low variability in TRW can also be related to the seasonally fluctuating rate of xylogenesis [55]. The largest part of which (earlywood) is formed during the first part of the period [38] when soil moisture is abundant [13]. As the growing season advances and the weather gets warmer and drier, the effects of moisture availability on growth increases, yet only a part of TRW is affected [38,55], thus explaining the limited interannual variability (gini of TRW; Table 1). The negative effects of water shortage on TRW might also be partially offset by the positive effects of warming [6,56]. Accordingly, the moderate expression of a common sensitivity of growth can still be considered informative for the sustainability of genotypes to weather conditions and, particularly, moisture deficiency, which can have devastating legacy effects on Norway spruce [3,41]. The plasticity of increment and its responses to environmental fluctuation, however, can be considered an adaptation to an increasing range of environmental conditions [4,9,11], as the trees grow productively. Alternatively, this might be an artefact of the relatively small age of the trees, when the radial increment is largely affected by competition rather than weather conditions [54].

4.2. Weather Controls of Increment

Under the temperate climate in the hemiboreal zone, weather controls of tree growth are complex, as the effects of cold damage by low temperatures, vegetation, and/or growing season length and intensifying droughts interact [36,37]. Accordingly, the effects of weather conditions previously identified as the regional drivers of spruce growth [2,39] were associated with the PY and annual changes in the TRW of the studied genotypes (Figure 3, Table 2). These relationships confirmed the spatiotemporal complexity of meteorological effects [39]. The effects of intensifying hot droughts, which are emerging threats to forest growth [14,37,42], were indicated by the weather anomalies in summer moisture availability underlying the PY, particularly in the latter part of the analyzed period (Table 2). The correlations with summer temperature (negative) and precipitation (positive; Figure 3) particularly supported the sensitivity of local genotypes to hot droughts [42]. Considering Norway spruce as a boreal species that is adapted to cold climates [7,57], the differences in climatic forcing of TRW showed an explicit thermal cline in sensitivity of increment among the trials, such as on the distribution limit [2]. In turn, the positive correlations in the KLN trial with moisture availability in the previous vegetation season (Figure 3) might be explained by the assimilation of additional nutrients and, hence, the facilitated early growth in cases when moisture was not limited [36,38].

Considering the location in a cool temperate climate, the weather conditions during the dormancy period had a carry-over effect on increment. The effects of winter temperature and precipitation (Figure 3), which are usually in the form of snow, might be explained by the insulating effects of snow cover, which affect root dynamics and water relations in the following vegetation season [40]. Additionally, the amount of winter precipitation that replenishes groundwater determines the moisture availability during the first part of the vegetation period [13]. This was particularly so under the warmer and drier climate of the BKT trials. Contrastingly, the negative correlations with May precipitation under the cooler climate of the KLN trial might be related to reduced assimilation in the case of rainy and hence, cloudy conditions [36,58].

The positive PY (Figure 2) tended to coincide with a cool or moist summer (Table 2) implying that increased moisture availability boosted increment at a juvenile age, when drought tolerance of spruce is high [59]. As the trees age, their drought sensitivity increases due to changes in their hydraulic architecture and increasing maintenance costs [60]; accordingly, the limiting effect of moisture deficiency emerged. However, at the maturing age, the incidence of a negative PY with increased winter temperatures (warmth records might be related to interrupted dormancy and respiratory nutrient loss [58,61], and hence, reduced formation of earlywood [38]. Such a shift in responsiveness to anomalies implicitly suggests that 20–25 years is the age when the drought sensitivity of trees increases [60].

Although spruce populations from the two local provenance regions show explicit differences in productivity [43,44], only minor differences were estimated for the resilience components of growth (Figure 4), suggesting a comparable tolerance to weather anomalies. Still, the resistance of TRW in the negative PYs, which were drought-related (Table 2), was higher for genotypes from the western local provenance region, indicating some degree of local adaptation [2,7,24]. Although resistance is inverse to plasticity, it has been related to the productivity of the genotypes of Scots pine [62]. For the studied Norway spruce genotypes, resilience was similar, implying even susceptibility of growth to weather extremes [7,36], although recovery appeared sufficient [15]. The clonal differences in resilience components (Table 3), though, suggested the presence of adaptations to microsite conditions [9,45]. This was also supported by the variability in resilience in response to the combinations of weather anomalies underlying PY (Table 3). The weather-growth relationships were estimated for progenies of plus trees, and hence, might be biased for general populations [63]. Nevertheless, adaptive management implies that improved forest regenerative material is used, thus emphasizing the superior genotypes [21,26,28]

4.3. Genetic Controls over Sensitivity of Growth

Trees possess large genomes, which is an adaptation allowing them to survive a range of environmental conditions and their extremes [64], for which the rare alleles can be crucial [65]. Accordingly, the genetic control of the increment peaked during the PY, indicating immediate genetic responses to weather disturbances and highlighting small-scale local genetic adaptations [31,45,57,66]. Still, meteorological conditions apparently had uneven adaptive significance on increment [8,66], as the peaks in heritability differed by PY (Figure 2, Table 4), thus highlighting the complexity of genotype and environment interactions [10,11,18]. Genetic controls over the sensitivity of the TRW, particularly regarding moisture availability, have been observed between populations of Scots pine from the same region [32], highlighting the growing relevance of local adaptations regarding the sustainability of genotypes [26,31].

The adaptability of populations to environmental changes depends on local genetic adaptation and phenotypic plasticity [8,12,57], which are reflected by the variance components of genotype and genotype-by-environment interaction, respectively [11,18]. The differences in the strength of peaks in genetic (clonal) effect represented by H^2 between the BKT/TUK and KLN trials (Figure 2), which had small differences in climate (Supplementary Material, Figure S1), suggested negative effects of warming [2,4,6]. This was also supported phenotypically by the higher proportion of clones, which represented two local provenances, showing common correlations (Figure 3). Accordingly, the diminishing genetic effects under a warmer climate project a reducing adaptability for the local populations that are used to cool and moist climates [7,67], supporting the hypothesis of this study. Genetic differences in weather-growth relationships at a small scale have been attributed to the near-marginal parts of the distribution of widespread species [45], while the actual genetic differences diminish [2]. Accordingly, the estimated structure of the variance component of a clone (Figure 2; Table 4) supports the rapidly increasing marginality of climatic conditions for the native populations of Norway spruce, also within the species range [4]. Hence, the marginality of growing conditions apparently also applies to the population level, supporting regional differences in the adaptability of native stands [28,37].

When the heritability estimates were generalized for the reference periods across the trials (Table 4), the negligible variance of clone indicated a limited adaptability to extending climatic gradients and intensifying droughts [7,57,68]. Still, the variance of the clone-by-year, as well as by year and trial interactions, which comprised up to 33% of the total for resilience components (Table 4), implied moderate phenotypic plasticity, and hence, some mid-term adaptability of the populations [8,10,18,26]. Accordingly, the moderate, yet diminishing genetic controls over the weather-growth responses (Figure 2 Table 4) supported the projected decline in the abundance of the local population [1,4] and indicated a weakening potential for improvements by conservative breeding [7,19,23]. Still the location of the Baltics within the range of spruce suggests that assisted gene transfer of genotypes adapted to warmer and drier conditions could still contribute to the sustainability of Norway spruce. Accordingly, supplementation of local breeding populations with north transferred genotypes appears compulsory for sustaining commercial spruce forests in the Baltics [24,57]. For the screening of the most adapted genotypes, the sensitivity of increment particularly in response to the intensifying weather anomalies (Table 2, Figure 2), appears as a crucial and informative trait [30,35].

5. Conclusions

The increment of studied clones of Norway spruce plus trees under a temperate climate in the near marginal tailing part of the species distribution was complexly controlled by summer moisture availability and winter thermal regime, which have been identified as the regional drivers of growth. The increment of the tree, however, appeared quite robust to meteorological conditions, as only coinciding anomalies in both winter thermal and summer moisture regimes were associated with abrupt (yet moderate) changes in growth. The strength of these environmental effects showed clonal differences that suggest genetic controls over the sensitivity of increment. The genetic controls, however, were weaker under a warmer climate, implying an increasing marginality of growing conditions and limited adaptability of local populations to it. Still, the weather-growth relationships showed moderate genotype-by-environment interactions and, hence, phenotypical plasticity, thus highlighting the ability to cope with moderate environmental changes. This suggests that the increase in marginality also applies to local populations. Considering the diminishing genetic effects and moderate plasticity of native genotypes/populations regarding the weather-growth relationships, supplementation of breeding populations with genotypes that are adapted to warmer and drier climates via assisted gene transfer is encouraged.

Supplementary Materials: The following supporting information can be downloaded at: https://www. mdpi.com/article/10.3390/f15010015/s1, Figure S1: Climatic description of the trials. Mean monthly temperature (line) and precipitation (bars) with their standard deviations (whiskers and polygon) for the period 1988–2017 are shown. Figure S2: The proportion of clones showing "significant" pointer year in TRW in the studied clonal plantations of local plus-trees of Norway spruce in Latvia (bars). Lines indicate broad sense clonal heritability (H2) of the resilience components (resistance, recovery, resilience and relative resilience) of TRW calculated on annual basis. Heritability coefficients are calculated for the periods represented by ≥5 clones and ≥50 observations.

Author Contributions: Conceptualization, R.M. and Ā.J.; Methodology, R.M. and P.Z.; Software, R.M. and D.J.; Validation, J.K.; Formal analysis, R.M. and J.K.; Investigation, J.K., P.Z. and D.J.; Resources, Ā.J.; Data curation, J.K., P.Z. and D.J.; Writing—original draft, R.M., P.Z. and Ā.J.; Writing—review and editing, R.M., P.Z., D.J. and Ā.J.; Visualization, R.M.; Supervision, R.M., P.Z. and Ā.J.; Project administration, Ā.J.; Funding acquisition, Ā.J. All authors have read and agreed to the published version of the manuscript.

Funding: This study was funded by the European Regional Development Fund project 'Decision support tool for increased forest productivity via efficient climate-adjusted transfer of genetic gain' (no 1.1.1.1/19/A/111).

Data Availability Statement: Data are contained within the article.

Conflicts of Interest: The authors declare no conflict of interest.

References

1. Buras, A.; Menzel, A. Projecting tree species composition changes of European forests for 2061–2090 under RCP 4.5 and RCP 8.5 Scenarios. *Front. Plant Sci.* **2019**, *9*, 1986. [CrossRef] [PubMed]
2. Klisz, M.; Buras, A.; Sass-Klaassen, U.; Puchałka, R.; Koprowski, M.; Ukalska, J. Limitations at the limit? Diminishing of genetic effects in Norway spruce provenance trials. *Front. Plant Sci.* **2019**, *10*, 306. [CrossRef] [PubMed]
3. Knoke, T.; Gosling, E.; Thom, D.; Chreptun, C.; Rammig, A.; Seidl, R. Economic losses from natural disturbances in Norway spruce forests—A quantification using Monte-Carlo simulations. *Ecol. Econ.* **2021**, *185*, 107046. [CrossRef]
4. Garzón, M.B.; Robson, T.M.; Hampe, A. Δ Trait SDMs: Species distribution models that account for local adaptation and phenotypic plasticity. *New Phytol.* **2019**, *222*, 1757–1765. [CrossRef] [PubMed]
5. Valladares, F.; Matesanz, S.; Guilhaumon, F.; Araújo, M.B.; Balaguer, L.; Benito-Garzón, M.; Cornwell, W.; Gianoli, E.; van Kleunen, M.; Naya, D.E.; et al. The effects of phenotypic plasticity and local adaptation on forecasts of species range shifts under climate change. *Ecol. Lett.* **2014**, *17*, 1351–1364. [CrossRef] [PubMed]
6. Reyer, C.; Lasch-Born, P.; Suckow, F.; Gutsch, M.; Murawski, A.; Pilz, T. Projections of regional changes in forest net primary productivity for different tree species in Europe driven by climate change and carbon dioxide. *Ann. For. Sci.* **2014**, *71*, 211–225. [CrossRef]
7. Liepe, K.J.; van der Maaten, E.; van der Maaten-Theunissen, M.; Liesebach, M. High phenotypic plasticity, but low signals of local adaptation to climate in a large-scale transplant experiment of *Picea abies* (L.) Karst. in Europe. *Front. For. Glob. Chang.* **2022**, *5*, 804857. [CrossRef]
8. Moran, E.; Lauder, J.; Musser, C.; Stathos, A.; Shu, M. The genetics of drought tolerance in conifers. *New Phytol.* **2017**, *216*, 1034–1048. [CrossRef]
9. Chevin, L.-M.; Hoffmann, A.A. Evolution of phenotypic plasticity in extreme environments. *Philos. Trans. R. Soc. B Biol. Sci.* **2017**, *372*, 20160138. [CrossRef]
10. Li, Y.; Suontama, M.; Burdon, R.D.; Dungey, H.S. Genotype by environment interactions in forest tree breeding: Review of methodology and perspectives on research and application. *Tree Genet. Genomes* **2017**, *13*, 60. [CrossRef]
11. Chmura, D.J.; Barzdajn, W.; Kowalkowski, W.; Guzicka, M.; Rożkowski, R. Analysis of genotype-by-environment interaction in a multisite progeny test with Scots pine for supporting selection decisions. *Eur. J. For. Res.* **2021**, *140*, 1457–1467. [CrossRef]
12. Leites, L.P.; Rehfeldt, G.E.; Robinson, A.P.; Crookston, N.L.; Jaquish, B. Possibilities and limitations of using historic provenance tests to infer forest species growth responses to climate change. *Nat. Resour. Model.* **2012**, *25*, 409–433. [CrossRef]
13. Meier, H.E.M.; Kniebusch, M.; Dieterich, C.; Gröger, M.; Zorita, E.; Elmgren, R.; Myrberg, K.; Ahola, M.P.; Bartosova, A.; Bonsdorff, E.; et al. Climate change in the Baltic Sea region: A summary. *Earth Syst. Dyn.* **2022**, *13*, 457–593. [CrossRef]
14. Reyer, C.P.O.; Bathgate, S.; Blennow, K.; Borges, J.G.; Bugmann, H.; Delzon, S.; Faias, S.P.; Garcia-Gonzalo, J.; Gardiner, B.; Gonzalez-Olabarria, J.R.; et al. Are forest disturbances amplifying or canceling out climate change-induced productivity changes in European forests? *Environ. Res. Lett.* **2017**, *12*, 034027. [CrossRef] [PubMed]
15. Lloret, F.; Keeling, E.G.; Sala, A. Components of tree resilience: Effects of successive low-growth episodes in old ponderosa pine forests. *Oikos* **2011**, *120*, 1909–1920. [CrossRef]
16. Schwarz, J.; Skiadaresis, G.; Kohler, M.; Kunz, J.; Schnabel, F.; Vitali, V.; Bauhus, J. Quantifying growth responses of trees to drought—A critique of commonly used resilience indices and recommendations for future studies. *Curr. For. Rep.* **2020**, *6*, 185–200. [CrossRef]
17. Nabuurs, G.J.; Verkerk, P.J.; Schelhaas, M.J.; González Olabarria, J.R.; Trasobares, A.; Cienciala, E. Climate-Smart Forestry: Mitigation Impacts in Three European Regions. 2018. Available online: https://www.efi.int/sites/default/files/files/publication-bank/2018/efi_fstp_6_2018.pdf (accessed on 14 July 2023).
18. Ansarifar, J.; Akhavizadegan, F.; Wang, L. Performance prediction of crosses in plant breeding through genotype by environment interactions. *Sci. Rep.* **2020**, *10*, 11533. [CrossRef]
19. Baliuckas, V.; Pliūra, A.; Eriksson, G. Forest tree breeding strategies in Nordic and Baltic countries and the possible implications on Lithuanian tree breeding strategy. *Balt. For.* **2004**, *10*, 95–103.
20. Jansons, Ā.; Donis, J.; Danusevičius, D.; Baumanis, I. Differential analysis for next breeding cycle for Norway spruce in Latvia. *Balt. For.* **2015**, *21*, 285–297.
21. Jansson, G.; Hansen, J.K.; Haapanen, M.; Kvaalen, H.; Steffenrem, A. The genetic and economic gains from forest tree breeding programmes in Scandinavia and Finland. *Scand. J. For. Res.* **2017**, *32*, 273–286. [CrossRef]
22. Burdon, R.D.; Klápště, J. Alternative selection methods and explicit or implied economic-worth functions for different traits in tree breeding. *Tree Genet. Genomes* **2019**, *15*, 79. [CrossRef]
23. Hong, Z.; Fries, A.; Wu, H.X. High negative genetic correlations between growth traits and wood properties suggest incorporating multiple traits selection including economic weights for the future Scots pine breeding programs. *Ann. For. Sci.* **2014**, *71*, 463–472. [CrossRef]
24. Trujillo-Moya, C.; George, J.-P.; Fluch, S.; Geburek, T.; Grabner, M.; Karanitsch-Ackerl, S.; Konrad, H.; Mayer, K.; Sehr, E.M.; Wischnitzki, E.; et al. Drought sensitivity of Norway spruce at the species' warmest fringe: Quantitative and molecular analysis reveals high genetic variation among and within provenances. *Genes Genom. Genet.* **2018**, *8*, 1225–1245. [CrossRef] [PubMed]
25. Goude, M.; Nilsson, U.; Mason, E.; Vico, G. Using hybrid modelling to predict basal area and evaluate effects of climate change on growth of Norway spruce and Scots pine stands. *Scand. J. For. Res.* **2022**, *37*, 59–73. [CrossRef]

26. MacLachlan, I.R.; Wang, T.; Hamann, A.; Smets, P.; Aitken, S.N. Selective breeding of lodgepole pine increases growth and maintains climatic adaptation. *For. Ecol. Manag.* **2017**, *391*, 404–416. [CrossRef]
27. Breed, M.F.; Stead, M.G.; Ottewell, K.M.; Gardner, M.G.; Lowe, A.J. Priority actions to improve provenance decision-making. *BioScience* **2018**, *68*, 510–516. [CrossRef]
28. O'neill, G.A.; Stoehr, M.; Jaquish, B. Quantifying safe seed transfer distance and impacts of tree breeding on adaptation. *For. Ecol. Manag.* **2014**, *328*, 122–130. [CrossRef]
29. Xu, K.; Wang, X.; Liang, P.; An, H.; Sun, H.; Han, W.; Li, Q. Tree-ring widths are good proxies of annual variation in forest productivity in temperate forests. *Sci. Rep.* **2017**, *7*, 1945. [CrossRef]
30. Zhang, Z.; Babst, F.; Bellassen, V.; Frank, D.; Launois, T.; Tan, K.; Ciais, P.; Poulter, B. Converging climate sensitivities of European forests between observed radial tree growth and vegetation models. *Ecosystems* **2018**, *21*, 410–425. [CrossRef]
31. Housset, J.M.; Nadeau, S.; Isabel, N.; Depardieu, C.; Duchesne, I.; Lenz, P.; Girardin, M.P. Tree rings provide a new class of phenotypes for genetic associations that foster insights into adaptation of conifers to climate change. *New Phytol.* **2018**, *218*, 630–645. [CrossRef]
32. Matisons, R.; Schneck, V.; Jansone, D.; Bāders, E.; Dubra, S.; Zeltiņš, P.; Jansons, Ā. South-eastern Baltic provenances of Scots pine show heritable weather-growth relationships. *Forests* **2021**, *12*, 1101. [CrossRef]
33. McCullough, I.M.; Davis, F.W.; Williams, A.P. A range of possibilities: Assessing geographic variation in climate sensitivity of ponderosa pine using tree rings. *For. Ecol. Manag.* **2017**, *402*, 223–233. [CrossRef]
34. Jetschke, G.; van der Maaten, E.; van der Maaten-Theunissen, M. Towards the extremes: A critical analysis of pointer year detection methods. *Dendrochronologia* **2019**, *53*, 55–62. [CrossRef]
35. Bräuning, A.; Bolte, A.; Nabais, C.; Rossi, S.; Sass-Klaassen, U. Studying tree responses to extreme events. *Front. Plant Sci.* **2017**, *8*, N506. [CrossRef]
36. Jönsson, A.M.; Lagergren, F. Effects of climate and soil conditions on the productivity and defence capacity of *Picea abies* in Sweden—An ecosystem model assessment. *Ecol. Model.* **2018**, *384*, 154–167. [CrossRef]
37. Bosela, M.; Tumajer, J.; Cienciala, E.; Dobor, L.; Kulla, L.; Marčiš, P.; Popa, I.; Sedmák, R.; Sedmáková, D.; Sitko, R.; et al. Climate warming induced synchronous growth decline in Norway spruce populations across biogeographical gradients since 2000. *Sci Total Environ.* **2021**, *752*, 141794. [CrossRef]
38. Jyske, T.; Mäkinen, H.; Kalliokoski, T.; Nöjd, P. Intra-annual tracheid production of Norway spruce and Scots pine across a latitudinal gradient in Finland. *Agric. For. Meteorol.* **2014**, *194*, 241–254. [CrossRef]
39. Matisons, R.; Elferts, D.; Krišāns, O.; Schneck, V.; Gärtner, H.; Wojda, T.; Kowalczyk, J.; Jansons, Ā. Nonlinear weather-growth relationships suggest disproportional growth changes of Norway spruce in the eastern Baltic region. *Forests* **2021**, *12*, 661 [CrossRef]
40. Tierney, G.L.; Fahey, T.J.; Groffman, P.M.; Hardy, J.P.; Fitzhugh, R.D.; Driscoll, C.T. Soil freezing alters fine root dynamics in a northern hardwood forest. *Biogeochemistry* **2001**, *56*, 175–190. [CrossRef]
41. Seidl, R.; Rammer, W. Climate change amplifies the interactions between wind and bark beetle disturbances in forest landscapes. *Landsc. Ecol.* **2017**, *32*, 1485–1498. [CrossRef]
42. Allen, C.D.; Breshears, D.D.; McDowell, N.G. On underestimation of global vulnerability to tree mortality and forest die-off from hotter drought in the Anthropocene. *Ecosphere* **2015**, *6*, 1–55. [CrossRef]
43. Gailis, A. Norway spruce provenances in Latvia. In Proceedings of the Norway Spruce Provenances and Breeding: Proceedings of IUFRO (S2.2–11) Symposium, Riga, Latvia, 14–18 September 1993; pp. 44–49.
44. Zeltiņš, P.; Gailis, A.; Zariņa, I. Long-term performance of Norway spruce in two provenance trials in Latvia. *Balt. For.* **2021** *27*, 195. [CrossRef]
45. Cavin, L.; Jump, A.S. Highest drought sensitivity and lowest resistance to growth suppression are found in the range core of the tree *Fagus sylvatica* L. not the equatorial range edge. *Glob. Chang. Biol.* **2017**, *23*, 362–379. [CrossRef]
46. van der Maaten-Theunissen, M.; Trouillier, M.; Schwarz, J.; Skiadaresis, G.; Thurm, E.A.; van der Maaten, E. pointRes 2.0: New functions to describe tree resilience. *Dendrochronologia* **2021**, *70*, 125899. [CrossRef]
47. Harris, I.; Osborn, T.J.; Jones, P.; Lister, D. Version 4 of the CRU TS monthly high-resolution gridded multivariate climate dataset *Sci. Data* **2020**, *7*, 1–18. [CrossRef]
48. Vicente-Serrano, S.M.; Beguería, S.; López-Moreno, J.I. A multiscalar drought index sensitive to global warming: The standardized precipitation evapotranspiration index—SPEI. *J. Clim.* **2010**, *23*, 1696–1718. [CrossRef]
49. Bates, D.; Mächler, M.; Bolker, B.; Walker, S. Fitting linear mixed-effects models using lme4. *J. Stat. Softw.* **2015**, *67*, 1–48. [CrossRef]
50. Falconer, D.S.; Mackay, T.F.C. *Introduction to Quantitative Genetics*, 4th ed.; Longmans Green: Harlow, UK, 1996; 635p.
51. Loha, A.; Tigabu, M.; Teketay, D.; Lundkvist, K.; Fries, A. Provenance variation in seed morphometric traits, germination, and seedling growth of *Cordia africana* Lam. *New For.* **2006**, *32*, 71–86. [CrossRef]
52. R Core Team. *R: A Language and Environment for Statistical Computing*; R Foundation for Statistical Computing: Vienna, Austria 2022. Available online: https://www.R-project.org/ (accessed on 17 December 2023).
53. Bunn, A.G. A dendrochronology program library in R (dplR). *Dendrochronologia* **2008**, *26*, 115–124. [CrossRef]
54. Speer, J.H. *Fundamentals of Tree-Ring Research*; The University of Arizona Press: Tucson, AZ, USA, 2010; 333p.
55. Cuny, H.E.; Fonti, P.; Rathgeber, C.B.K.; Von Arx, G.; Peters, R.L.; Frank, D.C. Couplings in cell differentiation kinetics mitigate air temperature influence on conifer wood anatomy. *Plant Cell Environ.* **2019**, *42*, 1222–1232. [CrossRef]

56. Socha, J.; Solberg, S.; Tymińska-Czabańska, L.; Tompalski, P.; Vallet, P. Height growth rate of Scots pine in Central Europe increased by 29% between 1900 and 2000 due to changes in site productivity. *For. Ecol. Manag.* **2021**, *490*, 119102. [CrossRef]
57. Milesi, P.; Berlin, M.; Chen, J.; Orsucci, M.; Li, L.; Jansson, G.; Karlsson, B.; Lascoux, M. Assessing the potential for assisted gene flow using past introduction of Norway spruce in southern Sweden: Local adaptation and genetic basis of quantitative traits in trees. *Evol. Appl.* **2019**, *12*, 1946–1959. [CrossRef]
58. Strand, M.; Löfvenius, M.O.; Bergsten, U.; Lundmark, T.; Rosvall, O. Height growth of planted conifer seedlings in relation to solar radiation and position in Scots pine shelterwood. *For. Ecol. Manag.* **2006**, *224*, 258–265. [CrossRef]
59. Matisons, R.; Krišāns, O.; Jansons, Ā.; Kondratovičs, T.; Elferts, D.; Ievinsh, G. Norway spruce seedlings from an Eastern Baltic provenance show tolerance to simulated drought. *Forests* **2021**, *12*, 82. [CrossRef]
60. Prendin, A.L.; Mayr, S.; Beikircher, B.; von Arx, G.; Petit, G. Xylem anatomical adjustments prioritize hydraulic efficiency over safety as Norway spruce trees grow taller. *Tree Physiol.* **2018**, *38*, 1088–1097. [CrossRef] [PubMed]
61. Ögren, E. Relationship between temperature, respiratory loss of sugar and premature dehardening in dormant Scots pine seedlings. *Tree Physiol.* **1997**, *17*, 47–51. [CrossRef]
62. Matisons, R.; Jansone, D.; Elferts, D.; Adamovičs, A.; Schneck, V.; Jansons, Ā. Plasticity of response of tree-ring width of Scots pine provenances to weather extremes in Latvia. *Dendrochronologia* **2019**, *54*, 1–10. [CrossRef]
63. Klesse, S.; DeRose, R.J.; Guiterman, C.H.; Lynch, A.M.; O'connor, C.D.; Shaw, J.D.; Evans, M.E.K. Sampling bias overestimates climate change impacts on forest growth in the southwestern United States. *Nat. Commun.* **2018**, *9*, 5336. [CrossRef] [PubMed]
64. Neale, D.B.; Martínez-García, P.J.; De La Torre, A.R.; Montanari, S.; Wei, X.-X. Novel insights into tree biology and genome evolution as revealed through genomics. *Annu. Rev. Plant Biol.* **2017**, *68*, 457–483. [CrossRef] [PubMed]
65. Schaberg, P.G.; DeHayes, D.H.; Hawley, G.J.; Nijensohn, S.E. Anthropogenic alterations of genetic diversity within tree populations: Implications for forest ecosystem resilience. *For. Ecol. Manag.* **2008**, *256*, 855–862. [CrossRef]
66. de la Mata, R.; Zas, R.; Bustingorri, G.; Sampedro, L.; Rust, M.; Hernandez-Serrano, A.; Sala, A. Drivers of population differentiation in phenotypic plasticity in a temperate conifer: A 27-year study. *Evol. Appl.* **2022**, *15*, 1945–1962. [CrossRef] [PubMed]
67. Isaac-Renton, M.; Montwé, D.; Hamann, A.; Spiecker, H.; Cherubini, P.; Treydte, K. Northern forest tree populations are physiologically maladapted to drought. *Nat. Commun.* **2018**, *9*, 5254. [CrossRef] [PubMed]
68. Zeltiņš, P.; Katrevičs, J.; Gailis, A.; Maaten, T.; Desaine, I.; Jansons, Ā. Adaptation capacity of Norway spruce provenances in Western Latvia. *Forests* **2019**, *10*, 840. [CrossRef]

forests

MDPI

Article

Factors Driving Unexpected Drought-Induced *Nothofagus dombeyi* Mortality in a Valdivian Temperate Rainforest, Argentina

María Laura Suarez *, Yamila Sasal and Loreta Facciano

Laboratorio de Investigaciones en Ecología de Bosques-LIEB, Instituto de Investigaciones en Biodiversidad y Medioambiente (INIBIOMA, CONICET-UNCO), Pasaje Gutiérrez 1415, San Carlos de Bariloche CP8400, Rio Negro, Argentina
* Correspondence: mlsuarez@comahue-conicet.gob.ar

Abstract: Understanding the drivers of drought-induced tree mortality remains a significant scientific challenge. Here, we investigated an unexpected mortality event of *Nothofagus dombeyi* (Mirb.) Oerst. following the 2014–15 drought in a Valdivian rainforest, Argentina. Our focus was on long-term growth trend differences between vital and dead trees, and how the mixing of species in tree neighbourhoods drives tree growth during drought. The inter-annual variation of basal area increments of vital and 2014–15-dead *N. dombeyi* trees showed a similar pattern through the 1930–2015 period, while the climate–growth relationships indicated that precipitation during the growing season promoted growth in both vitality classes, regardless of whether they were in the wettest location. For the period 1990–2015, both vitality classes showed similar estimated growth regardless of competition level, whereas species mingling in the neighbourhood significantly affected the dead tree growth. Network analysis revealed that drought performance covaried positively with a neighbourhood dominated by species functionally different from the focal species only in vital trees. These findings suggest a nuanced response of *N. dombeyi* to drought, shaped by multifaceted interactions at both the individual tree and neighbourhood levels. This research underscores that species-specific relationships under different mixtures imply different tree responses within a stand and add complexity to understanding drought response at the individual level.

Keywords: neighbourhood; climate–tree growth relationships; competition; mingling; tree-ring grow

Citation: Suarez, M.L.; Sasal, Y.; Facciano, L. Factors Driving Unexpected Drought-Induced *Nothofagus dombeyi* Mortality in a Valdivian Temperate Rainforest, Argentina. *Forests* **2024**, *15*, 1355. https://doi.org/10.3390/f15081355

Academic Editors: Yassine Messaoud, Jan Světlík and Giorgio Alberti

Received: 28 June 2024
Revised: 30 July 2024
Accepted: 31 July 2024
Published: 2 August 2024

1. Introduction

A report by the IPCC warns that the average global temperature will reach 1.5 °C warmer than pre-industrial times due to human activity, and the frequency and severity of heatwaves and droughts are projected to increase in many regions worldwide [1]. In this context, the southern Andean region has proved to be no exception, as the region has shown an abrupt trend toward warmer and drier conditions since the mid-twentieth century [2]. These already occurring and expected changes in temperature and precipitation may have important consequences for the southern ecosystems with a significant level of endemism and endangered species. The temperate forests of southern Patagonia, among other Andean ecosystems, have begun to show clear evidence of a negative impact from the changes in climate parameter trends, as widespread tree mortality events following droughts are being recorded in Chile and Argentina [3–5]. However, understanding and predicting tree mortality during or after droughts remains a scientific challenge worldwide, as it is still hard to predict when and where it will occur, and which tree will die as a consequence of a drought [6].

Tree rings store information on climate, site, and local environmental effects on tree growth, and are a powerful tool for studying drought impacts on multi-year tree performance. Tree ring patterns can reveal not only short- and long-term trends in climate influence on tree productivity, but also mid- and long-term trends in nutrient uptake [7], water use efficiency [8], and competition pressure at the local neighbourhood scale during dry spells [9]. This retrospective quantification of drought impact on long-term growth and vigour highlights the declining trend that prevails after droughts in many species worldwide [10]. Successive drought events can compound the stress that trees experience, leading to more pronounced decreases in growth rates [11], which finally erodes trees' resilience capacity (i.e., reduces their ability to regain pre-drought growth rates [12,13]).

Although tree physiological traits remain an important factor in understanding drought-driven death [14], the accuracy of explanations has been limited as trees' drought performance covaries with factors acting at individual, site, and community levels. In this regard, tree resilience components, in terms of growth performance, depend among others factors like soil characteristics [15], stand density [16], and species composition [17]. This creates a complex network that ultimately influences individual tree performance and resilience to dry spells [18], which makes it very difficult to rank individual responses in a stand. Species mixing implies the presence of different compensation strategies to cope with droughts in relation to the species' different functional traits [19]. These strategies can operate at the underground level, with root systems occupying different soil layers to access water resources, thereby creating spatial complementarity that can benefit other species [20]. Additionally, the development of canopies with varied architecture and phenology allows for the efficient use of light. Shade-tolerant species can establish beneath the canopy, while light-dependent species thrive in direct sunlight [21]. While stand diversity improves the average stand productivity due to complementary, the positive effects cannot prevail over negative effects (e.g., competition) during stressful conditions [22,23]. However, additional factors could unbalance this relationship, leading to disparate results of the effects of stand diversity on tree performance in the face of drought [24,25]. Different climate–growth sensitivity along gradients drives different patterns in tree drought response across species' distribution range [26]. Thus, in dry sites where water availability is a major constraint, competition could override any complementarity advantage of stand diversity, whereas in humid sites, the complementarity effect could determine drought response [22]. However, species-specific relationships under different mixtures in the immediate tree neighbourhood add complexity to understanding drought response at the individual level [27]. While the influence of climatic stressors on tree growth responses is recognized, the specific impact of other ecological traits on these relationships remains a topic of ongoing research.

As we stated above, Andean temperate forests have not been an exception to the context of forest vulnerability to climate change. Over the last few decades, there has been a significant negative impact of droughts on *Nothofagus dombeyi* (Mirb.) Oerst. (Nothofagaceae), a dominant tree species of the temperate forest in Patagonia, Argentina, with strong pervasive effects on species growth, leading to mortality [5,7,28,29]. *N. dombeyi* is one of the ten *Nothofagus* species in southern South America, and is considered a shade- and drought-intolerant species, conditions found in several environments of the Patagonian Andes. As a broadleaf evergreen species, reaching 30–40 m tall, it has suffered leaf browning and shedding during past droughts, which implies crown dieback and sudden tree mortality [28]. All the recorded mortality events in *N. dombeyi* have been observed towards the eastern limit of the species distribution, suggesting that the negative impacts of drought are more pronounced on the leeward side of the Andes, which typically receives less moisture. This spatial dimension in the discussion of *N. dombeyi's* response to drought and mortality led us to suggest a gradient of vulnerability to drought impacts [30]. However, following the 2014–15 dry period, an unexpected event of sudden *N. dombeyi* mortality occurred in a portion of the Valdivian rainforests, a region with ~3500 mm year^{-1} in precipitation [31].

Motivated by this unique and rare event, we started a dendroecological study to uncover the factors that led to *N. dombeyi* mortality in this specific location, which also has a high level of tree biodiversity. We specifically (i) reconstructed and analysed the growth patterns and climate sensitivity in vital and dead *N. dombeyi* trees, and (ii) related these growth patterns and drought responses to several factors associated with tree neighbourhood characteristics in order to determine their role as drivers of tree mortality. Our main hypothesis posited that an increased proportion of species diversity within the tree's vicinity, or a less dense neighbourhood, positively influences both individual growth patterns and a tree's response to drought, enhancing its resilience. The study aims to unravel these factors and shed light on why some trees were affected more than others, ultimately contributing to a better understanding of the species response to environmental stressors in this specific context. Although the main areas affected by *N. dombeyi* forests are included in national parks and no management actions are planned, the identification of mortality factors that can be reduced or reversed, and the synthesis of responses, can be brought back to a set of general management goals relevant for national park policies worldwide.

2. Materials and Methods

2.1. Study Site and Species

The study area is located near Puerto Blest (41°02′ S, 71° 49′ W; 750 m.a.s.l., Figure 1), and belongs to one of the easternmost relicts of the Valdivian rainforest on the leeward side of the Andes. In Argentina, Valdivian rainforest extends in small fragmented areas, from 40° S to 42° S, and in Nahuel Huapi National Park, Río Negro Province, it comprises a total area of 3.6 km^2 (Figure 1) distributed in different small fragments across the National Park. Valdivian rainforest is the westernmost end of a striking west-to-east gradient of precipitation and associated vegetation that characterize the southern Andes, with precipitation ranging from ~3500 mm year^{-1} (in Valdivian rainforest, Figure S1) to ~500 mm year^{-1} (on Patagonian steppe) only 100 km to the east [31]. Average annual precipitation in the study area is estimated at ~3500 mm year^{-1} (Figure S1) and is seasonally distributed with approximately 60% falling between May and August (winter in the Southern Hemisphere), and the annual mean temperature is 9 °C (Figure S1). Between 750 and 900 m. a.s.l., the rainforest is dominated by the shade-intolerant *Nothofagus dombeyi* along with the shade-tolerant broadleaf evergreen tree species *Saxegothaea conspicua* Lindl (Podocarpaceae), *Archidasyphyllum diacanthoides* (Less.) P.L. Ferreira, Saavedra & Groppo (Asteraceae), and *Weismania tricosperma* Cav. (Cunoniaceae). In addition, this relict of the Valdivian rainforests holds the wettest margin population of the conifer *Austrocedrus chilensis* (D. Don) Florin & Boutelje (Cupressaceae). *N. dombeyi* establishes immediately following fires and for a period lasting c. 30 years, while the shade-tolerant co-occurring species continue the establishment period under the closed canopy [32]. The understory is shaded by a dense forest canopy and consists of a great diversity of plants [33]. Layers of volcanic ash cover the glacial topography, and the soil throughout the region is derived from these parent materials (Andosol type soils). At the study site, the mean slope was c. 20%, with a northern orientation.

The study area came to our attention when, in the summer of 2015–2016 (September to March in the Southern Hemisphere), a wide portion of the Valdivian rainforests near Puerto Blest showed browned crowns of *N. dombeyi* trees (Figure 1). This event constitutes the first recorded mortality episode in a Valdivian rainforest of Argentina, but it was not isolated, as we documented other mortality locations in Lago Puelo National Park during the same summer (Suarez, personal record).

2.2. Climate Data, Drought Index, and Selected Dry Spells

Due to the lack of a meteorological station network in northern Patagonia that provides long-term records of climatic data, we downloaded climate data from the Climatic Research Unit, UK (http://www.cru.uea.ac.uk/, accessed on 18 August 2022). We used monthly temperature and precipitation grid-box data for 41.25 S and 71.75 W (grid cell 0.5° × 0.5° latitude–longitude) from the CRUTS v4.05 dataset over the period 1930–2020 [34] to characterize drought severity and climate–growth performance. To do this, we used the climatic data to calculate the standardized precipitation evapotranspiration index (SPEI) for the period 1930–2014 using the SPEI R package [35]. SPEI is a standardized multi-scalar drought index based on precipitation and potential evapotranspiration, in which negative values indicate a negative cumulative water balance [36]. The bootstrapped correlation between residual tree ring chronologies and SPEI showed the strongest correlation during the summer season ($SPEI3_{Jan}$; Figure S2). We used the R-package treeclim [37] to assess the climate–growth relationships using the correction for the Southern Hemisphere. We focused on the dry spells that caused *N. dombeyi* mortality (2014–15) and the two main droughts that happened during the last thirty years, 1998–99 and 2008–09, taking into account the calendar year when the growing process starts in the Southern Hemisphere. We defined dry years under the criterion that SPEI values were higher than −1.5 SD (or values were within 10% of the lower values during the same period).

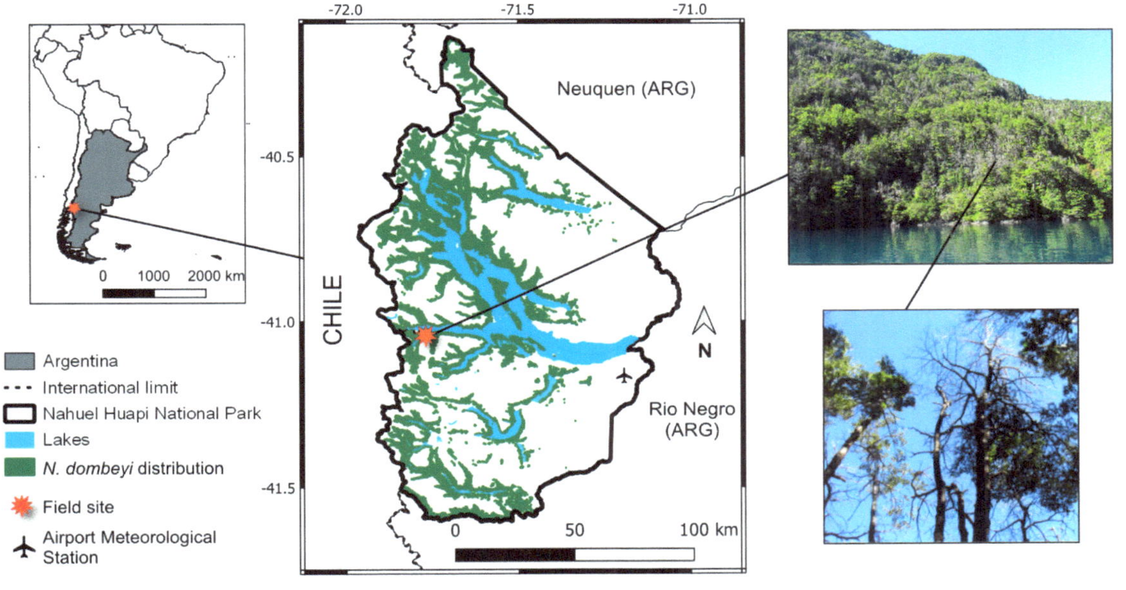

Figure 1. Map showing *N. dombeyi* distribution (green) in Nahuel Huapi National Park, northern Patagonia (Argentina), and the location of the field sampling site (red star). Coordinates on the axes are given in the World Geodetic System (WGS 84). To the right, illustrative views of the site in 2019 (top photograph), and trees that died in 2015 (bottom photograph).

2.3. Field Sampling and Dendrochronological Methods

Across the study area, we sampled 30 dominants that were vital *N. dombeyi* trees, and 30 dominants that were defoliated (dead) *N. dombeyi* individuals. The vitality state of the selected trees (focal trees) was assessed based on the presence of non-defoliated crowns. Focal trees were randomly selected across the affected area, with an aspect from north to west. Each focal tree was cored twice at a height of 1.30 m (diameter at breast height (DBH)) using a Pressler increment borer, and perpendicular to the slope. Cores were taken back to the pith in order to estimate tree age at coring height. We recorded the DBH, height, and crown dimensions of the focal tree. Additionally, we characterized its neighbourhood environmental pressure through the following variables: the distance to each neighbour, and the DBH, identity and height of each tree in the focal tree's vicinity. The area of the neighbourhood was variable as any tree crown that was in contact with the focal tree's crown was considered as a neighbour. In the lab, the tree cores were air-dried and polished with sandpaper until the tree rings were clearly visible. Afterwards, the tree ring samples were visually cross-dated, scanned at 2400–3200 dpi resolution, and measured to the nearest 0.01 mm using CooRecorder software v9.6. The quality of the cross-dating was examined using COFECHA v 6.06P [38], and we followed the convention for the Southern Hemisphere to date cores (assigning to each tree ring the date of the year in which growth started). The mortality dates of the dead trees (the date of the outermost tree ring) were defined using the chronology developed for healthy trees, and were stated to the summer of 2014–15.

2.4. Long-Term Growth Trend of Vital and Dead Trees

The tree ring width series were transformed into basal area increments (BAIs), which account for decreasing ring widths with increasing tree size [39]. For the BAI calculations we used the formula

$$\mathrm{BAI} = \pi\,(\mathrm{r^2}_{\mathrm{t}} - \mathrm{r^2}_{\mathrm{t-1}}), \tag{1}$$

where $\mathrm{r_t}$ and $\mathrm{r_{t-1}}$ is the stem radius corresponding to years t and t − 1, respectively. Each BAI chronology (vital and dead) was obtained by averaging the individual series year-by-year using a biweight robust mean. To detrend each individual tree ring width series (for climate–growth correlation inspection), we applied a cubic regression spline method to remove any influence of age-related trends, with a 50% frequency response cut off at 2/3 of the ring width series length. Afterwards, we applied an autoregressive model to each detrended series to remove the first-order autocorrelation, therefore building residual pre-whitened ring width indexes. The BAI and ring width index (RWI) chronologies were developed for each tree vitality class (vital and dead). The BAI calculations, series detrending, and chronology computation were performed using the dplR R-package [40] in R software v4.3.3 [41].

To quantify *N. dombeyi* long-term growth trends according to vitality for the period 1925–2015 and as a function of the tree characteristics and response to climate variables relevant to growth, we built a generalized additive mixed model (GAMM; [42]) by adjusting the individual BAI as a function of linear and smooth predictors. We considered tree vitality (vital or dead) and DBH (to control for ontogenetic factors) as linear predictors, and calendar year, $\mathrm{SPEI3_{Jan}}$, and their interaction as thin plant spline regression with 4 degrees of freedom or tensor products [42]. As the BAI data represent repeated measures of the same individual, we included tree identity as a random factor. Finally, we included a first-order autocorrelation structure (AR1) to account for dependency of the BAI in year t on the BAI of the previous year, t−1. To achieve normality assumptions, the BAI was log-transformed [log (BAI + 1)] prior to the analyses, and all explanatory variables were standardized by subtracting their means and dividing by their standard deviation to provide comparable coefficients. We used the mgcv R-package [43] to build the generalized additive mixed models.

2.5. Effect of Species Diversity on Tree Growth and Drought Performance

We quantified the immediate neighbourhood influence by a tree-based size–distance dependent competition index (CI) and a species mingling index (M). The CI of each focal tree *i* was calculated taking into account the DBH of the focal tree and the number, DBH and distance to the neighbouring trees as follows:

$$\mathrm{CI}_i = \sum\nolimits_{j=1}^{N} \frac{\mathrm{DBH}_j}{\mathrm{DBH}_i \times \mathrm{dist}_{ij}}, \tag{2}$$

where DBH_i and DBH_j are the DBH measurements for the *i* focal and *j* neighbouring trees, respectively, and dist_{ij} is the distance between them [44]. The mingling index describes the species variety (v_{ij}) in the vicinity of the focal tree *i* and is defined as the proportion of the *n* nearest neighbours *j* that do not belong to the same species [45], as follows:

$$\mathrm{M}_i = \frac{1}{n} \times \sum\nolimits_{j=1}^{n} v_{ij}, \tag{3}$$

The mingling values range between zero and one ($0 \leq M \leq 1$), with M close to zero meaning that the neighbour j belongs to the same species as the focal tree i, while M values close to one indicate the opposite. The advantage of mingling over common diversity indices (e.g., Shannon) is the ability to distinguish a neighbourhood dominated by *N. dombeyi* from one dominated by a species other than the focal species, both being of low diversity.

To evaluate the effect of neighbourhood influence on tree growth, we built a linear mixed-effect model for the period 1990–2014. We selected a linear model instead of an additive one, because we shortened the period to avoid uncertainty regarding the stand structure >25 years ago (although no evidence of past logging, recent fire, or mortality was recorded, and the trees are long-lived species). We included the effect of $SPEI3_{Jan}$ (standardized drought index over November, December, and January), competition (CI), neighbourhood mingling (M), and the vitality condition of the *N. dombeyi* trees as fixed effects. As above, since the BAI measurements represent repeated measures of the same individuals, tree identity was regarded as a random factor; and we included a first-order autocorrelation structure (AR1) to account for the yearly dependence on the BAI.

In addition, we quantified short-term growth responses to each selected drought and according to tree vitality. The short-term growth responses were quantified by applying a resilience index [25,46,47], considering consistent pre-drought and post-drought periods of three years. We used a period of 3 years to calculate the resilience components, as a good compromise with low bias in the calculation of resilience (Figure S3, Table S1). In addition, we set a maximum recovery period of 6 years, as a good compromise with not including subsequent droughts. However, due to the sampling date, we were limited to including short-term responses following the last drought. All the conventional indexes (resistance (Rt), recovery (Rc), resilience (Rs)) were calculated following the formulas in Lloret et al. [46], and based on the ratios of the pre-drought, drought, and post-drought BAI values. We quantified mid-term responses following the 1998–99 drought event to account for the direction of potential legacy effects as a consequence of the most severe drought in the region, by considering the ratio between the 10 years before and after the dry spell (legacy98). We applied linear mixed-effect models to compare the short-term responses (Rt, Rc, Rs) as a function of the fixed factors, i.e., vitality, magnitude of the drought event ($SPEI3_{Jan}$), DBH, CI, and M, as well as the interactions between them. Again, we included tree identity as a random factor and an autocorrelation structure as explained above. All the linear mixed-effects models were built using the nlme R-package [48] in R software, and the goodness-of-fit of the models was assessed using conditional and marginal R^2.

2.6. Network Analysis for Vital and Dead Trees

We evaluated the covariations between all the considered variables and their effect on the health status of each *N. dombeyi* tree by correlational network analysis, which is flexible, allows for the untargeted exploration of the data without prior known relationships, and accommodates many data types [49]. We linked all the tree and neighbourhood variables by tree vitality status, using a set of variables grouped in four classes: tree characteristics (BAI25, DBH, height (ht), and crown spread (W)); climate–growth relationship (with temperature (T) and precipitation (P)); drought–growth performance (Rt, Rc, Rs, and legacy98), and neighbourhood pressure (CI, % *N. dombeyi* (%Nd), % *A. chilensis* (%Au), % broadleaf codominant species (%Bl), % subdominant species (%sub), and proportion of dead *N. dombeyi* following 2014–15 drought (%NDT)). We decided not to include mingling (M), as we characterized mingling values with the proportion of species in the neighbourhood.

In the correlation network analysis, selected variables are represented as nodes in the network, and significant relationships between them are represented as edges. To test the significance of the correlation coefficients between the variables, we used Pearson correlations, where only significant coefficients ($p < 0.05$) were considered as connections between nodes (edges). We used the igraph R-package [50] in R environment [41]. The *degree* function, which counts the number of connections of each node (significant correlations), was used to analyse the importance of the nodes within a network (centrality). Network analysis provides valuable insights into the structure of systems with interconnected elements, and is a powerful tool to understand and optimize complex relationships.

3. Results

3.1. Long-Term Growth Trend Differed between Vital and Dead Trees

The inter-annual variation of the basal area increment (BAI) of the vital and 2014–15-dead *N. dombeyi* trees showed a similar pattern through the analysed period, and a common growth reduction in response to the main droughts happened in the last decades (Figure 2a). However, the vital trees presented higher BAI values than the trees that died during the last drought, for almost all the analysed period (Figure 2a grey rectangles). In addition, the vital trees belonged to a group of larger trees compared to the dead ones (Figure A1), although the larger structure (height, diameter) of the vital trees did not lead to a larger mean canopy spread (Figure A1).

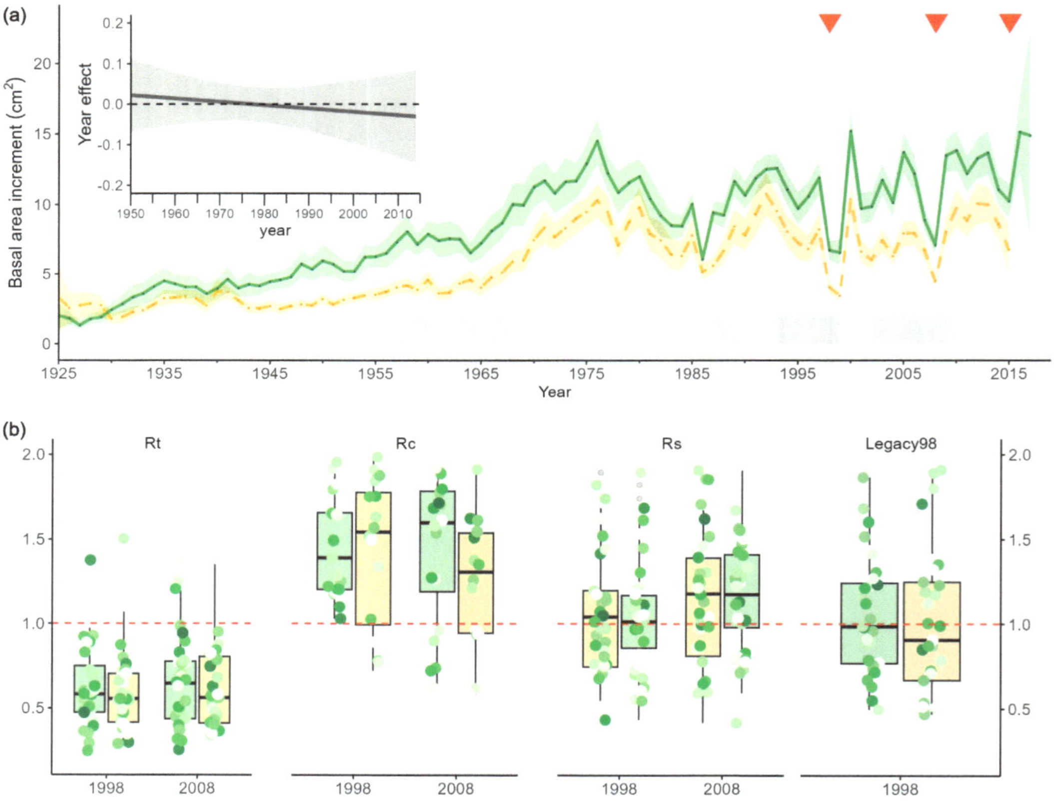

Figure 2. (**a**) Inter-annual variation of basal area increment (BAI) of vital (solid green lines) and dead (dashed orange lines). Solid lines represent the means and shaded areas around them the standard error of the mean. The grey rectangles indicate the periods when BAI significantly ($p < 0.05$) differed between vigour classes (Wilcoxon rank-sum tests). Downward triangles in red represent the main drought years during the last decades (1998, 2008 and 2015). The plot inset in figure represents predicted smooth differences in the effect of calendar year on growth trends according to GAMMs. (**b**) Resistance (Rt), recovery (Rc), resilience (Rs) and Legacy98 (pre/post 10 years) against the driest spells (1998 and 2008). The mingling index (M) was used as colour scale (0 = light green to 1 = dark green) for individual resilience components.

Consistent with similar BAI variability between the vital and dead trees, we found comparable responses to climate in both vitality groups that modelled tree growth. The correlation coefficients between the vital and dead residual RWI chronologies and climate variables indicated that precipitation during December and January in the current growing season promoted growth, whereas higher temperatures during the same period inhibited tree growth. Only dead trees showed an additional dependence on temperature during the previous growing season (October and February). Although the ranges were often similar, the correlation between climate and individual detrended and prewhitened series mostly indicated stronger correlations than would be expected based on the mean group RWI chronology (Figure 3).

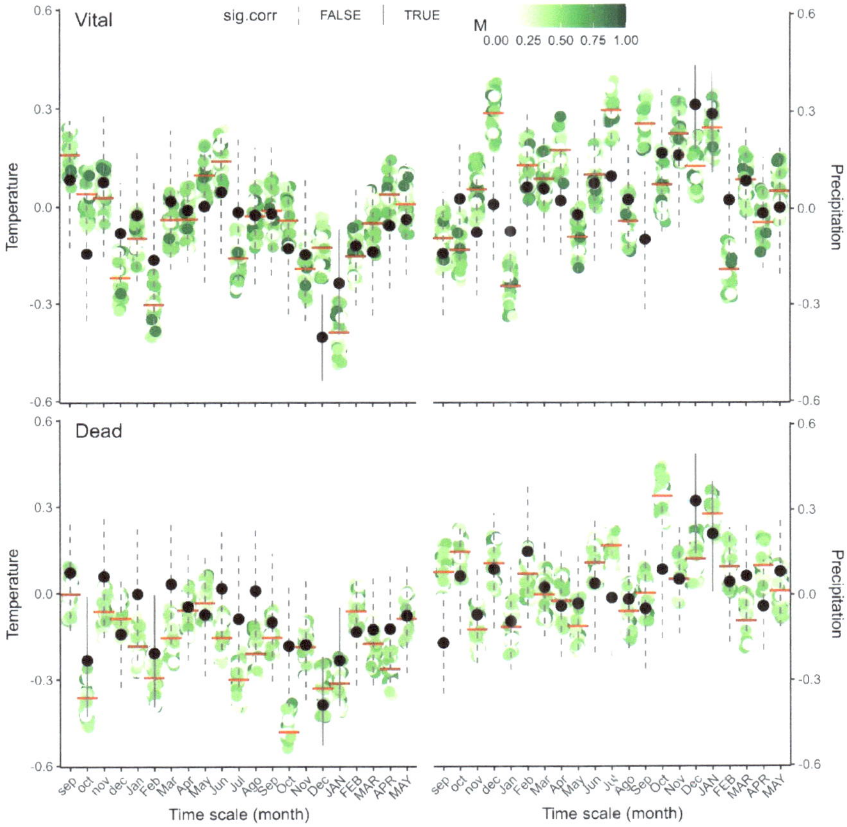

Figure 3. Pearson correlation coefficients using detrended prewhitened individual tree ring index (coloured jitter points with mean (red line)) and group residual chronology (black point range) with mean monthly temperature and total monthly precipitation for vital (**top**) and dead (**bottom**) *N. dombeyi* trees. Solid black point range depicts significant correlation according to bootstrap. The mingling index (M) was used as colour scale (0 = light green to 1 = dark green) for individual climate–growth tree coefficients.

According to the GAMM, the effect of year did not differ between the vitality classes (Table 1; Figure 2a, inset in figure). Further, for the vital and dead trees, the spline described a positive effect of year starting from the mid-1970s, pointing to a growth enhancement in both vitality classes since the late 20th–early 21st centuries (Figure A2). Changes in the climate variables relevant to growth, with the integer as the 3-month January SPEI ($SPEI3_{Jan}$; including November, December, and January) significantly influenced the BAI (Table 1) as the resultant growth was lowest under dry conditions and significantly increased during moister conditions. However, the vital trees did not show a higher growth performance during moister conditions, and neither did the dead trees show their lowest growth rates during drier ones (Table 1, Figure A3).

Table 1. Summary of the fitted generalized additive mixed model (GAMM) explaining changes in BAI (log BAI + 1) of vital and dead *N. dombeyi* trees over the period 1925–2015. Smooth terms correspond to the difference in the smooth terms between vital and dead trees, as the model was set with vital group as the reference level of vitality. Estimate and standard error (Std. Error) is shown for the linear term, whereas estimated degree of freedom (Edf) and the F-value are given for the smooth terms. Significant p-values are in bold.

Predictor Variables	Estimate/Edf	Std. Error/ F-Value	p
Linear Terms			
Height	0.037	0.027	0.164
DBH	0.377	0.025	**<0.001**
Vitality (dead)	0.069	0.049	0.159
Smooth terms			
$SPEI3_{Jan}$	1.494	9.286	**0.007**
Year	3.910	70.40	**<0.001**
Year × Vitality (dead)	1.000	0.305	0.581
$SPEI3_{Jan}$ × Vitality (dead)	1.000	0.140	0.708
Year × $SPEI3_{Jan}$	10.985	13.759	**<0.001**
Year × $SPEI3_{Jan}$ × Vitality (dead)	1.028	0.178	0.072
Observations		4533	
Adjusted R^2		0.554	

3.2. Role Played by Neighbourhood in Growth and Resilience

For the period 1990–2015, the linear mixed-effect model detected no significant differential effect of competition on the BAI between the vital and dead trees (Table 2). Thus, the dead and vital trees showed similar estimated BAI values regardless of the competition level exerted by the immediate neighbourhood. In contrast, species mingling in the neighbourhood (M) had a significant effect on the 1990–2015 BAI in the dead trees (Table 2), as the greater the species mingling in the neighbourhood, higher the estimated BAI values of the trees.

Table 2. Summary of the fitted linear mixed-effect model estimating the effect of $SPEI3_{Jan}$, competition, and species mingling on BAI (log BAI + 1) of vital and dead *N. dombeyi* trees over the period 1990–2015. $SPEI3_{Jan}$ corresponds to standardized drought index over November, December, and January, CI indicates the competition index, and M is the mingling index. Significant *p*-values are in bold.

Variables	Estimates	Std. Error	*p*
Fixed effects			
Intercept	−3.39	4.03	0.360
Vitality (vital)	0.71	0.37	0.060
Year	0.002	0.002	0.181
$SPEI3_{Jan}$	0.03	0.01	**<0.001**
CI	−0.03	0.05	0.466
M	1.16	0.40	**0.005**
Vitality (vital) × CI	0.03	0.08	0.719
Vitality (vital) × M	−1.35	0.60	**0.027**
Random Effects			
σ_i^2 *		0.14	
N		59_{ID}	
Observations		1457	
Marginal R^2/Conditional R^2		0.176/0.644	

* σ_i^2 represents the mean random effect variance of the model. R-squared values are based on Nakagawa et al. [51].

In agreement with similar climate–growth sensitivity and BAI patterns, no significant differences were observed in the post-drought responses (1998 and 2008) between the vitality classes of the trees. Likewise, there were no differences in the magnitude of the legacy effect following the 1998 drought (Figure 2b, Table 3). During the adverse climatic conditions prevailing in the summer of the 1998 and 2008 droughts, both vitality classes showed resistance values of around 0.5, although recovered their previous growth in the following 3 years (Figure 2b). Moreover, we obtained unexpected findings regarding the impact of the competition index and mingling values on growth response during droughts. The linear mixed-effect models did not reveal any significant influence of competition or mingling on driving resistance or recovery, nor did they indicate a potential buffering effect of neighbourhood on post-drought responses (Table 3). We found only an effect of DBH on the resilience indices for the dead trees, where bigger dead trees showed a more significant drop (lower resistance) during the previous drought, but also showed better recovery following the dry spells (Table 3). Finally, $SPEI_{Jan}$ during the analysed droughts did not affect the resilience indices, neither as simple predictors nor as an interactions with the vitality class.

Table 3. Summary of the fitted linear mixed-effect model estimating the effect of $SPEI3_{Jan}$ (standardized drought index over November, December, and January), competition (CI), and neighbourhood mingling (M) on resilience component (resistance, recovery, and resilience) of vital and dead *N. dombeyi* trees during 1998 and 2008 dry spells. Significant *p*-values are in bold.

Variables	Log (Resistance)		Log (Recovery)		Log (Resilience)	
	Est. (Std. Error)	*p*	Est. (Std. Error)	*p*	Est. (Std. Error)	*p*
Fixed effects						
Intercept	−0.33 (0.38)	0.39	0.39 (0.45)	0.39	0.21 (0.29)	0.47
$SPEI3_{Jan}$	0.08 (0.13)	0.55	0.04 (0.13)	0.75	0.10 (0.12)	0.43
DBH	−0.001 (0.01)	0.84	0.003 (0.01)	0.76	−0.002(0.01)	0.72
Vitality (dead)	0.87 (0.50)	0.08	−0.76 (0.59)	0.21	−0.16 (0.45)	0.73
CI	0.03 (0.04)	0.50	0.03 (0.06)	0.59	0.02 (0.03)	0.50
M	−0.24 (0.28)	0.40	0.29 (0.36)	0.42	0.15 (0.24)	0.53
DBH × Vitality (dead)	−0.03 (0.01)	**0.006**	0.03 (0.01)	**0.05**	0.004 (0.01)	0.68
CI × Vitality (dead)	−0.05 (0.05)	0.37	0.01 (0.07)	0.91	0.04 (0.05)	0.38
M × Vitality (dead)	0.24 (0.39)	0.55	−0.13 (0.50)	0.79	−0.10 (0.35)	0.76
$SPEI3_{Jan}$ × Vitality (dead)	0.07 (0.19)	0.72	−0.10 (0.18)	0.58	0.02 (0.17)	0.88
Random Effects						
σ_i^2 *	0.16		0.15		0.12	
N	61		61		61	
Observations	122		120		121	
Marginal R^2/Conditional R^2	0.150/0.154		0.108/0.333		0.044/0.044	

* σ_i^2 represents the mean random effect variance of the model. R-squared values are based on Nakagawa et al. [51].

3.3. Integrating Predictors of Tree Mortality

The visual representation of the covariation among the predictors highlighted variables with greater centrality (i.e., number of significant correlations with other predictors) for the dead trees (Table 4). However, the expected covariation between the drought performance and neighbourhood characteristics was only evidenced for the vital trees (Figure 4a).

Despite the predictable significant correlations of variables within the groups (particularly within drought response and tree characteristic group), network analysis showed that the neighbour pressure predictors were related to drought response in the vital trees. The percentage of subcanopy species (%sub), *A. chilensis* (%Au), or dead *N. dombeyi* (%NDT) in the neighbourhood showed a positive relationship with *N. dombeyi* drought performance (Figure 4a). Furthermore, the vital trees experienced greater sensitivity to precipitation under increased competitive pressure, while if the neighbourhood was dominated by *N dombeyi*, temperature sensitivity become more relevant (more negative). Thus, drought performance following droughts (Rc and Rs) covaried positively with a neighbourhood dominated by species taxonomically and functionally different from the focal species (i.e. %Au; Figure 4a).

Table 4. Studied variables and values for degree centrality (number of edges per node) in the established network.

Variables Tree Vitality	Group	Centrality Vital	 Dead
Diameter at breast height (DBH)	Tree characteristic	2	4
25-year mean BAI (BAI25)	Tree characteristic	1	5
Tree height (ht)	Tree characteristic	2	-
Crown spread (W)	Tree characteristic	1	1
Temperature relationship (T)	Climate–growth	1	2
Precipitation relationship (P	Climate–growth	2	1
Resistance (Rt)	Drought response indices	2	3
Recovery (Rc)	Drought response indices	4	4
Resilience (Rs)	Drought response indices	2	1
10-year post-drought response (leg98)	Drought response indices	2	1
Competition index (CI)	Neighbour pressure	1	1
Percentage of *N. dombeyi* (%Nd)	Neighbour pressure	3	5
Percentage of co-dominant broadleaf (%Bl)	Neighbour pressure	1	1
Percentage of *A. chilensis* (%Au)	Neighbour pressure	3	1
Percentage of subcanopy species (%sub) *	Neighbour pressure	2	2
Percentage of dead *N. dombeyi* (%NDT)	Neighbour pressure	4	1

* *Raukaua laetevirens* (Gay) Frodin, *Luma apiculate* (DC.) Burret, *Maytenus magellanica* (Lam.) Hook. f., *Lomatia hirsuta* (Lam.) Diels.

For dead trees, network analysis showed that BAI25 and the percentage of *N. dombeyi* in the neighbourhood (%Nd) were the variables with the highest centrality (Table 4). As expected, BAI25 was positively related to DBH; and the drought performance predictors covaried with each other (Figure 4b). The variables with higher centrality were negatively related, indicating that the dead trees that had grown less (lower BAI25) were surrounded by a higher proportion of *N. dombeyi* and the neighbourhood experienced higher levels of 2014–15 tree mortality. Also, smaller trees (low DBH and low BAI25) experienced lower Rt values (values far from 1) during the 1998 and 2008 droughts, and had higher dependence on growing season precipitation (Figure 4b). In contrast, bigger dead trees had endured previous droughts and demonstrated higher recoveries, and, indirectly, higher resilience values. Finally, the dead trees that experienced higher competition pressure were related to more negative temperature sensitivity. However, these predictors did not depict covariation with any other variable in the network.

(a) Vital

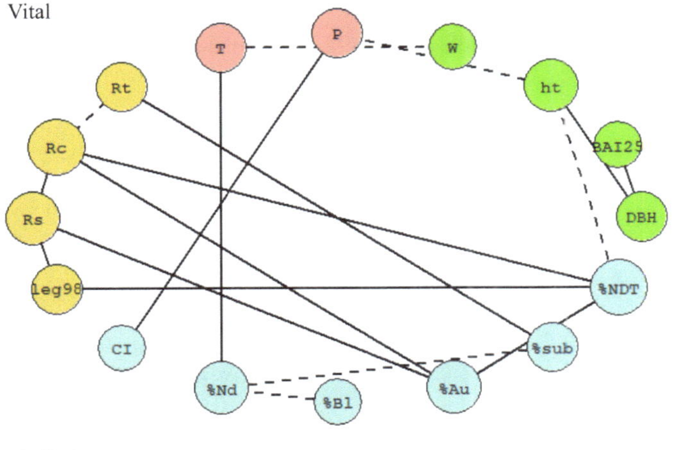

Figure 4. *Cont.*

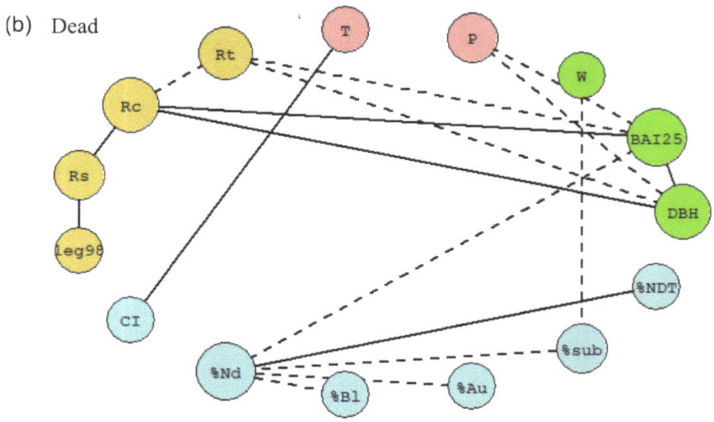

Figure 4. Visual representation of the significant correlation between tree level characteristics (diameter at breast height = DBH, 25-year mean BAI = BAI25, tree height = ht, crown spread = W green circles), climate–growth relationships (P = with precipitation, T = with temperature, light pink circles), drought response indices (resistance = Rt, recovery = Rc, resilience = Rs, 10-year post-drought response following 1998 drought = leg98, brown circles), and neighbour pressure (competition index = CI, percentage of *N. dombeyi* = %Nd, percentage of co-dominant broadleaf = %Bl, percentage of *A. chilensis* = %Au, percentage of subcanopy species = %sub, percentage of dead *N. dombeyi* = %NDT light blue circles). Solid lines (edges) represent positive correlations and dashed lines negative correlations. Circle (nodes) size represents degree of centrality. The top panel (**a**) corresponds to network representation of vital trees, and the bottom panel (**b**) to dead ones. Please refer to the web version of this article for an interpretation of the reference colours.

4. Discussion

In this study, we have observed how long-term growth trends, climate sensitivity, response to previous droughts, and neighbourhood structure constitute a complex network of predictors of *N. dombeyi* growth when the species develops in highly diverse and wet environments. *N. dombeyi* is a species with high safety margins, and, like other *Nothofagus* species, its leaves are more vulnerable to embolism than its stems [52]. Thus, death appears to be a consequence of the total loss of leaf hydraulic conductance, which leads to sudden leaf dehydration and leaf drop under stress conditions [53]. This mechanism seems a plausible explanation for what happened during the 2014–15 drought, as dead trees were characterized by entirely desiccated crowns (brownish leaves) in the late summer of 2015. However, rather than elucidate which potential predictors could help us to estimate death following 2014–15 drought, the vital and dead co-occurring tree differences were mainly associated with the size-related variables (BAI, DBH, height). Contrary to our expectations, we did not find divergent growth trajectories in the co-occurring vital and dead trees. Thus, the trees impacted by the 2014–15 drought had not experienced previous growth decline or crown dieback, highlighting the sudden nature of the mortality event and its relationship with other driving factors rather than previous vigour loss. In addition, the drought response, in terms of the resilience components during the previous droughts, was also not related to tree death during 2014–15 drought. Finally, the dead trees grew in neighbourhoods that were neither denser nor less diverse than those of the vital trees, factors which could have been expected to contribute to the probability of the trees dying.

4.1. Tree Size and Long-Term Growth in Relation to Drought Response

Several studies have predicted that taller and dominant trees will be at greater risk of dying following droughts due to their inherent vulnerability to hydraulic stress, higher radiation, and evaporative demand experienced by their exposed crowns [54–56]. This idea coupled with greater leaf size and stomata amount in *N. dombeyi* genotypes from humid environments [57], led us to predict a greater risk for bigger *N. dombeyi* individuals in the Puerto Blest mortality event. However, our results were in agreement with others [58,59] who found a greater survivorship of bigger trees. Despite a low number of studies recreating in detail the leaf–soil moisture dynamics during droughts [60], we may speculate that lower growth (and lower size) in dead trees could probably be a consequence of less favourable microsite conditions that were acting to limit growth [28]. This coincides with the fact that, although the vital and dead trees showed comparable responses to precipitation and temperature, the trees that died during the 2014–15 drought had experienced more dependence on the previous year's climatic conditions. Mortality tended to be higher in small-sized *N. dombeyi* individuals [61] with smaller basal area increments and more dependence on the previous year's carbon reserves, probably due to smaller rooting systems related to shallow soils or a higher degree of rockiness at the microsite level [59]. The potentially less favourable microsite conditions prevailing since the tree establishment may have led the now-dead *N. dombeyi* trees to exhibit lower growth performance during their lifetime, although the extremely wet climate did not exacerbate the growth-limiting factors enough to drive differences in noticeable climatic sensitivity between the vital and dead trees. Moreover, less favourable microsite conditions, serving as an underlying inciting factor at the tree level, are consistent with the observation that the dead *N. dombeyi* trees had a larger number of 2014–15 dead trees in their vicinity.

Even though the dead trees were smaller compared to the vital trees, our results strongly highlight the similar long-term growth trend and variability between the co-occurring vital and dead trees. A long-term reduction in radial growth (vigour loss) is a common growth pattern of trees that died following severe droughts [4,62,63], pointing to the so-called 'legacy effect' as a framework for explaining tree death [64–66]. Physiological changes can lead to higher drought impact and reduce growth recovery (loss of resilience capacity) in subsequent droughts, ultimately resulting in tree death [11]. This legacy effect has been identified in *N. dombeyi* trees that died following severe droughts in northern Patagonia [28]. Moreover, a study considering highly to fully defoliated *N. dombeyi* trees following the 1998–99 drought affecting the easternmost populations exhibited a significant growth decline that was related to reduced growth and a negative post-drought response during previous droughts (Suarez ML submitted manuscript). This increasing loss in resilience points to an irreversible increase in drought vulnerability in the more recent droughts, indicating a state of change that denotes a negative legacy effect, although the dead *N. dombeyi* trees growing in the relict Valdivian rainforest near Puerto Blest neither showed previous growth decline nor a long-term loss of growth resilience. Furthermore, the vital and 2014–15 dead trees showed similar resistance to drought, and rapid growth recovery upon rewatering when previous droughts ended. This represented an unexpected result considering the behaviour of the species in the eastern populations. Nevertheless, it would be simplistic to assume that the loss of resilience alone can explain tree death probability.

4.2. Neighbour Effect on Tree Growth and Drought Performance

Bigger and taller surviving *N. dombeyi* trees appear to be surrounded by a more species-mixed environment (Figure 2). Besides other factors, species mingling in the tree vicinity exhibiting different functional traits is assumable to imply different resource utilization during drought, attenuating negative growth reactions [9,27,67–69]. Inter-specific differences in physiology (e.g., isohydric vs. anisohydric) and morphology, as well as intra-specific differences resulting from intra- and inter-specific interactions (e.g., vertical positioning within the canopy or phenotypes), contribute to a positive effect on radial growth and performance in the face of drought [17,27,67], while the opposite could be attributed to a less complementary neighbourhood. In the studied stand, better drought performance during the previous dry spells was shaped by a neighbourhood dominated by a shade-tolerant conifer species with different physiological and morphological traits compared to

N. dombeyi [70], and/or species with different maximum heights as a species trait, leading to differences in resource consumption strategies.

However, the relationship between neighbourhood pressure (competition and mingling) and the previous drought performance of the now-dead *N. dombeyi* trees was less straightforward. Furthermore, competition levels neither explained part of the recent growth trends at local scales, nor implied negative growth trends in the dead trees. This result was in agreement with studies suggesting that the current competition levels (last decades) have a low effect on growth, potentially due to underestimated past competition [62,71] or due to pressure from neighbours, not considering the critical role played by neighbours' identity in plant performance [72]. The beneficial effects of mingling have been shown to strongly influence tree growth [25], though its role in lessening the severity of drought impacts and survival is still controversial [24,27]. Here, we use the basal area increment as an indicator of the tree's stress during its lifetime. Thereby, we can assume that the current species mingling arrangement in the neighbourhood of the now-dead trees had enhanced the tree growth, as the trees with greater species mingling in their neighbourhood showed higher estimated BAI values during the last 25 years. However, bigger trees, which are assumed to be benefited by the species mixture in their surroundings, displayed lower resistance (closer to 0.5) and even higher recovery following past droughts, highlighting that the net effects of neighbourhood pressure can vary, resulting in a different outcome in drought performance. It is noteworthy that the strength of the correlation between the factors could be constrained by the number of analysed individuals.

Finally, it is of particular interest that higher competitive pressure (measured as size–distance competition index) or fewer species mingling in the environment (measured as a neighbourhood dominated by *N. dombeyi*) covaried with the climatic sensitivity of the vital and dead trees. By considering the intra-population variability in the growth responses to the climate, we can capture the range of responses to drought, recognizing which local neighbours could be more resilient or vulnerable than others, responses that are often averaged or masked when we focus on competition or mixing influences at plot or stand level [73,74]. At first glance, the climate–growth responses varied among the trees, with more climate sensitivity as the neighbourhood pressure increased in both vitality classes (neighbourhoods dominated by *N. dombeyi* or greater size–distance competition levels) The intra-population variability in the *N. dombeyi* climate–growth relationship could be attributed to xylem anatomy variability or root development, modulated by the balance between the intra- and inter-specific relationships in the neighbourhood, soil conditions, and genetic uniqueness. Therefore, our findings agree with studies pointing out that trees with high competition in the close vicinity show greater climate sensitivity and could potentially be more prone to drought-induced mortality [18,71,75]. This suggests that the local context, in terms of the micro-environmental factors and species arrangement in the vicinity, interact in a complex way, impacting the resilience and adaptability of tree populations. Understanding the dynamics of this network can be crucial for effective forest management and conservation efforts.

5. Conclusions

Our findings shed light on the possible drivers acting at the tree-to-tree level in driving growth, resilience, and death following the 2014–15 drought in the northern Patagonian rainforest. Despite differences in factor covariation between the two vitality classes, here we have demonstrated that smaller trees were more susceptible to drought-induced mortality which is likely to be due to the worse microsite conditions they experienced during their lifetime. However, contrary to our expectations, based on the potential vigour loss in the dead trees, both vitality classes showed similar long-term growth trends, with no significant prior growth declines characterizing the dead trees. This highlights the sudden nature of mortality events, rather than a gradual loss of vigour, and the role played by other driving factors. Furthermore, our hypothesis posited increased neighbourhood diversity or a less dense vicinity as driving factors influencing both individual growth patterns and a tree's response to drought, enhancing tree resilience. However, partially agreeing with our initial predictions, the results showed that neighbour pressure was related to drought response in the vital trees. It is worth mentioning that the ecological relationships between individual trees and their neighbours are more complex than those we evidenced, and the positive effect of a less uniform neighbourhood on the drought performance of the vital trees helps us to forecast *N. dombeyi* drought responses under the prevailing climate in humid diverse stands. These results constitute a useful tool in planning effective forest management and conservation efforts, considering the complex interplay of tree size, growth history, neighbourhood composition, and micro-environmental factors to enhance the resilience and adaptability of tree populations in future dry spells.

Supplementary Materials: The following supporting information can be downloaded at: https://www.mdpi.com/article/10.3390/f15081355/s1, Figure S1: Puerto Blest Climograph; Figure S2: Bootstrapped climate–growth correlation coefficients; Figure S3: Integration period of resilience values; Table S1: ANOVA table of integration period model.

Author Contributions: Conceptualization and methodology, M.L.S. and Y.S.; tree ring measurements and dendroecological and network analysis, M.L.S.; data analysis, M.L.S. and Y.S.; writing—original draft preparation, M.L.S.; review and editing, M.L.S., Y.S., and L.F.; funding acquisition, M.L.S. and Y.S. All authors have read and agreed to the published version of the manuscript.

Funding: The study has been funded by Fondo para la Investigación Científica y Tecnológica (AGENCIA I+D+i) projects PICT 2017-3843 and PICT 2021-0187; and by CONICET project PIP 2017-0484. M.L.S. is grateful to Idea Wild (Fort Collins, CO) for providing essential equipment.

Data Availability Statement: The data in this study are available from the authors upon reasonable request.

Acknowledgments: We acknowledge logistical support from the personnel of the Nahuel Huapi National Park. The authors are also grateful to the technical staff of INIBIOMA for their commitment and dedication to the fieldwork.

Conflicts of Interest: The authors declare no conflicts of interest.

Appendix A. Vital and Dead Tree Characteristics

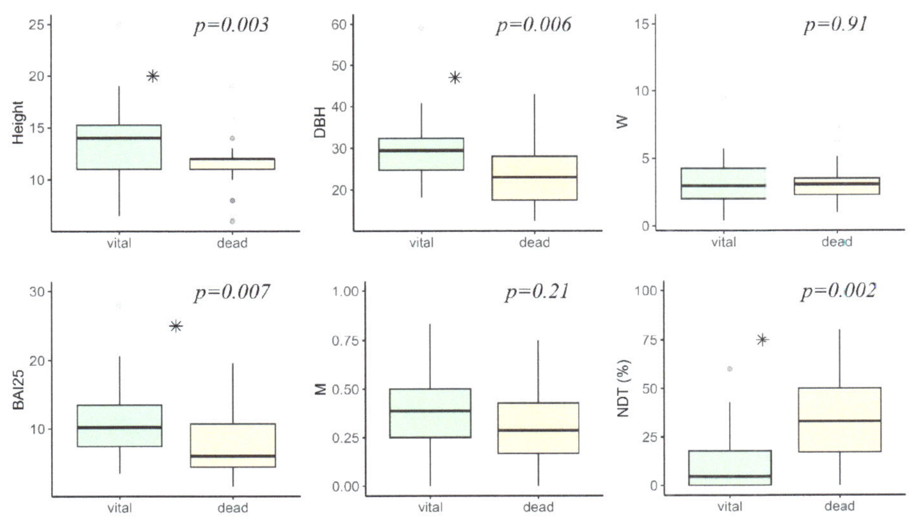

Figure A1. Differences in tree characteristics between vital (light green) and dead (light orange) *N. dombeyi* trees. W depicts average crown spread, M is the mingling index, and NDT is the proportion of dead trees in the focal neighbourhood. Asterisks and *p*-values depict differences according to Wilcoxon rank sum test.

Appendix B. Partial Effect of the General Additive Linear Model

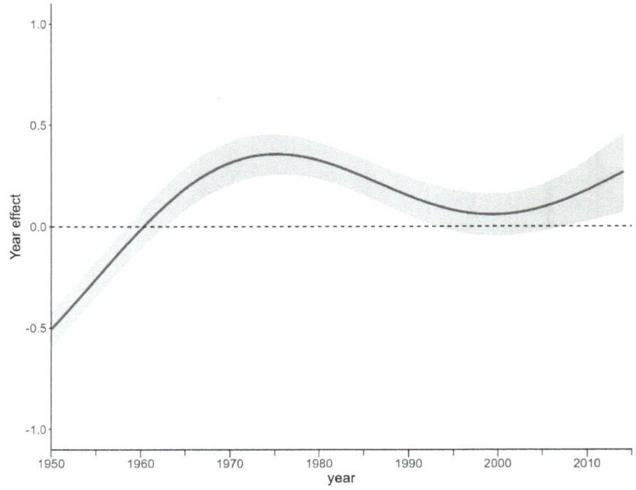

Figure A2. Effect of calendar year on growth trends (BAI) of *N. dombeyi* regardless of vital conditions

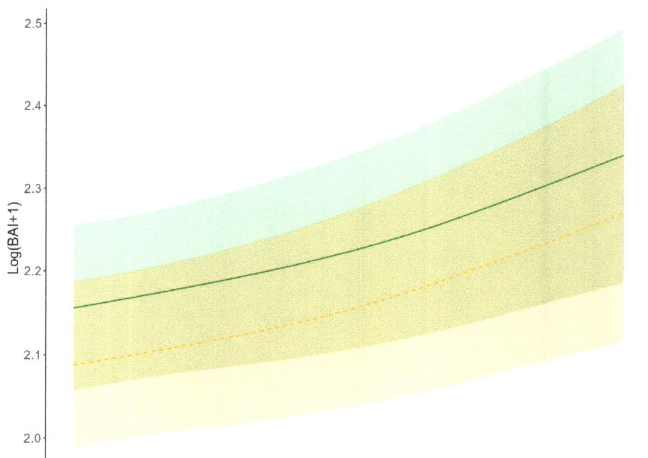

Figure A3. Predicted log (BAI + 1) for vital (solid green lines) and dead (dashed orange lines *N. dombeyi* trees in function of changes in $SPEI3_{Jan}$ according to GAMM model. The shaded area depict the 95% confidence intervals.

References

1. Lee, H.; Calvin, K.; Dasgupta, D.; Krinner, G.; Mukherji, A.; Thorne, P.; Trisos, C.; Romero, J.; Aldunce, P.; Barret, K. *IPCC, 2023: Cl mate Change 2023: Synthesis Report, Summary for Policymakers. Contribution of Working Groups I, II and III to the Sixth Assessment Repor of the Intergovernmental Panel on Climate Change*; Core Writing Team, Lee, H., Romero, J., Eds.; IPCC: Geneva, Switzerland, 2023
2. Aguayo, R.; León-Muñoz, J.; Garreaud, R.; Montecinos, A. Hydrological Droughts in the Southern Andes (40–45 S) from a Ensemble Experiment Using CMIP5 and CMIP6 Models. *Sci. Rep.* **2021**, *11*, 5530. [CrossRef] [PubMed]
3. Urrutia-Jalabert, R.; Barichivich, J.; Rozas, V.; Lara, A.; Rojas, Y.; Bahamondez, C.; Rojas-Badilla, M.; Gipoulou-Zuñiga, T Cuq, E. Climate Response and Drought Resilience of Nothofagus Obliqua Secondary Forests across a Latitudinal Gradient i South-Central Chile. *For. Ecol. Manag.* **2021**, *485*, 118962. [CrossRef]

4. Rodríguez-Catón, M.; Villalba, R.; Srur, A.; Williams, A.P. Radial Growth Patterns Associated with Tree Mortality in Nothofagus Pumilio Forest. *Forests* **2019**, *10*, 489. [CrossRef]
5. Fasanella, M.; Suarez, M.L.; Hasbun, R.; Premoli, A.C. Individual-Based Dendrogenomic Analysis of Forest Dieback Driven by Extreme Droughts. *Can. J. For. Res.* **2021**, *51*, 420–432. [CrossRef]
6. Trugman, A.T.; Anderegg, L.D.L.; Anderegg, W.R.L.; Das, A.J.; Stephenson, N.L. Why Is Tree Drought Mortality so Hard to Predict? *Trends Ecol. Evol.* **2021**, *36*, 520–532. [CrossRef] [PubMed]
7. González de Andrés, E.; Suárez, M.L.; Querejeta, J.I.; Camarero, J.J. Chronically Low Nutrient Concentrations in Tree Rings Are Linked to Greater Tree Vulnerability to Drought in Nothofagus Dombeyi. *Forests* **2021**, *12*, 1180. [CrossRef]
8. Mathias, J.M.; Thomas, R.B. Global Tree Intrinsic Water Use Efficiency Is Enhanced by Increased Atmospheric CO_2 and Modulated by Climate and Plant Functional Types. *Proc. Natl. Acad. Sci. USA* **2021**, *118*, e2014286118. [CrossRef] [PubMed]
9. Castagneri, D.; Vacchiano, G.; Hacket-Pain, A.; DeRose, R.J.; Klein, T.; Bottero, A. Meta-Analysis Reveals Different Competition Effects on Tree Growth Resistance and Resilience to Drought. *Ecosystems* **2022**, *25*, 30–43. [CrossRef]
10. Cailleret, M.; Dakos, V.; Jansen, S.; Robert, E.M.; Aakala, T.; Amoroso, M.M.; Antos, J.A.; Bigler, C.; Bugmann, H.; Caccianiga, M. Early-Warning Signals of Individual Tree Mortality Based on Annual Radial Growth. *Front. Plant Sci.* **2019**, *9*, 1964. [CrossRef]
11. Peltier, D.M.P.; Ogle, K. Legacies of More Frequent Drought in Ponderosa Pine across the Western United States. *Glob. Change Biol.* **2019**, *25*, 3803–3816. [CrossRef]
12. Serra-Maluquer, X.; Granda, E.; Camarero, J.J.; Vilà-Cabrera, A.; Jump, A.S.; Sánchez-Salguero, R.; Sangüesa-Barreda, G.; Imbert, J.B.; Gazol, A. Impacts of Recurrent Dry and Wet Years Alter Long-term Tree Growth Trajectories. *J. Ecol.* **2021**, *109*, 1561–1574. [CrossRef]
13. Cabon, A.; DeRose, R.J.; Shaw, J.D.; Anderegg, W.R.L. Declining Tree Growth Resilience Mediates Subsequent Forest Mortality in the US Mountain West. *Glob. Change Biol.* **2023**, *29*, 4826–4841. [CrossRef] [PubMed]
14. Serra-Maluquer, X.; Gazol, A.; Anderegg, W.R.L.; Martínez-Vilalta, J.; Mencuccini, M.; Camarero, J.J. Wood Density and Hydraulic Traits Influence Species' Growth Response to Drought across Biomes. *Glob. Change Biol.* **2022**, *28*, 3871–3882. [CrossRef] [PubMed]
15. Pretzsch, H.; Dieler, J. The Dependency of the Size-Growth Relationship of Norway Spruce (*Picea abies* [L.] Karst.) and European Beech (*Fagus sylvatica* [L.]) in Forest Stands on Long-Term Site Conditions, Drought Events, and Ozone Stress. *Trees* **2011**, *25*, 355–369. [CrossRef]
16. Schmitt, A.; Trouvé, R.; Seynave, I.; Lebourgeois, F. Decreasing Stand Density Favors Resistance, Resilience, and Recovery of *Quercus petraea* Trees to a Severe Drought, Particularly on Dry Sites. *Ann. For. Sci.* **2020**, *77*, 52. [CrossRef]
17. Bello, J.; Vallet, P.; Perot, T.; Balandier, P.; Seigner, V.; Perret, S.; Couteau, C.; Korboulewsky, N. How Do Mixing Tree Species and Stand Density Affect Seasonal Radial Growth during Drought Events? *For. Ecol. Manag.* **2019**, *432*, 436–445. [CrossRef]
18. Serra-Maluquer, X.; Gazol, A.; Igual, J.M.; Camarero, J.J. Silver Fir Growth Responses to Drought Depend on Interactions between Tree Characteristics, Soil and Neighbourhood Features. *For. Ecol. Manag.* **2021**, *480*, 118625. [CrossRef]
19. Anderegg, L.D.L.; HilleRisLambers, J. Drought Stress Limits the Geographic Ranges of Two Tree Species via Different Physiological Mechanisms. *Glob. Change Biol.* **2016**, *22*, 1029–1045. [CrossRef] [PubMed]
20. Zapater, M.; Hossann, C.; Bréda, N.; Bréchet, C.; Bonal, D.; Granier, A. Evidence of Hydraulic Lift in a Young Beech and Oak Mixed Forest Using 18O Soil Water Labelling. *Trees* **2011**, *25*, 885–894. [CrossRef]
21. Forrester, D.I. Linking Forest Growth with Stand Structure: Tree Size Inequality, Tree Growth or Resource Partitioning and the Asymmetry of Competition. *For. Ecol. Manag.* **2019**, *447*, 139–157. [CrossRef]
22. González De Andrés, E.; Seely, B.; Blanco, J.A.; Imbert, J.B.; Lo, Y.; Castillo, F.J. Increased Complementarity in Water-limited Environments in Scots Pine and European Beech Mixtures under Climate Change. *Ecohydrol. Hydrobiol.* **2017**, *10*, e1810. [CrossRef]
23. Liu, C.L.C.; Kuchma, O.; Krutovsky, K.V. Mixed-Species versus Monocultures in Plantation Forestry: Development, Benefits, Ecosystem Services and Perspectives for the Future. *Glob. Ecol. Conserv.* **2018**, *15*, e00419. [CrossRef]
24. Gillerot, L.; Forrester, D.I.; Bottero, A.; Rigling, A.; Lévesque, M. Tree Neighbourhood Diversity Has Negligible Effects on Drought Resilience of European Beech, Silver Fir and Norway Spruce. *Ecosystems* **2021**, *24*, 20–36. [CrossRef]
25. Schwarz, J.A.; Bauhus, J. Benefits of Mixtures on Growth Performance of Silver Fir (*Abies alba*) and European Beech (*Fagus sylvatica*) Increase with Tree Size without Reducing Drought Tolerance. *Front. For. Glob. Change* **2019**, *2*, 79. [CrossRef]
26. Sánchez-Salguero, R.; Camarero, J.J.; Rozas, V.; Génova, M.; Olano, J.M.; Arzac, A.; Gazol, A.; Caminero, L.; Tejedor, E.; De Luis, M.; et al. Resist, Recover or Both? Growth Plasticity in Response to Drought Is Geographically Structured and Linked to Intraspecific Variability in *Pinus pinaster*. *J. Biogeogr.* **2018**, *45*, 1126–1139. [CrossRef]
27. Vitali, V.; Forrester, D.I.; Bauhus, J. Know Your Neighbours: Drought Response of Norway Spruce, Silver Fir and Douglas Fir in Mixed Forests Depends on Species Identity and Diversity of Tree Neighbourhoods. *Ecosystems* **2018**, *21*, 1215–1229. [CrossRef]
28. Suarez, M.L.; Ghermandi, L.; Kitzberger, T. Factors Predisposing Episodic Drought-Induced Tree Mortality in Nothofagus– Site, Climatic Sensitivity and Growth Trends. *J. Ecol.* **2004**, *92*, 954–966. [CrossRef]
29. Suarez, M.L.; Kitzberger, T. Differential Effects of Climate Variability on Forest Dynamics along a Precipitation Gradient in Northern Patagonia. *J. Ecol.* **2010**, *98*, 1023–1034. [CrossRef]
30. Suarez, M.L.; Villalba, R.; Mundo, I.A.; Schroeder, N. Sensitivity of Nothofagus Dombeyi Tree Growth to Climate Changes along a Precipitation Gradient in Northern Patagonia, Argentina. *Trees* **2015**, *29*, 1053–1067. [CrossRef]
31. Perotti, M.G.; Diéguez, M.C.; Jara, F.G. State of the Knowledge of Moist Soils of the Patagónico North (Argentina): Excellent Aspects and Importance for the Conservation of the Regional Biodiversity. *Rev. Chil. Hist. Nat.* **2005**, *78*, 723–737.

32. Kitzberger, T. *Fire Regime Variation along a Northern Patagonian Forest-Steppe Gradient: Stand and Landscape Response*; University of Colorado, Department of Geography: Boulder, CO, USA, 1994.
33. Ezcurra, C.; Brion, C. *Plantas Del Nahuel Huapi Catálogo de La Flora Vascular Del Parque Nacional Nahuel Huapi, Argentina*; Universidad Nacional del Comahue, Red Latinoamericana de Botánica: Bariloche, Argentina, 2005.
34. Harris, I.; Osborn, T.J.; Jones, P.; Lister, D. Version 4 of the CRU TS Monthly High-Resolution Gridded Multivariate Climate Dataset. *Sci. Data* **2020**, *7*, 109. [CrossRef]
35. Beguería, S.; Vicente-Serrano, S.M. SPEI: Calculation of the Standardised Precipitation-Evapotranspiration Index. R Package Version 1.7. 2017. Available online: https://CRAN.R-project.org/package=SPEI (accessed on 12 October 2022).
36. Vicente-Serrano, S.M.; Beguería, S.; López-Moreno, J.I. A Multiscalar Drought Index Sensitive to Global Warming: The Standardized Precipitation Evapotranspiration Index. *J. Clim.* **2010**, *23*, 1696–1718. [CrossRef]
37. Zang, C.; Biondi, F. Treeclim: An R Package for the Numerical Calibration of Proxy-Climate Relationships. *Ecography* **2015**, *38*, 431–436. [CrossRef]
38. Holmes, R.L. Computer-Assisted Quality Control in Tree-Ring Dating and Measurement. *Tree-Ring Bull.* **1983**, *43*, 69–75.
39. Biondi, F.; Qeadan, F. A Theory-Driven Approach to Tree-Ring Standardization: Defining the Biological Trend from Expected Basal Area Increment. *Tree* **2008**, *64*, 81–96. [CrossRef]
40. Bunn, A.; Korpela, M.; Biondi, F.; Campelo, F.; Mérian, P.; Qeadan, F.; Zang, C. dplR: Dendrochronology Program Library in R. R Package Version 1.7.6. 2020. Available online: https://CRAN.R-project.org/package=dplR (accessed on 23 June 2022).
41. R Core Team. *R: A Language and Environment for Statistical Computing*; R Core Team: Vienna, Austria, 2021.
42. Wood, S.N. *Generalized Additive Models: An Introduction with R*; CRC Press: Boca Raton, FL, USA, 2017.
43. Wood, S.; Wood, M.S. Package 'Mgcv'. *R Package Version* **2015**, *1*, 29.
44. Hui, G.; Wang, Y.; Zhang, G.; Zhao, Z.; Bai, C.; Liu, W. A Novel Approach for Assessing the Neighborhood Competition in Two Different Aged Forests. *For. Ecol. Manag.* **2018**, *422*, 49–58. [CrossRef]
45. Von Gadow, K.; Hui, G.Y. Characterizing Forest Spatial Structure and Diversity. In *Sustainable Forestry in Temperate Regions*; Björk, L., Ed.; SUFOR, University of Lund: Lund, Sweden, 2002; pp. 20–30.
46. Lloret, F.; Keeling, E.G.; Sala, A. Components of Tree Resilience: Effects of Successive Low-Growth Episodes in Old Ponderosa Pine Forests. *Oikos* **2011**, *120*, 1909–1920. [CrossRef]
47. Thurm, E.A.; Uhl, E.; Pretzsch, H. Mixture Reduces Climate Sensitivity of Douglas-Fir Stem Growth. *For. Ecol. Manag.* **2016**, *376*, 205–220. [CrossRef]
48. Pinheiro, J.; Bates, D.; DebRoy, S.; Sarkar, D.; Team, R.C. Nlme: Linear and Nonlinear Mixed Effects Models. *R Package Version* **2013**, *3*, 111.
49. Ramirez, K.S.; Geisen, S.; Morriën, E.; Snoek, B.L.; van der Putten, W.H. Network Analyses Can Advance Above-Belowground Ecology. *Trends Plant Sci.* **2018**, *23*, 759–768. [CrossRef]
50. Csardi, G.; Nepusz, T. The Igraph Software Package for Complex Network Research. *Int. J. Complex Syst.* **2006**, *1695*, 1–9.
51. Nakagawa, S.; Johnson, P.C.; Schielzeth, H. The Coefficient of Determination R 2 and Intra-Class Correlation Coefficient from Generalized Linear Mixed-Effects Models Revisited and Expanded. *J. R. Soc. Interface* **2017**, *14*, 20170213. [CrossRef] [PubMed]
52. Bucci, S.J.; Scholz, F.G.; Campanello, P.I.; Montti, L.; Jimenez-Castillo, M.; Rockwell, F.A.; Manna, L.L.; Guerra, P.; Bernal, P.L.; Troncoso, O. Hydraulic Differences along the Water Transport System of South American Nothofagus Species: Do Leaves Protect the Stem Functionality? *Tree Physiol.* **2012**, *32*, 880–893. [CrossRef] [PubMed]
53. Scholz, F.G.; Bucci, S.J.; Goldstein, G. Strong Hydraulic Segmentation and Leaf Senescence Due to Dehydration May Trigger Die-Back in Nothofagus Dombeyi under Severe Droughts: A Comparison with the Co-Occurring Austrocedrus Chilensis. *Trees* **2014**, *28*, 1475–1487. [CrossRef]
54. Bennett, A.C.; McDowell, N.G.; Allen, C.D.; Anderson-Teixeira, K.J. Larger Trees Suffer Most during Drought in Forests Worldwide. *Nat. Plants* **2015**, *1*, 15139. [CrossRef] [PubMed]
55. McDowell, N.G.; Allen, C.D. Darcy's Law Predicts Widespread Forest Mortality under Climate Warming. *Nat. Clim. Change* **2015**, *5*, 669–672. [CrossRef]
56. Leifsson, C.; Buras, A.; Klesse, S.; Baittinger, C.; Bat-Enerel, B.; Battipaglia, G.; Biondi, F.; Stajić, B.; Budeanu, M.; Čada, V.; et al. Identifying Drivers of Non-Stationary Climate-Growth Relationships of European Beech. *Sci. Total Environ.* **2024**, *937*, 173321. [CrossRef] [PubMed]
57. Diaz, D.G.; Ignazi, G.; Mathiasen, P.; Premoli, A.C. Climate-Driven Adaptive Responses to Drought of Dominant Tree Species from Patagonia. *New For.* **2022**, *53*, 57–80. [CrossRef]
58. Colangelo, M.; Camarero, J.J.; Borghetti, M.; Gazol, A.; Gentilesca, T.; Ripullone, F. Size Matters a Lot: Drought-Affected Italian Oaks Are Smaller and Show Lower Growth Prior to Tree Death. *Front. Plant Sci.* **2017**, *8*, 135. [CrossRef]
59. Galiano, L.; Martínez-Vilalta, J.; Sabaté, S.; Lloret, F. Determinants of Drought Effects on Crown Condition and Their Relationship with Depletion of Carbon Reserves in a Mediterranean Holm Oak Forest. *Tree Physiol.* **2012**, *32*, 478–489. [CrossRef]
60. Cranston, B.M.; Powers, B.F.; Macinnis-Ng, C. Inexpensive Throughfall Exclusion Experiment for Single Large Trees. *Appl. Plant Sci.* **2020**, *8*, e11325. [CrossRef]
61. Molowny-Horas, R.; Suarez, M.L.; Lloret, F. Changes in the Natural Dynamics of Nothofagus Dombeyi Forests: Population Modeling with Increasing Drought Frequencies. *Ecosphere* **2017**, *8*, e01708. [CrossRef]

62. Neycken, A.; Scheggia, M.; Bigler, C.; Lévesque, M. Long-Term Growth Decline Precedes Sudden Crown Dieback of European Beech. *Agric. For. Meteorol.* **2022**, *324*, 109103. [CrossRef]
63. Cailleret, M.; Jansen, S.; Robert, E.M.; Desoto, L.; Aakala, T.; Antos, J.A.; Beikircher, B.; Bigler, C.; Bugmann, H.; Caccianiga, M. A Synthesis of Radial Growth Patterns Preceding Tree Mortality. *Glob. Change Biol.* **2017**, *23*, 1675–1690. [CrossRef]
64. Kannenberg, S.A.; Novick, K.A.; Alexander, M.R.; Maxwell, J.T.; Moore, D.J.; Phillips, R.P.; Anderegg, W.R. Linking Drought Legacy Effects across Scales: From Leaves to Tree Rings to Ecosystems. *Glob. Change Biol.* **2019**, *25*, 2978–2992. [CrossRef]
65. Gazol, A.; Camarero, J.J.; Sánchez-Salguero, R.; Vicente-Serrano, S.M.; Serra-Maluquer, X.; Gutiérrez, E.; de Luis, M.; Sangüesa-Barreda, G.; Novak, K.; Rozas, V. Drought Legacies Are Short, Prevail in Dry Conifer Forests and Depend on Growth Variability. *J. Ecol.* **2020**, *108*, 2473–2484. [CrossRef]
66. Vilonen, L.; Ross, M.; Smith, M.D. What Happens after Drought Ends: Synthesizing Terms and Definitions. *New Phytol.* **2022**, *235*, 420–431. [CrossRef]
67. Chen, Y.; Wright, S.J.; Muller-Landau, H.C.; Hubbell, S.P.; Wang, Y.; Yu, S. Positive Effects of Neighborhood Complementarity on Tree Growth in a Neotropical Forest. *Ecology* **2016**, *97*, 776–785. [CrossRef]
68. Fichtner, A.; Härdtle, W.; Bruelheide, H.; Kunz, M.; Li, Y.; von Oheimb, G. Neighbourhood Interactions Drive Overyielding in Mixed-Species Tree Communities. *Nat. Commun.* **2018**, *9*, 1144. [CrossRef]
69. Pretzsch, H.; Schütze, G.; Uhl, E. Resistance of European Tree Species to Drought Stress in Mixed versus Pure Forests: Evidence of Stress Release by Inter-specific Facilitation. *Plant Biol.* **2013**, *15*, 483–495. [CrossRef]
70. Gyenge, J.E.; Fernández, M.E.; Dalla Salda, G.; Schlichter, T. Leaf and Whole-Plant Water Relations of the Patagonian Conifer Austrocedrus Chilensis (D. Don) Pic. Ser. et Bizzarri: Implications on Its Drought Resistance Capacity. *Ann. For. Sci.* **2005**, *62*, 297–302. [CrossRef]
71. Sangüesa-Barreda, G.; Camarero, J.J.; Oliva, J.; Montes, F.; Gazol, A. Past Logging, Drought and Pathogens Interact and Contribute to Forest Dieback. *Agric. For. Meteorol.* **2015**, *208*, 85–94. [CrossRef]
72. Grossiord, C. Having the Right Neighbors: How Tree Species Diversity Modulates Drought Impacts on Forests. *New Phytol.* **2020**, *228*, 42–49. [CrossRef]
73. Ford, K.R.; Breckheimer, I.K.; Franklin, J.F.; Freund, J.A.; Kroiss, S.J.; Larson, A.J.; Theobald, E.J.; HilleRisLambers, J. Competition Alters Tree Growth Responses to Climate at Individual and Stand Scales. *Can. J. For. Res.* **2017**, *47*, 53–62. [CrossRef]
74. Versace, S.; Gianelle, D.; Garfì, V.; Battipaglia, G.; Lombardi, F.; Marchetti, M.; Tognetti, R. Interannual Radial Growth Sensitivity to Climatic Variations and Extreme Events in Mixed-Species and Pure Forest Stands of Silver Fir and European Beech in the Italian Peninsula. *Eur. J. For. Res.* **2020**, *139*, 627–645. [CrossRef]
75. González De Andrés, E.; Camarero, J.J.; Blanco, J.A.; Imbert, J.B.; Lo, Y.; Sangüesa-Barreda, G.; Castillo, F.J. Tree-to-tree Competition in Mixed European Beech–Scots Pine Forests Has Different Impacts on Growth and Water-use Efficiency Depending on Site Conditions. *J. Ecol.* **2018**, *106*, 59–75. [CrossRef]

forests

MDPI

Article

Climate Adaptation in White Oak (*Quercus alba*, L.): A Forty-Year Study of Growth and Phenology

Austin M. Thomas [1,*], Mark V. Coggeshall [2,3], Philip A. O'Connor [4] and C. Dana Nelson [5,6,*]

1 Oak Ridge Institute for Science and Education (ORISE), Forest Health Research & Education Center, Southern Research Station, USDA Forest Service, Lexington, KY 40546, USA
2 School of Natural Resources, University of Missouri, Columbia, MO 65211, USA; coggeshallm@missouri.edu
3 Hardwood Tree Improvement and Regeneration Center, Northern Research Station, USDA Forest Service, West Lafayette, IN 47907, USA
4 Vallonia Tree Seedling Nursery, Indiana Division of Forestry, Vallonia, IN 47281, USA; poconnor@dnr.in.gov
5 Forest Health Research & Education Center, Southern Research Station, USDA Forest Service, Lexington, KY 40546, USA
6 Southern Institute of Forest Genetics, Southern Research Station, USDA Forest Service, Saucier, MS 39574, USA
* Correspondence: austin.thomas@uky.edu (A.M.T.); charles.d.nelson@usda.gov (C.D.N.)

Abstract: Climate change poses a significant threat to the resilience and sustainability of forest ecosystems. This study examines the performance of white oak (*Quercus alba*, L.) across a range of provenances in a common garden planting, focusing on the species' response to climatic variables and the potential role of assisted migration in forest management. We evaluated the survival and growth rates of white oak provenances originating from various points along a latitudinal gradient over a period of 40 years. These provenances were planted in a common garden situated near the midpoint of this latitudinal gradient, where we also monitored their phenological traits, such as budburst and leaf senescence. The results revealed substantial variation in phenological responses and growth patterns among the provenances, with southern provenances demonstrating faster growth and later senescence relative to local sources, with limited impact on survival. In contrast the northern provenances demonstrated slower growth, resulting in later-aged competition-induced mortality. The findings highlight the necessity of incorporating genetic diversity into white oak reforestation and conservation strategies, as the local provenance may no longer be the most suitable option for current and future conditions. We advocate for a nuanced approach to forest management that leverages genetic insights to optimize seed source selection for reforestation, fostering resilient forest landscapes in the face of ongoing climate shifts.

Keywords: white oak; *Quercus alba*; provenance test; common garden; phenology; assisted migration; climate change; tree growth

Citation: Thomas, A.M.; Coggeshall, M.V.; O'Connor, P.A.; Nelson, C.D. Climate Adaptation in White Oak (*Quercus alba*, L.): A Forty-Year Study of Growth and Phenology. *Forests* **2024**, *15*, 520. https://doi.org/10.3390/f15030520

Academic Editors: Yassine Messaoud, Jan Světlík and Giorgio Alberti

Received: 30 December 2023
Revised: 4 March 2024
Accepted: 4 March 2024
Published: 12 March 2024

1. Introduction

White oak (*Quercus alba*, L.) is an iconic deciduous tree prevalent throughout the hardwood forests of the eastern United States, extending from Maine to Florida and from Minnesota to Texas [1,2]. Although capable of thriving in diverse soils, it exhibits a particular preference for well-drained loamy soils that offer moderate moisture and fertility [2,3]. Throughout its range, this species is a key component of various forest types including oak-hickory, oak-pine, and mixed hardwoods [4,5]. As a keystone species, white oak provides essential ecosystem services, playing a pivotal role in carbon sequestration, oxygen generation, and water purification, all while forming the basis of intricate multi-trophic systems [5]. The resilience of white oak is exemplified by its longevity [6]. However, recent evidence suggests a concerning decline in white oak populations within eastern forests attributed to overharvesting, high grading, inconsistent regeneration, and poor recruitment [7–9].

The genus *Quercus* includes approximately 435–500 oak species native to temperate and subtropical regions [10,11]. Among these, white oak stands out as the most preferred of only three oak species widely used in liquid goods cooperage [11]. Its robust and durable heartwood is particularly prized for crafting bourbon barrels, which are subsequently repurposed in the production of other whiskeys and wine [12]. Given the commercial and ecological value of white oak, coupled with its regional decline [7], there is growing interest among federal, state, and private entities in white oak reforestation and tree improvement [13–15]. Despite the genus *Quercus* being known for its adaptability to environmental variation [16], such plasticity in white oak, specifically, has not been thoroughly investigated.

In light of ongoing global climate change, the consideration of both phenotypic and genetic factors influencing white oak performance is crucial. Assisted migration, and more specifically assisted population migration, i.e., the transfer of seeds from the southern part of a species' geographical range to the north (in the northern hemisphere) [17], is regarded as one way to assist a species' adaptation to global warming. Traditional forestry practices advocate for the use of local seed sources that are presumed to be well adapted to local conditions. However, these notions are challenged by ongoing rapid environmental changes, suggesting that southern genotypes might be better suited for northern locations in the future due to their adaptation to warmer climates [18]. Economically important forest tree species, especially those with large geographical ranges, often have ecotypes or substantive clines adapted to local site conditions. The phenotypic and genetic intraspecific variation of forest trees in the context of assisted migration have been less studied than interspecific variation and competition [19].

A comprehensive understanding of white oak's response to climatic variables over its lifespan is necessary for delineating seed zones and guiding assisted migration strategies in the context of climate change. The majority of eastern hardwood species are primarily experiencing a southern range retraction with evidence of a significant lag between temperature shifts and northward migration [20]. Dispersal models of white oak indicate that unassisted, the species' natural distribution will lag significantly behind the northward shift of the species' suitable climate envelope [21]. If white oak lacks the plasticity to adapt to rapid climate warming, then the assisted northward range expansion and population migration of provenances northward may be necessary to avoid substantial range retraction exacerbated by the species' particularly long life and limited natural dispersal abilities [22].

In this study, we investigate the survival, growth, and phenological traits of white oak provenances growing in a common garden planting. While provenance performance is ideally assessed at multiple planting sites with varying environmental conditions, this study's scope is confined to a single common garden near the center of the species range. We acknowledge conclusions drawn from this research must be made with caution but believe there is a pressing need for the quantitative assessment of white oak provenance performance. This common garden study represents, to the best of our knowledge, the only long-term investigation incorporating white oak genetics representative of most of the latitudinal range of the species, the foremost requirement of a provenance study [23]. Our analysis spans 40 years of tree growth measurements and increment core records complemented by spring leaf budburst and fall leaf senescence data collected in years 39 and 40. We used these phenology data to estimate the length of the growing season for each provenance under the contemporary climate conditions at the planting site.

We expected that the white oak provenances in this study would exhibit distinct growth, survival, and phenological responses to the environmental conditions at our common garden site. We predicted high mortality in the Mississippi provenance due to late spring and early fall freezes, and potential slower growth in northern provenances. These hypotheses reflected our anticipation of latitudinal influences on provenance performance. Some of our hypotheses were confirmed by the study outcomes, while others were contradicted by the data. Analyzing the growth and phenological traits of white oak seed source origins spanning a range of climatic conditions, this study underscores the critical role of adaptive genetic management in forest conservation, providing invaluable insights for tree breeders and forest managers as they navigate the challenges posed by climate change.

2. Materials and Methods

Two-year-old bare root seedlings grown at the Vallonia Nursery (Indiana Department of Natural Resources, Vallonia, IN, USA) were planted in 1983 at the Starve Hollow State Recreation Area on a site 2 km from the nursery. The common garden experiment was planted on level terrain, situated 180 meters north of Starve Hollow Lake, in the well-drained sandy-loam soils characteristic of the Bloomfield-Alvin complex [24]. The seedlings represented six distinct provenances, including southern Minnesota (MN), southern Wisconsin (WI), central Ohio (OH), southern Indiana (IN), southern Illinois (IL), and northern Mississippi (MS). Detailed coordinates, seed source information, and climate summaries for each provenance are provided in Table 1 and Figure 1. The selection of seed source locations was guided by two primary considerations. Firstly, an effort was made to ensure that these locations spanned the latitudinal range of white oak, with the intent of capturing most of the species' genetic diversity as it relates to climate adaptation represented by plant hardiness zones [25]. However, the predominant consideration, given the nature of white oak's recalcitrant seeds, was the location of accessible acorns in the fall of 1981 and the availability of a suitable planting site.

The common garden planting was arranged in a 12 row by 18 column grid (n = 216) with 2.5 meter tree spacing. This grid was divided into six randomized complete blocks, each containing nine four-tree row plots. Some provenances contained more than one plot (see Table 1). The southern Indiana seed sources were collected from two locations 100 km apart, each of which were planted in their own plot. Given their similar latitude and proximity, both Indiana sources were treated as a single provenance (IN) in our analysis. The southern Illinois seed sources consisted of three open pollinated families, coming from a single location, each of which were planted in their own plots. We also treated these Illinois sourced plots as a single provenance (IL) for the purposes of our analysis. A one-tree border row was planted around the whole experiment with a local seed source The experiment was thinned in year 12, halving the number of trees in each plot (n = 108 after thinning) to reduce tree-tree competition within and between plots. Tree mortality (natural and via thinning) was tracked throughout the study. Local tree density (number of orthogonal and diagonal neighbors) was included as a cofactor in all statistical models to account for the effects of neighbor-tree competition and canopy gaps left by tree mortality or thinning.

Table 1. Coordinates in decimal degrees and the USDA Plant Hardiness Zones [25] of provenances planted in the common garden study located near Vallonia, Indiana (latitude = 38.82 longitude = −86.08, elevation = 174.04 m).

Prov.	Lat.	Long.	Elev. (m)	USDA Zone 1990	USDA Zone 2023	Mother Trees per Prov.	Plots per Block	State	County	Place Name
MN	45.00	−93.10	256.34	4a	5a	10	1	Minnesota	Ramsey	University of Minnesota campus
WI	43.00	−89.30	269.44	4b	5a	10	1	Wisconsin	Dane	Nevin State Fish Hatchery
OH	40.28	−84.10	272.00	5a	6a	6	1	Ohio	Auglaize	Wapakoneta Fairgrounds
IN	39.06	−86.59	176.78	5a	6a–6b	6	2	Indiana	Greene, Jackson	Shakamak State Park, Vallonia Tree Seedling Nursery
IL	37.40	−89.13	183.25	6a	7a	3	3	Illinois	Jackson	Private residences in Makanda
MS	34.37	−88.36	120.40	7b	8a	8	1	Mississippi	Prentiss	Private residences south of Booneville

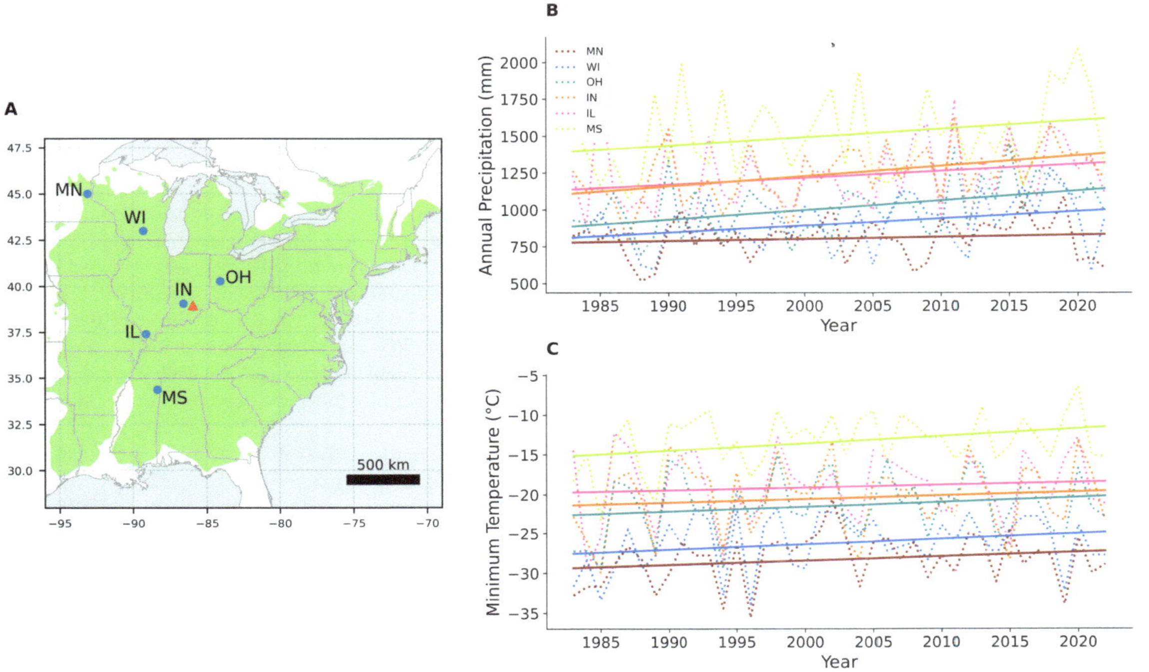

Figure 1. Locations of seed source origins and relevant climate information for each provenance. (**A**) Map of provenance locations. Solid gray lines indicate US state and Canadian provincial borders. Green shading indicates the natural range of white oak (*Quercus alba*, L.) [1]. Provenances are represented as labeled blue dots. The location of the common garden in southern Indiana is represented as a red triangle. (**B**) Annual precipitation (mm) of each provenance [26]. Solid lines represent the linear trend of each provenance with the corresponding color. (**C**) Annual mean absolute minimum temperatures (°C) of each provenance [26]. Again, solid lines represent the linear trend of each provenance with the corresponding color. The Indiana provenance and test location data are based on regional climate data covering both seed source locations and the common garden site.

Diameter at breast height (DBH, approximately 1.3 m) was measured to the nearest 0.1 inches (2.54 mm) in the 11th, 20th, and 40th years using a diameter tape. Annual DBH estimates were derived from two 5.15 mm diameter increment cores extracted from each tree in early spring 2022, oriented east-west and north-south, respectively, and taken as close to breast height as possible while avoiding any epicormic branching. Increment cores were mounted on poplar blocks and sanded smooth before precision dendrochronological measurements were made using a Velmex TA measurement system (Velmex, Inc., Bloomfield, NY, USA). The average radial growth from these two cores was calculated, adjusted for average bark thickness, and multiplied by two to represent annual DBH growth. These measurements were analyzed using nonlinear modeling to evaluate growth differences between provenances, employing the 'nlme' package in R [27,28]. Cumulative DBH growth curves were fit using the Schumacher growth model [29,30] (Equation (1)), where a = the growth asymptote, b = the growth scale factor, and x = the year.

$$y = a * e^{-b\frac{1}{x}} \quad (1)$$

The height of each tree was measured at the time of planting (year 0) as well as in the fall of years 1–5, year 20, and year 40. The height of trees in years 0–5 was measured via a measuring rod to the nearest cm, and in years 20 and 40 via a Forestry Pro II hypsometer

(Nikon, Minato, Tokyo) to the nearest 0.1 ft (3.048 cm). Again, a nonlinear model was used to discern statistical differences across provenances. Height growth curves were fit via the 3P Gompertz growth model (Equation (2)), where a = the growth asymptote, b = x-axis displacement (translation along the x-axis), c = the growth rate, and x = the year.

$$y = a * e^{-bc^{x}} \tag{2}$$

Phenological observations of spring leaf budburst were recorded in 2022 (year 39) and 2023 (year 40). Canopies were visually sectioned into thirds (lower, middle, and upper), and leaf budburst was assessed for each section using a 0–6 point scale, adapted from Cole and Sheldon's seven-stage key for oak bud development [31] on a roughly weekly basis. Leaf scores grouped by provenance were then analyzed via a linear mixed model. The date at which half of the trees reached a score of $\geq$3 was established as the onset of the growing season for each provenance (50% of trees at 50% budburst or greater). Celsius growing degree days (GDDs (°C)) corresponding to the date of season start of each provenance were calculated using a base temperature of 10 °C and method one described by McMaster and Wilhelm [32,33].

Fall leaf senescence observations were made in the fall of 2022. Estimates of the percentage of brown (marcescent) or abscised (senescent) leaves in each tree canopy were made on a roughly weekly basis. A nonlinear model was used to discern statistical differences across provenances, fitting each provenance to a modified 4P logistic model (Equation (3)), where a = the point of inflection, b = Hill's slope, and x = the date.

$$y = \frac{1}{1 + e^{\frac{a-x}{b}}} \tag{3}$$

Peak senescence intensity, defined by the PhenObs initiative as 50% of leaves fully brown or abscised [34], was established as the end of the growing season for each provenance. The length of growing season was calculated by summing the number of days between growing season start and growing season end.

3. Results

The DBH growth model revealed significant differences among the six white oak provenances planted in the common garden experiment ($p < 0.01$, Figure 2). In general, maximum DBH was greater in the southern provenances, with the Mississippi provenance (MS = 42.98 cm) exhibiting the highest growth asymptote (Figure 2B). The Wisconsin provenance appeared to be an outlier, underperforming compared to even the Minnesota provenance (WI = 29.55 cm, MN = 37.69 cm), which originated further north (Table 1). However, several Wisconsin trees suppressed in the understory are likely to die in the near future, whereas the Minnesota-sourced trees that were suppressed had all died by year 40, persisting only as partially decayed snags or stumps. As all surviving trees were used in the analysis, this may account for the outlying low DBH growth performance of the Wisconsin provenance. Minnesota and Ohio sourced trees had a notably greater Schumacher scale factor than the other provenances, indicating slower early growth than other provenances in the study (Figure 2C).

The height growth model also revealed significant differences among the six white oak provenances ($p < 0.01$, Figure 3). The Indiana and Illinois provenances reached the greatest heights (IN = 26.04 m, IL = 26.03 m), although the Mississippi-sourced trees were nearly as tall (24.68 m). There were no significant differences in the x-axis displacement parameter between provenances, and height growth started immediately for all sources, indicating that there was no delay due to transplant shock or unfavorable site factors.

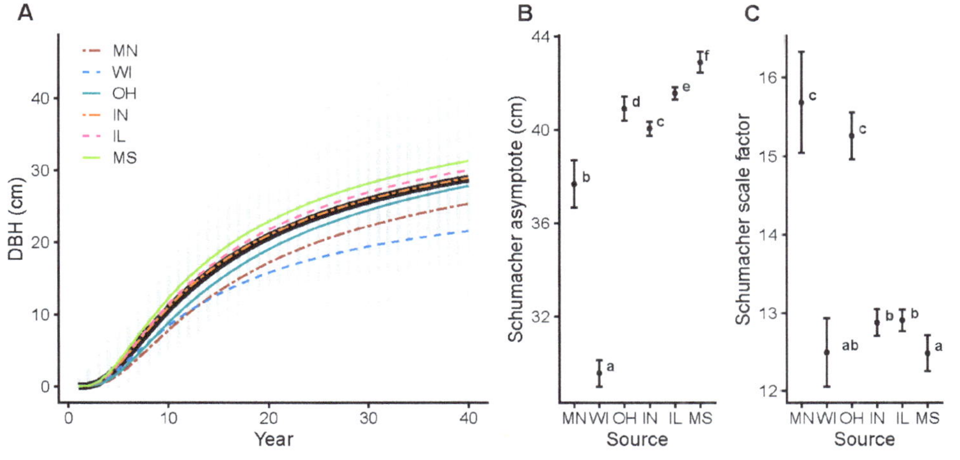

Figure 2. Summary of DBH growth (cm) of the six provenances planted in the southern Indiana common garden experiment (n = 81). DBH was reconstructed from increment cores and regular DBH measurements over the 40-year study period. (**A**) Schumacher growth model fits representing the mean DBH growth of each seed source. The solid black line represents the total population fit. (**B**) Connecting letters diagram (CLD) of the Schumacher asymptote parameter. Gray dots represent DBH measurement values. (**C**) CLD of the Schumacher scale factor parameter. For all CLDs, means not sharing any letter are significantly different at α = 0.05 according to a Holm-Bonferroni corrected post-hoc test.

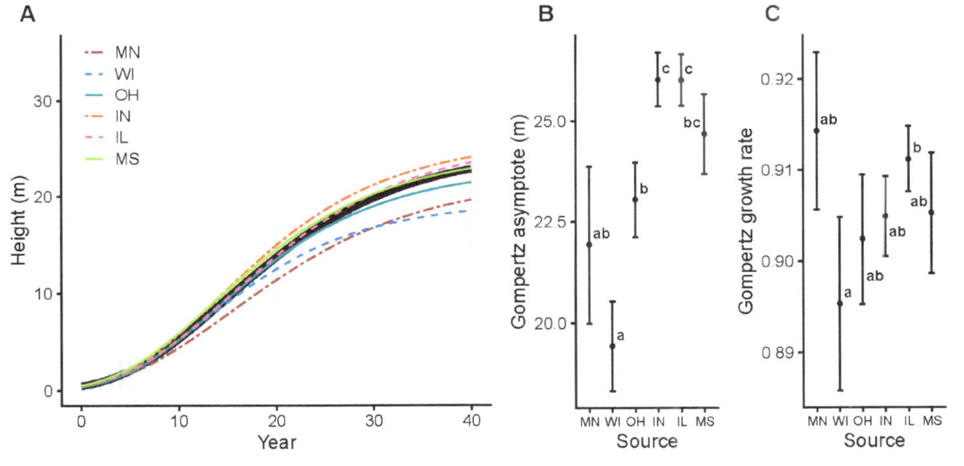

Figure 3. Summary of height growth (m) of the six white oak provenances planted in the southern Indiana common garden experiment based on regular height measurements over the 40-year study period (n = 81). (**A**) 3P Gompertz growth model fits representing the mean height growth of each provenance. The solid black line represents the total population fit. Gray dots represent height measurement values. (**B**) Connecting letters diagram (CLD) of the 3P Gompertz asymptote parameter. (**C**) CLD of the 3P Gompertz growth rate (y-scaling) parameter. For all CLDs, means not sharing any letter are significantly different at α = 0.05 according to a Holm-Bonferroni corrected post-hoc test. Note that the post-hoc test for parameter b (x-axis displacement) is not shown as there were no significant differences between provenances.

The Gompertz growth rate parameter tracks inversely with growth, meaning a lower Gompertz growth rate parameter corresponds to a relatively steeper growth curve, although this is attenuated by the asymptote parameter. Growth rate parameters varied, with the Illinois and Minnesota provenances having the highest growth rate values (IL= 0.91, MN = 0.91). Thus, these provenances experienced relatively slower early height growth rates, but appeared to maintain their initial growth rate for longer (year ~15 to ~25) than the other provenances.

Linear mixed model analysis indicated a uniform onset of budburst across all provenances in the spring of 2022, as depicted by the closely clustered provenance mean scores in Figure 4A, and the lack of significant differences between provenances in the post-hoc comparisons in Figure 4B. This suggests a synchronous phenological response to springtime cues. However, spring conditions in 2022 were very atypical, with colder than average April temperatures (2.11 °C below 1980–2023 mean temperatures) and a very wet May (5.9 cm above average precipitation, ~150% of the norm; also more than 5.0 cm above the average May rainfall of all provenance seed origins) [35,36]. Due to the abnormal spring weather conditions in 2022, spring budburst was measured again in the spring of 2023, when seasonal conditions were much closer to the area's long-term averages.

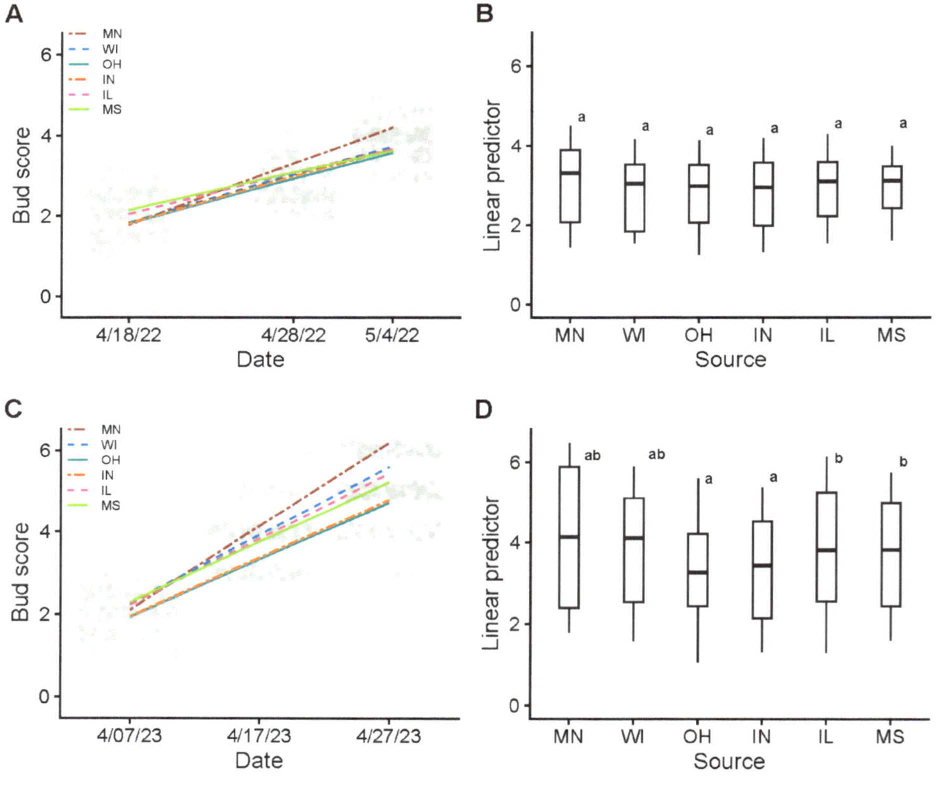

Figure 4. White oak budburst for each provenance planted in the southern Indiana common garden experiment. (**A**,**C**) Budburst scores over time for the six provenances during the spring seasons of 2022 and 2023, respectively (n = 81). Each gray point represents an individual tree's budburst score and the lines indicate the average trend for each provenance. (**B**,**D**) Post-hoc analysis of the linear mixed models for the two years, presented as Tukey's HSD tests. For all connecting letters diagrams, means not sharing any letter are significantly different at α = 0.05.

The 2023 budburst score data illustrates a departure from the previous year's uniformity. Figures 4C,D reveal discernible variations in the budburst scores among the provenances. The Indiana and Ohio provenances, the sources closest in latitude to the common garden site, were more conservative than the more northern and the more southern provenances. Despite a statistically significant difference in budburst progression among provenances, this difference only accounts for a four-day difference (when rounding to the nearest 24-h period) in growing season start date (Table 2).

Table 2. Survival rates and estimated growing season length of white oak seed sources planted in the southern Indiana common garden experiment based on 2022/2023 bud scores and 2022 leaf senescence. Note that 11-year survival rates were measured prior to stand thinning at the end of year 12, while 40-year survival rates were measured after stand thinning.

State	Latitude	11 yr. Survival	40 yr. Survival	Season Start	Growing Degree Days (°C)	Season End	Season Length (Days)
MN	45.00	100%	25%	4/11	83.9	10/6	178
WI	43.00	96%	50%	4/11	83.9	10/13	185
OH	40.28	96%	92%	4/15	111.4	10/20	188
IN	39.06	94%	96%	4/14	101.9	10/20	189
IL	37.40	94%	89%	4/11	83.9	10/18	190
MS	34.37	100%	83%	4/10	81.9	10/23	196

Differences in the progression of fall leaf senescence were much more pronounced between seed sources (Figure 5A) than differences in spring budburst scores. Trees of more northern provenance lost their leaves significantly earlier than southern provenance trees (Figure 5B). Also notable was the timespan over which senescence occurred, with the most locally sourced trees (Indiana and Ohio) having lost their leaves over a much shorter period compared to more southern and northern provenances (Figure 5C).

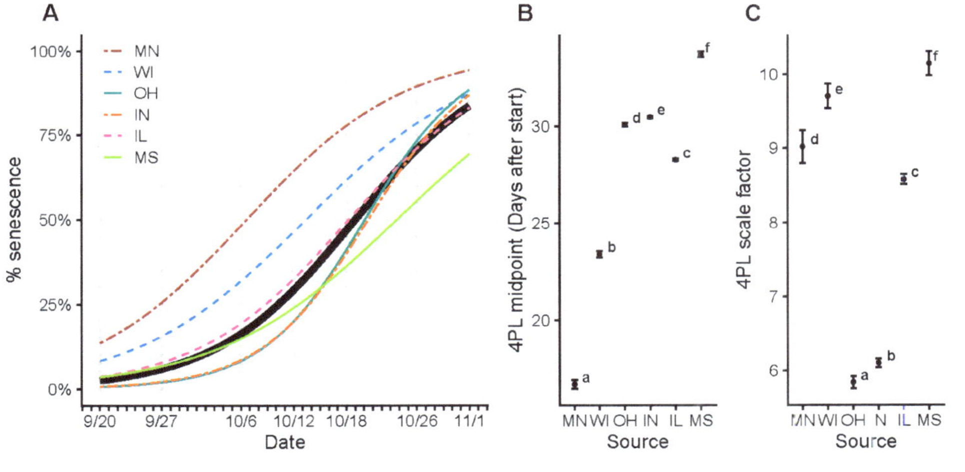

Figure 5. Summary of fall percent senescence, measured in fall 2022, for each of the six white oak seed sources growing in the southern Indiana common garden experiment (n = 81). Although based on the 4P logistic growth model, the start parameter was fixed at 0%, and the asymptote parameter was set at 100% for all provenances. (**A**) Model fits. The solid black line represents the total population fit. Gray dots represent the senescence score of individual trees. (**B**) Connecting letters diagram (CLD) of the 4P logistic midpoint parameter. (**C**) CLD of the 4P logistic scale factor parameter. For all CLDs, means not sharing any letter are significantly different at α = 0.05 according to a Holm-Bonferroni corrected post-hoc test.

Differences between the estimated growing season length for each provenance were driven primarily by fall senescence due to the greater variability in fall phenology: a spread of 17 days vs. only four days in the spring (Table 2). Thus, the total estimated season length was greatest in the southernmost provenance of Mississippi (196 days) and shortest in the northernmost provenance of Minnesota (178 days), with a difference in length of 18 days. Over the course of the 40-year study, that equates to an 'extra' 720 days (or about four growing seasons) worth of growth in Mississippi provenance trees compared to Minnesota-sourced trees.

The provenance survival rate prior to thinning the stand at 11 years was close to 100% across all seed sources (Table 2). However, after 40 years of growth, a 50% and a 75% mortality rate (not including thinned trees) were observed in Wisconsin and Minnesota provenances, respectively (Table 2). This may be due in part to their slower growth and lesser competitive ability at this relatively southern site compared to the more southern provenances caused by their shorter effective growing season, among other potential maladaptive traits. Perhaps most surprising is the relatively low mortality (<20%) of Mississippi provenance trees after 40 years despite being planted in a climate that experiences regular hard freezes, which are relatively uncommon in Mississippi.

4. Discussion

Overall, white oak seedlings responded well to the plantation setting in southern Indiana, with no evidence of transplant shock or delayed growth. White oak provenances significantly influence phenological responses such as leaf budburst and leaf senescence which are critical to the adaptive capacity of the species. The variation in growth performance and survival rates among provenances, particularly the superior growth observed in southern provenances like Mississippi, aligns with the concept that genetic selection for climate resilience could bolster forest health and productivity [37]. Furthermore, the low mortality of the Mississippi and southern Illinois provenances should assuage fears of 'overshooting' the climate envelope as these southern seed sources are already resilient to freezing conditions.

Some aspects of local provenances remain distinct from both more northern and more southern seed sources, but it is not clear if these traits impart a selective advantage. The spring phenology of the Indiana and Ohio provenances was slightly more conservative than both the northern and the southern provenances in 2023, although budburst was only postponed by roughly four days (2.1 percent of the overall average growing season length). Despite this relatively small effect size, middle latitudes (around 41° N) experienced the greatest number of growing season frost days (GSFDs) in the spring and are expected to continue to experience even more GSFDs as the climate warms [38]. In addition, these four days represent up to 27.5 GDDs (°C), which may be a better indicator of adaptation than average date of spring budburst at specific locations. The later onset of budburst among these middle latitude provenances could be an evolutionary adaptation to increased regional GSFDs. The faster and more synchronous fall senescence of the Indiana and Ohio provenances compared to other seed sources was much more pronounced than the differences is spring budburst. However, the date of peak senescence was clearly intermediate between the northern and southern provenances.

The differential growth responses documented among the six provenances in this study can be attributed to a complex interplay between genetic adaptation to local climates and the immediate environmental conditions of the common garden, which cannot be fully accounted for in this single-site study. Nevertheless, these findings resonate with recent studies that highlight the potential for preexisting mismatches between tree species genetic adaptations and rapidly changing climates [6]. The significant mortality rate observed in northern seed sources (Minnesota and Wisconsin) post-thinning could be indicative of the adaptive limitations of these provenances under the warmer climatic conditions of the common garden site, which mirror broader climate warming trends [9]. Northern provenance trees simply did not compete well with their faster growing southern seed-sourced neighbors. In addition, as the climate continues to warm, local white oak provenances may fail to compete with both native and invasive tree species that are better adapted to the warmer and longer growing season.

Precipitation and temperature data representative of the six provenance origins indicate a consistent climatic gradient, with conditions transitioning from colder and drier in the northern locations to warmer and wetter towards the south. Over the course of the 40-year study, annual precipitation increased on average at all seed source locations (Figure 1B). Likewise, annual minimum temperatures also increased on average at all locations (Figure 1C). USDA plant hardiness zones have shifted northward during this period and now predict 2.8 °C to 5.6 °C warmer mean absolute minimum temperatures at all seed source origins [25]. Future climate projections based on the relatively conservative CCSM4 RCP 4.5 model project another temperature increase of 2.8 °C to 5.6 °C at these locations by 2070 [39]. While warming is expected to continue throughout the 21st century and the seasonal distribution of precipitation will vary, annual precipitation totals are not expected to significantly change in eastern North America over the latter half of the century [40]. Therefore, late 21st century conditions in southern Indiana may be similar to pre-1983 conditions in northern Mississippi. Likewise, growing conditions in southern Indiana over the last 40 years appear to be a reasonable analogue to late century conditions in southern Wisconsin and Minnesota [40].

While the 40-year performance of Minnesota and Wisconsin sourced white oaks in southern Indiana demonstrated the potential unsuitability of existing local genetics in a warmer future climate, the excellent performance of northern Mississippi and southern Illinois trees suggests the suitability of relatively southern genetics for the future southern Indiana climate. By sourcing trees from locations with mean absolute minimum temperatures 5.6 to 11.2 °C warmer [25], white oak plantings will be tolerant of current and likely future conditions. Overall, these findings clearly support proactive forest management actions, highlighting the importance of selecting provenances for reforestation and the expectation that assisted migration will help to ensure future forest ecosystem viability and productivity. Continued research is necessary to delve into the genetic underpinnings of climate adaptability, providing forest managers with the knowledge to bolster white oak forest resilience in a changing world.

5. Conclusions

The results of this study underscore the importance of assisted migration as a strategy for forest management and conservation in the context of climate change. In this study, we have comprehensively analyzed the growth and phenological responses of white oak provenances sourced from a wide range of climatic conditions relative to the common garden planting site in southern Indiana. Our findings reveal significant provenance differences in growth patterns and phenological responses, suggesting that white oak climate adaptation may be less plastic than expected. Rather, climate adaptation appears to be closely tied to genetic adaptation to the local environment prior to anthropogenic climate change. These results are critical for understanding the impact of climate change on forest tree species. For forest managers and tree breeders, this study provides essential insights into selecting and managing white oak provenances. This knowledge is crucial for developing effective reforestation and tree improvement programs that can mitigate the adverse effects of climate change and ensure the sustainability of future white oak forest ecosystems.

Author Contributions: Conceptualization, C.D.N. and M.V.C.; methodology, M.V.C., A.M.T. and C.D.N.; formal analysis, A.M.T.; investigation, A.M.T., P.A.O., M.V.C. and C.D.N.; resources, M.V.C. and C.D.N.; data curation, A.M.T., M.V.C. and C.D.N.; writing—original draft preparation, A.M.T.; writing—review and editing, C.D.N., A.M.T., P.A.O. and M.V.C.; visualization, A.M.T.; supervision, C.D.N. All authors have read and agreed to the published version of the manuscript.

Funding: This research was supported in part by an appointment to the United States Forest Service (USFS) Research Participation Program administered by the Oak Ridge Institute for Science and Education (ORISE) through an interagency agreement between the U.S. Department of Energy (DOE) and the U.S. Department of Agriculture (USDA). ORISE is managed by ORAU under DOE contract number DE-SC0014664. All opinions expressed in this paper are the author's and do not necessarily reflect the policies and views of USDA, DOE, or ORAU/ORISE.

Data Availability Statement: The data presented in this study are openly available in Dryad; DOI: https://doi.org/10.5061/dryad.m37pvmd94.

Acknowledgments: We thank Carolyn Pike, Jim Warren, Lee Grace, Maeve Draper, and Michael Rich for their help with fieldwork as well as Michael Rich and John Lhotka for their help with dendrochronology measurements. We also thank the Indiana Division of Forestry for providing access to the study planting and historical data related to the project, and the University of Kentucky Wood Utilization Center (Quicksand, KY) for producing the increment core mounting blocks.

Conflicts of Interest: The authors declare no conflicts of interest.

References

1. Little, E.L., Jr. *Atlas of United States Trees: Volume 1, Conifers and Important Hardwoods*, 1st ed.; U.S. Department of Agriculture, Forest Service: Washington, DC, USA, 1971; Volume 1.
2. Abrams, M.D. Distribution, Historical Development and Ecophysiological Attributes of Oak Species in the Eastern United States. *Ann. Sci. For.* **1996**, *53*, 487–512. [CrossRef]
3. Knapp, B.O.; Pallardy, S.G. Forty-Eight Years of Forest Succession: Tree Species Change across Four Forest Types in Mid-Missouri. *Forests* **2018**, *9*, 633. [CrossRef]
4. Abrams, M.D. Fire and the Development of Oak Forests. *BioScience* **1992**, *42*, 346–353. [CrossRef]
5. Fralish, J.S. The Keystone Role of Oak and Hickory in the Central Hardwood Forest. In Proceedings of the Upland Oak Ecology Symposium: History, Current Conditions, and Sustainability, Fayetteville, AK, USA, 7–10 October 2002; U.S. Department of Agriculture, Forest Service, Southern Research Station: Asheville, NC, USA, 2004; pp. 78–87.
6. Cavender-Bares, J. Diversification, Adaptation, and Community Assembly of the American Oaks (*Quercus*), a Model Clade for Integrating Ecology and Evolution. *New Phytol.* **2019**, *221*, 669–692. [CrossRef] [PubMed]
7. Abrams, M.D. Where Has All the White Oak Gone? *BioScience* **2003**, *53*, 927–939. [CrossRef]
8. Dey, D.C. Sustaining Oak Forests in Eastern North America: Regeneration and Recruitment, the Pillars of Sustainability. *For. Sci.* **2014**, *60*, 926–942. [CrossRef]
9. Dhungel, G.; Rossi, D.; Henderson, J.D.; Abt, R.C.; Sheffield, R.; Baker, J. Critical Market Tipping Points for High-Grade White Oak Inventory Decline in the Central Hardwood Region of the United States. *J. For.* **2023**, *121*, 224–234. [CrossRef]
10. Manos, P.S.; Hipp, A.L. An Updated Infrageneric Classification of the North American Oaks (*Quercus* Subgenus *Quercus*): Review of the Contribution of Phylogenomic Data to Biogeography and Species Diversity. *Forests* **2021**, *12*, 786. [CrossRef]
11. Gombau, J.; Cabanillas, P.; Mena, A.; Pérez-Navarro, J.; Ramos, J.; Torner, A.; Fort, F.; Gómez-Alonso, S.; García-Romero, E.; Canals, J.M.; et al. Comparative Study of Volatile Substances and Ellagitannins Released into Wine by *Quercus pyrenaica*, *Quercus petraea* and *Quercus alba* Barrels. *OENO One* **2022**, *56*, 243–255. [CrossRef]
12. Martínez-Gil, A.; Del Alamo-Sanza, M.; Sánchez-Gómez, R.; Nevares, I. Different Woods in Cooperage for Oenology: A Review. *Beverages* **2018**, *4*, 94. [CrossRef]
13. Dhungel, G.; Ochuodho, T.O.; Lhotka, J.M.; Stringer, J.W.; Poudel, K. Sustainability of White Oak (*Quercus alba*) Timber Supply in Kentucky. *J. For.* **2024**, *122*, 79–90. [CrossRef]
14. DeWald, L.E.; Hackworth, Z.J.; Nelson, C.D. White Oak (*Quercus alba*) Genetics and Tree Improvement Program: Range-Wide Collaborative Effort and Early Results. In Proceedings of the Proceedings of the 36th Southern Forest Tree Improvement Conference, Athens, Greece, 7–9 June 2021.
15. White Oak Initiative. Assessment & Conservation Plan. Available online: https://www.whiteoakinitiative.org/assessment-conservation-plan (accessed on 14 February 2024).
16. Saleh, D.; Chen, J.; Leplé, J.-C.; Leroy, T.; Truffaut, L.; Dencausse, B.; Lalanne, C.; Labadie, K.; Lesur, I.; Bert, D.; et al. Genome-Wide Evolutionary Response of European Oaks during the Anthropocene. *Evol. Lett.* **2022**, *6*, 4–20. [CrossRef]
17. Williams, M.I.; Dumroese, R.K. Preparing for Climate Change: Forestry and Assisted Migration. *J. For.* **2013**, *111*, 287–297. [CrossRef]
18. Poupon, V.; Chakraborty, D.; Stejskal, J.; Konrad, H.; Schueler, S.; Lstibůrek, M. Accelerating Adaptation of Forest Trees to Climate Change Using Individual Tree Response Functions. *Front. Plant Sci.* **2021**, *12*, 758221. [CrossRef] [PubMed]
19. Gustafson, E.J.; Kern, C.C.; Kabrick, J.M. Can Assisted Tree Migration Today Sustain Forest Ecosystem Goods and Services for the Future? *For. Ecol. Manag.* **2023**, *529*, 120723. [CrossRef]
20. Zhu, K.; Woodall, C.W.; Clark, J.S. Failure to Migrate: Lack of Tree Range Expansion in Response to Climate Change. *Glob. Change Biol.* **2012**, *18*, 1042–1052. [CrossRef]

21. Miller, K.M.; McGill, B.J. Land Use and Life History Limit Migration Capacity of Eastern Tree Species. *Glob. Ecol. Biogeogr.* **2017**, *27*, 57–67. [CrossRef]
22. McKenney, D.W.; Pedlar, J.H.; Lawrence, K.; Campbell, K.; Hutchinson, M.F. Potential Impacts of Climate Change on the Distribution of North American Trees. *BioScience* **2007**, *57*, 939–948. [CrossRef]
23. Leites, L.P.; Rehfeldt, G.E.; Robinson, A.P.; Crookston, N.L.; Jaquish, B. Possibilites and Limitations of Using Historic Provenance Tests to Infer Forest Species Growth Responses to Climate Change. *Nat. Resour. Model.* **2012**, *25*, 409–433. [CrossRef]
24. Soil Survey Staff—Natural Resources Conservation Service—United States Department of Agriculture. Web Soil Survey. Available online: https://websoilsurvey.nrcs.usda.gov/app/WebSoilSurvey.aspx (accessed on 3 February 2024).
25. Hanberry, B.B.; Fraser, J.S. *Current and Future Plant Hardiness Zones for the Conterminous United States*; U.S. Department of Agriculture: Madison, WI, USA, 2023.
26. National Centers for Environmental Information. Climate Data Online. Available online: https://www.ncdc.noaa.gov/cdo-web/ (accessed on 30 November 2023).
27. R Core Team. *R: A Language and Environment for Statistical Computing*; R Foundation for Statistical Computing: Vienna, Austria, 2023.
28. Pinheiro, J.; Bates, D. *R Core Team Nlme: Linear and Nonlinear Mixed Effects Models*; R Foundation for Statistical Computing: Vienna, Austria, 2023.
29. Canetti, A.; Braz, E.M.; de Mattos, P.P.; Basso, R.O.; Filho, A.F. A New Approach to Maximize the Wood Production in the Sustainable Management of Amazon Forest. *Ann. For. Sci.* **2021**, *78*, 67. [CrossRef]
30. Schumacher, F.X. A New Growth Curve and Its Applications to Timber Yield Studies. *J. For.* **1939**, *37*, 819–820.
31. Cole, E.F.; Sheldon, B.C. The Shifting Phenological Landscape: Within- and between-Species Variation in Leaf Emergence in a Mixed-Deciduous Woodland. *Ecol. Evol.* **2017**, *7*, 1135–1147. [CrossRef] [PubMed]
32. Fritts, H.C. The Relation of Radial Growth to Maximum and Minimum Temperatures in Three Tree Species. *Ecology* **1959**, *40*, 261–265. [CrossRef]
33. McMaster, G.S.; Wilhelm, W.W. Growing Degree-Days: One Equation, Two Interpretations. *Agric. For. Meteorol* **1997**, *87*, 291–300. [CrossRef]
34. Nordt, B.; Hensen, I.; Bucher, S.F.; Freiberg, M.; Primack, R.B.; Stevens, A.-D.; Bonn, A.; Wirth, C.; Jakubka, D.; Plos, C.; et al. The PhenObs Initiative: A Standardised Protocol for Monitoring Phenological Responses to Climate Change Using Herbaceous Plant Species in Botanical Gardens. *Funct. Ecol.* **2021**, *35*, 821–834. [CrossRef]
35. Global Modeling and Assimilation Office (GMAO). *MERRA-2: 2d,1-inst1_2d_asm_Nx, Hourly, Instantaneous, Single-Level, Assimilation, Single-Level Diagnostics*; National Aeronautics and Space Administration: Washington, DC, USA, 2015.
36. Lawrimore, J.H.; Ray, R.; Applequist, S.; Korzeniewski, B.; Menne, M.J. *Global Summary of the Month (GSOM). Version 1*; NOAA National Centers for Environmental Information: Jackson County, IN, USA, 2016.
37. Clair, B.S.; Howe, G. Genetic Options for Adapting Forests to Climate Change. *West. For.* **2009**, *54*, 9–11.
38. Liu, Q.; Piao, S.; Janssens, I.A.; Fu, Y.; Peng, S.; Lian, X.; Ciais, P.; Myneni, R.B.; Peñuelas, J.; Wang, T. Extension of the Growing Season Increases Vegetation Exposure to Frost. *Nat. Commun.* **2018**, *9*, 426. [CrossRef]
39. Matthews, S.N.; Iverson, L.R.; Peters, M.P.; Prasad, A.M. *Assessing Potential Climate Change Pressures across the Conterminous United States: Mapping Plant Hardiness Zones, Heat Zones, Growing Degree Days, and Cumulative Drought Severity throughout This Century*; U.S. Department of Agriculture: Madison, WI, USA, 2018; Volume 9, pp. 1–31. [CrossRef]
40. USGCRP. *Climate Science Special Report*; U.S. Global Change Research Program: Washington, DC, USA, 2017; pp. 1–470.

forests

MDPI

Article

Can Growth Increase of Small Trees after Drought Compensate for Large Trees' Growth Loss?

Mingqian Liu [1], Yihong Zhu [2], Rongrong Pang [1] and Lushuang Gao [1,*]

[1] Research Center of Forest Management Engineering of State Forestry and Grassland Administration, Beijing Forestry University, No. 35 Qinghua East Road, Beijing 100083, China; liu_mingqian@outlook.com (M.L.)

[2] Department of Environmental Science, Policy and Management, University of California, Berkeley, CA 94704, USA

* Correspondence: gaolushuang@bjfu.edu.cn

Abstract: Large trees dominate stand-level biomass but their growth suffers more from droughts while sheltering small trees during droughts. Under a warmer and drier climate, whether th growth decline of large trees could be compensated by prompted small trees' growth remain unknown. Based on the Standardized Precipitation Evapotranspiration Index (SPEI) series, drough characteristics were determined, and two drought events were selected. We reconstructed historica diameters at breast height (DBH) and the aboveground biomass of *Larix gmelinii* through tree ring dat allometric equations. To clarify the difference in the responses of tree size to drought, we calculate resistance, recovery, and resilience in each diameter class. We used a growth dominance coefficien (GDC) to exhibit the contributions of different-sized individuals to stand growth and demonstrate the growth dynamics of both the individual and stand level. The results proved that large trees wer more vulnerable to local droughts, the resilience of larch had a negative relationship with the DBH ($p < 0.05$), and small trees could recover to even exceed their pre-drought growth level. Most plot had a negative GDC and small trees contributed more to stand growth compared with their siz but their AGB growth was far less than that of large trees, which made it difficult to compensat for stand growth decline. Our results indicate that tree resilience has a negative relationship wit their pre-drought sizes, as large trees in the stand fail to regain their growth level before drough Even with a larger relative contribution and higher resilience, small trees cannot cover deficits i large trees' growth. Under more frequent droughts, the total aboveground biomass growth of larche would decline.

Keywords: *Larix gmelinii*; tree resilience; growth dynamics; drought events

Citation: Liu, M.; Zhu, Y.; Pang, R.; Gao, L. Can Growth Increase of Small Trees after Drought Compensate for Large Trees' Growth Loss? *Forests* **2024**, *15*, 448. https://doi.org/10.3390/f15030448

Academic Editor: Romà Ogaya

Received: 21 December 2023
Revised: 1 February 2024
Accepted: 8 February 2024
Published: 27 February 2024

1. Introduction

Boreal forests maintain the global terrestrial ecosystem carbon balance and the carbo sink stability varies under natural disturbances and forest management [1]. Climat change and large-scale disturbances, such as warming and drought, reduce forest biomas growth [2,3].

Diameter size also influences individual tree responses to climatic factors and dis turbance events. Dominant individuals are more sensitive to drought [4] but can recove better from disturbances [5]. Managements that adjust diameter class differences such a close-to-nature silviculture usually maintain the growth advantage of large trees better [6 Individuals with smaller initial diameters and faster growth rates before drought sho higher resilience, but this effect will change over time [7]. Small trees contribute mor steadily to forest growth and have a greater contribution to long-term carbon storag sequestration, and forest climate resilience [8]. Tree diameter classes may be more heterog neous as they respond to environmental conditions differently. We now should also focu on how smaller ones would perform under climate change.

Boreal forests in northern regions will face intensifying local drought conditions and accelerated climate warming under permafrost layer degradation [9] and increasing soil moisture evaporation [10]. In the northern part of China, the Greater Khingan Mountains region was considered as the south edge of boreal forests [11] and served as a crucial ecological barrier. Drought events occurred more frequently in the Greater Khingan Mountains from 1961 to 2012 [12]. Under such a background, the forests' net primary productivity will reduce, and forest decline will occur, leading to ecosystem shrinkage and severe degradation [13,14].

Larix gmelinii is the dominant tree species in northern forests in China [15]. The growth of *Larix gmelinii* is extremely sensitive to climate conditions [16], influenced by latitude [17], and has shown significant changes in growth trends under warmer and drier climatic conditions [18]. Under higher temperatures, growth patterns even vary within the same area. In high-latitude regions, the growth of larch is promoted, while in middle- and low-latitude regions, it was suppressed by drought events caused by high-temperature summers [17,19]. *Larix gmelinii* showed a positive correlation with the mean annual temperature [20], while with the increasing temperature leading to water stress, the growth of dominant larch individuals was suppressed [21]. With varied responses in different regions and sizes, it is important to look into whether the growth differences caused by a size difference could balance the slowing growth contribution of large trees to the forest.

To clarify the growth characteristics of *Larix gmelinii* across different diameter classes and how climate change affects the biomass dynamics, we focused on *Larix gmelinii* in the Greater Khingan Mountains. We reconstructed historical diameters based on tree rings and field data. Utilizing allometric growth equations, we derived historical biomass dynamics and the quantified growth dynamics of *L. gmelinii* through resilience in different diameter classes, and aimed to uncover the influence of climate change on the aboveground biomass growth contributions of various sizes. We hypothesized that:

1. In the southern part of the study area, droughts would suppress the growth of larches, while in the northern part, growth will be prompted by droughts.
2. Larger trees will suffer from droughts, but smaller trees could maintain stand growth after a drought event.
3. Multiple droughts will lead to larch aboveground biomass growth reduction in the study area.

2. Materials and Methods

2.1. Study Area

The research area locates in the Greater Khingan Mountains region in northeastern China (120°26′38.38″–126°30′22.12″ E, 47°31′34.67″–53°22′03.16″ N) (Figure 1). The soil type of this area is brown coniferous forest soil [22]. The vegetation community structure in this area is relatively uniform, with associated tree species including *Betula platyphylla* and *Pinus sylvestris*. We set twenty-six pure *Larix gmelinii* forest plots to analyze the impact of drought events. Each sample plot had a radius of 17.85 m and covered an area of 0.1 hectares. The elevation of these sample plots ranged from 300 to 1000 m above sea level, with the coldest monthly average temperature ranging from −29.5 to −22.6 °C and the warmest monthly average temperature ranging from 15.9 to 19.6 °C.

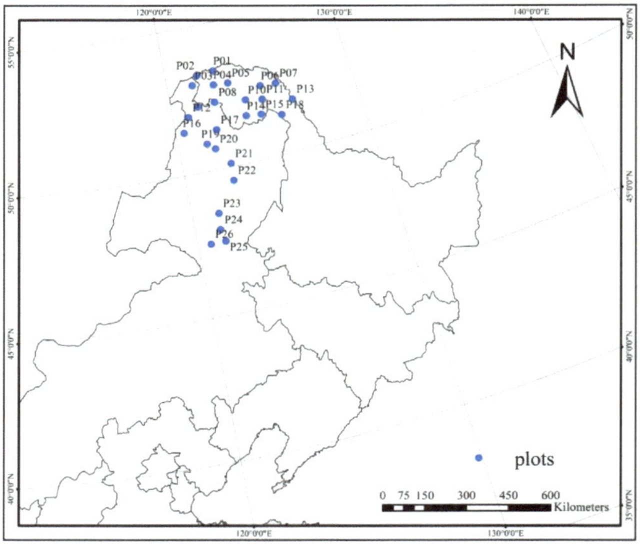

Figure 1. Locations of 26 pure *Larix gmelinii* forest sites in the Greater Khingan Range, China, and the area of each plot is 0.1 hectare. The numbering increases with latitude and longitude.

2.2. Field and Laboratory Measurements

Within the established circular sample plots, individuals with a diameter at breast height (DBH) that is greater than or equal to 5 cm were systematically measured clockwise. The following information was recorded for each individual: tree species, DBH, tree height, angle, and distance from the plot center. Additionally, at breast height (1.3 m above ground level), tree core samples were collected. These core samples were naturally air-dried in the laboratory, securely fastened to wooden mounts, and assigned unique identification numbers. The tree ring widths were measured using the LinTab 5 (RINNTECH, Heidelberg, DE, www.rinntech.com (accessed on 24 May 2021)) tree ring width measurement instrument, with a measurement accuracy of up to 0.01 mm. To ensure the accuracy of dating and measurements, the COFECHA software [23] was employed to perform correlation tests on 26 sample sequences within the study sites (Table 1). This process helped ensure precise dating, and only tree core samples with clear and reliable tree rings were selected for further analysis. We then used reconstructed historical DBH data as the initial DBH before the disturbance events for resilience analyses. Please refer to Table 2 for the specific characteristics of the tree core samples.

Table 1. Plots information.

Plots	Longitude (E)	Latitude (N)	Elevation/m	Slope/°	Slope Direction	Density/n·hm^{-2}	Mean DBH/cm	Mean Height/m
P01	122°41′28.88″	53°22′3.16″	522.6	3	southwest	590	18.5	24.7
P02	121°45′35.84″	53°19′20.97″	670.3	18	southwest	1060	15.5	22.9
P03	121°24′2.67″	53°1′22.07″	436.3	3	northeast	470	18.1	21.6
P04	122°33′16.06″	52°54′0.90″	498.5	8	northeast	790	19.4	26.6
P05	123°20′1.81″	52°51′25.90″	567.4	2	north	540	16.3	26.2
P06	125°0′58.02″	52°30′19.40″	368.3	7	southeast	720	18.7	18.6
P07	125°51′19.44″	52°27′27.90″	370.6	2	southeast	490	13.9	14.5
P08	122°23′32.41″	52°18′43.58″	741.5	22	south	880	16.4	22.3
P09	121°27′1.31″	52°15′57.03″	690.7	13	west	950	12.8	21.3
P10	124°3′3.28″	52°9′8.07″	506.8	2	northeast	640	9.1	10.1

Table 1. *Cont.*

Plots	Longitude (E)	Latitude (N)	Elevation/m	Slope/°	Slope Direction	Density/n·hm^{-2}	Mean DBH/cm	Mean Height/m
P11	124°56′20.70″	52°2′15.05″	481.3	2	north	710	13.5	13.6
P12	121°30′36.51″	50°56′22.80″	850.2	12	south	490	22.2	29
P13	126°30′22.12″	51°47′4.43″	283.3	3	southeast	690	16.3	15.9
P14	120°50′24.53″	51°58′0.16″	483.2	6	northeast	500	23.1	13.9
P15	123°52′56.99″	51°36′38.95″	721	3	northeast	490	15.6	12.2
P16	125°44′4.35″	51°21′7.70″	405.1	17	southwest	500	17	26.8
P17	120°26′38.38″	51°27′38.35″	589.7	10	southwest	530	25.4	17.8
P18	124°41′11.50″	51°32′28.00″	560.5	3	northeast	530	9.9	17.4
P19	121°53′29.55″	50°43′14.47″	822.5	9	southeast	640	21.2	20.8
P20	122°10′8.93″	51°21′31.38″	855.5	23	east	810	13	24.5
P21	122°31′0.17″	50°7′11.62″	572.6	1	south	300	25.3	29.3
P22	122°26′54.67″	49°31′15.08″	590.8	2	southeast	260	30.5	33.9
P23	121°20′10.55″	48°30′52.53″	997	10	west	340	31.2	14.8
P24	121°14′22.65″	47°57′11.19″	755.6	11	north	540	22.4	17.9
P25	121°22′43.57″	47°31′53.92″	610.8	18	northeast	400	25.4	15.6
P26	120°38′33.79″	47°31′34.67″	1087.7	12	northeast	670	20.5	17.6

Table 2. Tree ring characteristics.

Plots	Sample Depth	Common Interval	Mean Correlation Coefficient between Trees	Expressed Population Signal	Signal Noise Ratio	Mean Stand Age at Breast Height
P01	59	1954–2012	0.53	0.97	32.1	45
P02	106	1978–2015	0.44	0.98	43.4	44
P03	47	1988–2016	0.31	0.95	18.93	83
P04	79	1991–2016	0.24	0.93	13.06	49
P05	54	1995–2016	0.6	0.99	81.5	40
P06	72	1985–2016	0.31	0.88	7.08	24
P07	49	1969–2012	0.61	0.99	70.96	35
P08	88	1997–2016	0.11	0.74	2.88	49
P09	95	1992–2016	0.3	0.97	27.93	50
P10	64	1980–2016	0.15	0.84	5.08	21
P11	71	1982–2013	0.42	0.96	26.81	37
P12	50	1991–2013	0.2	0.82	4.6	31
P13	69	1958–2016	0.61	0.98	55.04	39
P14	49	1968–2016	0.36	0.93	14.07	129
P15	53	1984–2016	0.6	0.98	45.9	31
P16	53	1975–2016	0.37	0.97	27.75	37
P17	81	1991–2016	0.76	0.97	37.49	48
P18	50	1937–2016	0.65	0.95	13.8	47
P19	49	2000–2016	0.48	0.97	29.78	61
P20	64	1975–2014	0.68	0.98	47.06	49
P21	30	1977–2016	0.46	0.97	31.71	29
P22	26	1984–2015	0.71	0.98	60.62	37
P23	34	1904–2016	0.29	0.92	11.6	38
P24	54	1983–2014	0.19	0.84	5.1	45
P25	40	1976–2016	0.64	0.96	22.86	45
P26	67	1975–2016	0.72	0.98	55.14	48

2.3. Statistical Analysis Methods

2.3.1. Drought Events Selection and Region Partition

The growing season of *L. gmelinii* is from May to September [24]. We chose the historical average 3-month Standardized Precipitation Evapotranspiration Index (SPEI-3) [12,24,25] to measure drought intensity and determine the drought events of the study area. Drought data were downloaded from the Royal Netherlands Meteorological Institute (KNMI) website (https://climexp.knmi.nl (accessed on 26 September 2021)) and used Climatic Research Unit (CRU) Time-Series (TS) version 4.03 of high-resolution (0.5 × 0.5 degree) gridded data of month-by-month variation in the climate. Drought events are classified into five levels [25]. The threshold of drought was set to −0.5. We selected a period that started from the year 1984, when the local climate showed an increasing trend [26], to 2017. There are two main drought events that happened in 1999 and 2007, and we will discuss the resilience separately to avoid the first drought disrupting the plant responses during the subsequent drought occurrences. To explore the drought patterns in each plot, we calculated the characteristics based on the theory of runs in R (Version 4.2.3, Vienna, Austria) [17,27]. Scaled drought characteristics (drought number, duration, intensity, and peaks) were chosen to cluster plots in the study area. We chose the year that had a growth season SPEI below −0.5 as the drought year between 1984 and 2017 in each plot, and the time when SPEI dropped below −0.5 and then returned above −0.5 was the drought duration. During the drought duration, the absolute cumulative SPEI was the drought intensity, and the minimum absolute SPEI was the drought peak. We used factoextra package (version 1.07) to perform cluster analysis on the SPEI of each plot to divide the study area into 3 regions with different drought severities by the ward.D method (Table 3)

Table 3. Drought classification according to SPEI.

Drought Level	Drought Intensity	SPEI
1	Non-drought	$-0.5 < \text{SPEI}$
2	mild	$-1.0 < \text{SPEI} \leq -0.5$
3	moderate	$-1.5 < \text{SPEI} \leq -1.0$
4	severe	$-2.0 < \text{SPEI} \leq -1.5$
5	extreme	$\text{SPEI} \leq -2.0$

2.3.2. Reconstruction of Historical AGB

Based on the reconstructed historical diameter at breast height (DBH) data, we performed diameter class separation. This involved calculating the annual DBH of each tree and categorizing it into 2-cm diameter classes. We then integrated these data with the corresponding aboveground biomass increments to determine the specific growth performance of *L. gmelinii* within each diameter class annually. This approach allowed us to eliminate the influence of changes in growth rates during individual growth processes and investigate the dynamic contributions of different diameter classes to the total growth during the study period.

Using the historical DBH data, we employed the aboveground biomass formulas to calculate the aboveground biomass and growth increment of larches in the study plots [28]

Aboveground biomass calculation for *Larix gmelinii*:

$$M_L = 0.11270 \times D^{2.39582}\ (D \geq 5\ \text{cm}) \tag{1}$$

$$M_L = 0.18254 \times D^{2.09620}\ (D < 5\ \text{cm}) \tag{2}$$

M_L stands for the aboveground biomass of *Larix gmelinii* and D stands for the diameter.

2.3.3. Size-Different AGB Response to Drought Events

The response of *L. gmelinii* growth to disturbance events can be assessed by calculating resistance, recovery, and resilience, which reflect the trends in growth changes under extreme climatic disturbance events. A resistance value greater than 0.75 is considered a strong resistance to disturbance events, while a recovery value greater than 1.25 indicates strong recovery ability. When the resilience is less than 1.0, it is considered that the growth cannot recover to the pre-disturbance state [29].

$$\text{Resistance Rt} = \text{Dr}/\text{PreDr} \tag{3}$$

$$\text{Recovery Rc} = \text{PostDr}/\text{Dr} \tag{4}$$

$$\text{Resilience Rs} = \text{PostDr}/\text{PreDr} \tag{5}$$

Dr stands for the average AGB growth of *Larix gmelinii* during the disturbance period; PreDr stands for the average growth before the disturbance occurred; and PostDr represents the average growth afterwards. In this study, we chose the before and afterwards stage as the same length as the disturbance period.

To explore how trees in different sizes respond to drought events, we combined the initial DBH pre-drought and their resilience and used Pearson correlation to test the relationship between pre-drought size and resilience to corresponding drought events through the ggpubr package in R language (version 0.6.0) [30].

Large trees are believed to be more vulnerable to droughts and play a crucial role in forest growth; however, there are three common definitions for large trees [31]: the 99th percentile method (selecting the largest 1% of individuals in the population with a diameter at breast height of ≥1 cm), fixed diameter threshold (the specific threshold varies with tree species and forest types, with a common threshold being ≥20 cm in cold regions [32], and large diameter class threshold (individuals reaching this specific diameter class must collectively contribute to more than half of the live aboveground biomass). With resilience showing the growth potential of different diameter classes, the growth dynamics after the droughts of stands constituted by diverse individuals is uncertain. To investigate the contributions of large and small trees to the overall growth of *Larix gmelinii*, we selected the top 10% of individuals with the largest size from the historical population as the large tree group (Group L), and the bottom 10% of individuals with the smallest size as the small tree group (Group S). In this paper, Group L refers to individuals with a diameter class larger than 30 cm, while Group S refers to individuals with a diameter class smaller than 8 cm. The data analysis was completed in R version 4.2.3 [33]. We also completed ARIMA analysis using the astsa package (version 2.0) to forecast the AGB growth trend in the future and examine whether small trees could compensate for the loss of large trees.

The Growth Dominance Coefficient (GDC) reflects the relative contributions of different-sized individuals to the total growth in a stand [34,35]. It can represent the stage of forest development and can be calculated by the difference between the cumulative biomass increment Gini coefficient (GCis) and the cumulative biomass Gini coefficient (GCs).

$$\text{GDC} = \text{GCis} - \text{GCs} \tag{6}$$

$$\text{GC} = \frac{\sum_{i=1}^{n}\sum_{j=1}^{n}|x_i - x_j|}{2n(n-1)\overline{x}} \tag{7}$$

x_i and x_j represent the cumulative aboveground biomass or cumulative aboveground biomass increment of the i-th and j-th trees among the n trees in the stand.

3. Results

3.1. Stand Age of the Study Area

The EPS of 22 plots is above 0.85, which shows that the results are reliable. The larch forests in the study area were mainly young and middle-aged forests, with an average age of 60 years, and the lowest average age of the sample plots was 21 years and the longest average age of the sample plots reached 129 years (Table 2).

3.2. Size-Different Relationship between Aboveground Biomass Growth and Drought Severity

3.2.1. Drought Characteristics of Each Region

With a decreasing latitude, the drought level of the Greater Khingan Mountains increased, and the southern regions experienced even more severe aridity. According to the clustering result (Figure 2), plots located differently show three main drought patterns in the study period. The plots in Cluster 1 (referred to as Region I) went through the mildest drought in all three regions, Cluster 2 (Region II) showed a relatively moderate drought condition, and the plots in Cluster 3 (Region III) suffered the most under severe drought (Figure 2). The plots in the first cluster had the smallest drought number, shortest drought duration, mildest drought intensity, and lowest drought peaks. The plots in the second cluster had the highest drought number while the duration, severity, and peaks showed a moderate level. The plots in the third cluster had the longest duration, highest peaks, and the most severe drought events. The drought severity of Region I was 7.67, while in Region III the severity reached 12.35. The drought peaks in Region I indicated drought events which never reached a severe level, while in Region III, which had an average peak of 1.93, extreme drought happened. There were three clusters generated, sized 13, 7, and 6. The ratio of between the group sum of squares and the total sum of squares within is 80.4%, which shows a good fit.

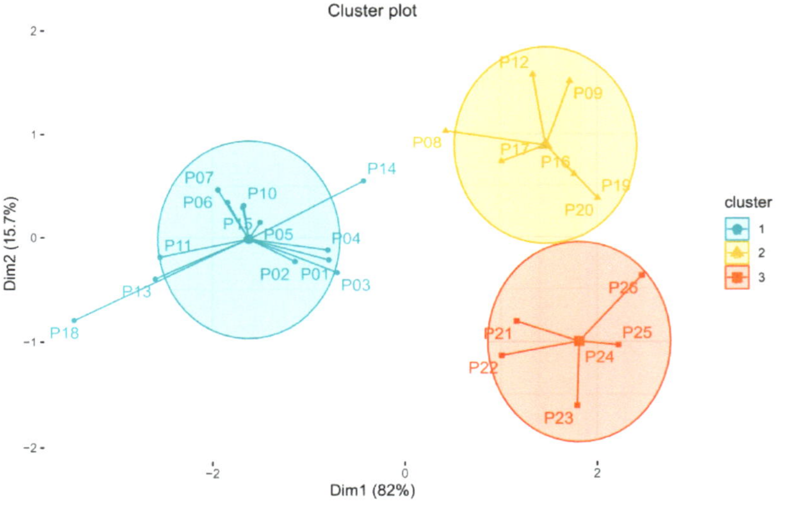

Figure 2. Plots clustering result according to drought characteristics.

3.2.2. Size Affects Relationships between AGB Growth and Local SPEI

The tree growth response to drought varied along with their size, and the relationship between the growth and drought severity of each size showed different trends in three regions. For trees with a DBH class smaller than 18 cm, there was no significant correlation between the growth and local SPEI. In Region I, trees with a size of 24 cm displayed a negative correlation with the local SPEI (see Figure 3), indicating that the growth of larch of this size is promoted under drier conditions. In Region II, the growth of trees with sizes of 18 cm and 28 cm even increased while the drought conditions turned worse. Conversely, in Region III where the drought duration lasted longer, trees that had a DBH at a 36 cm level grow faster when the drought has gone down, and the water condition helps with large tree growth in drier places. Trees that are relatively large are more sensitive to the local drought condition.

3.3. Size Affects Tree Resilience to Drought Events

With the growth response to drought severity varying in each diameter class, we further analyzed the relationships between the initial diameter class at the beginning of drought, and the resistance, recovery, and resilience under two drought events.

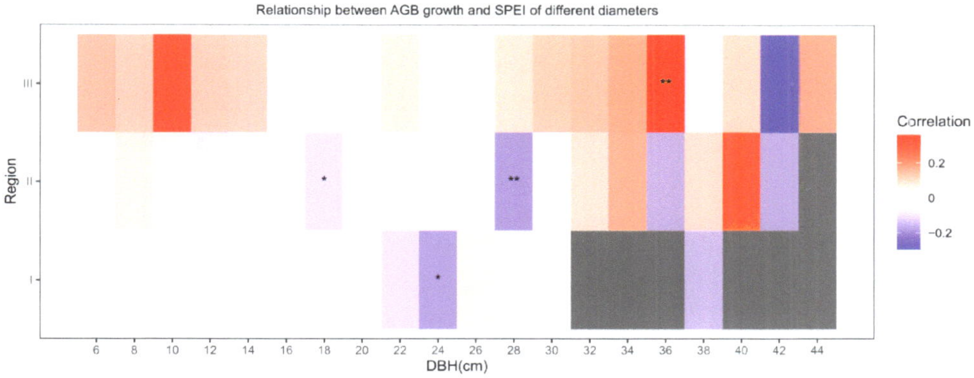

Figure 3. Relationships between AGB growth of different diameter classes (cm) (x-axis) and SPEI in each region (I, II, III). Red to blue stand for positive correlation to negative correlation. * stands for p-value less than 0.05, ** stands for p-value less than 0.01.

3.3.1. Resistance

The larch resistance was lowered when the size increased after the first drought, and the relationship disappeared after the second drought. During the first drought, both Region I and Region III exhibited a trend of decreasing resistance with the increasing diameter class ($p < 0.05$). Smaller-diameter individuals showed higher resistance. In Region III, individuals with a DBH above 34 cm displayed lower resistance and could not reach the 0.75 threshold (see Figure 4a(1st_drought)). After the second drought event, the correlation between the initial diameter class and resistance disappeared, but the resistance values remained above 0.75. All diameter classes in the three regions displayed strong resistance to drought, and the growth during the disturbance period was close to the pre-disturbance levels (see Figure 4a(2nd_drought)).

3.3.2. Recovery

Small trees recovered better after the first drought. Following an increase in the diameter class, both Region I and Region II showed a significant decrease in recovery ($p < 0.05$) after the first drought event. Regarding growth after the first drought, most diameter classes were unable to surpass their growth levels during the first drought. In Region III, recovery was not significantly related to the initial DBH, and most diameter ranges had strong recovery, with values exceeding 1.25 (see Figure 4b(1st_drought)). After the second drought event, only Region II exhibited a significant positive correlation between the initial DBH and recovery ($p < 0.05$) (see Figure 4b(2nd_drought)). Larger-diameter individuals with an initial DBH greater than 30 cm displayed stronger recovery. In Region I and III, there was no significant correlation between the initial DBH and recovery, but the recovery values remained above 1.0, although not reaching the 1.25 threshold. After the second drought, the post-disturbance growth approached the growth during the disturbance.

3.3.3. Resilience

After the first drought, large trees' growth decreased while the small tree growth level increased. Resilience fell with enlarged tree size, which exhibited a negative correlation with the initial DBH during the first drought event ($p < 0.05$). As the severity of the drought increased, smaller trees showed higher resilience. In Region I, only individuals with a DBH below 14 cm could recover their growth to pre-drought levels (see Figure 4c(1st_drought)). In Region III, individuals with a DBH below 30 cm were able to reach their pre-drought growth levels after drought. After the second drought event, most growth levels in all three regions achieved pre-disturbance levels, with resilience values exceeding 1.0. However, the correlation between the initial DBH and resilience weakened ($p > 0.05$) (see Figure 4c(2nd_drought)). Frequent droughts did not cause further growth decline.

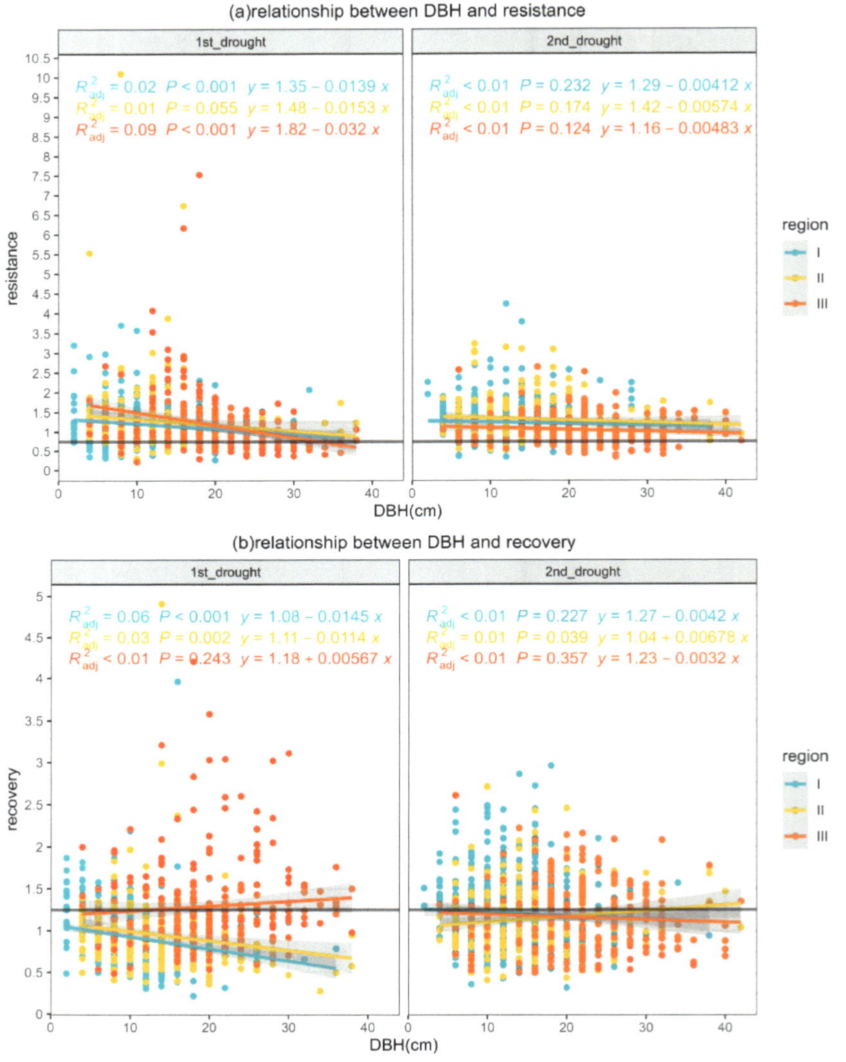

Figure 4. *Cont.*

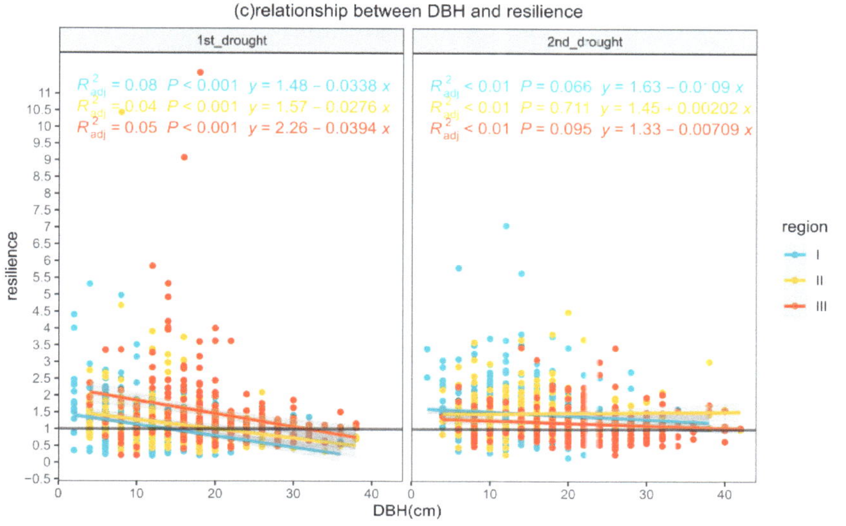

Figure 4. Relationships between DBH and growth resilience after two drought events: **1st_drought** stands for drought events that started at around 1999, **2nd_drought** stands for drought events that started at around 2007. (**a**) Relationships between DBH and resistance; (**b**) Relationships between DBH and recovery; (**c**) Relationships between DBH and resilience.

3.4. Growth of Small Trees Cannot Compensate for Large Trees' Growth Loss

With large trees failing to achieve their pre-drought growth level and small trees showing a resilience above 1.0, it is necessary to find out which group can decide forest growth. The Growth Dominance Coefficient showed the relative growth contributions of individuals compared to their size. The GDC in most plots was close to 0, indicating low competition and demonstrating that individual contributions to aboveground biomass growth were uniform. However, in specific plots like P03 and P14 with higher average ages, the GDC was less than −0.1, indicating that smaller trees contributed more due to a slowdown in large tree growth. The study area was primarily composed of young and middle-aged forests, and only eight plots followed the development pattern of large trees, which had a relatively greater growth contribution (GDC > 0). In contrast, 13 plots exhibited an early manifestation of low growth dominance (GDC < 0), whereas smaller-diameter trees contributed more to their biomass. Additionally, five plots initially showed low growth dominance and then reverted to GDC ≥ 0 (Figure 5). Plot P03, with the highest average age, consistently remained in the stage where smaller trees had a growth advantage. Small trees contributed more size-symmetrically in the study area.

To better connect tree growth actual contribution with their size in stands, instead of their relative contribution, we showed the rebuilt AGB growth dynamics of Group S (DBH class under 8 cm) and Group L (DBH class above 30 cm). Large trees had a more significant impact on the total growth of larch during the study period. Based on the predictions made by the ARIMA model, there was no apparent trend in the growth contribution of Group S and Group L in the next five years, as they remained stable (Figure 6a). The total aboveground biomass growth volume showed a different pattern of initially decreasing and then stabilizing at 3500 kg (Figure 6b). The increment in the aboveground biomass of Group S was considerably lower than that of individuals with a diameter class of above 8 cm, and the total growth during the latter period was less than 80 kg, which accounted for only less than 3% in the total stand AGB growth. The aboveground biomass growth of large-diameter individuals stayed at a relatively high level in the preceding five years, exceeding 1000 kg in the later period and contributing to more than 30% of the total stand

AGB growth. Throughout the research period, the total growth of Group S in the study area remained consistently low, while the individuals of other diameter classes tended to maintain relatively high levels of future growth, far surpassing the increments of small trees.

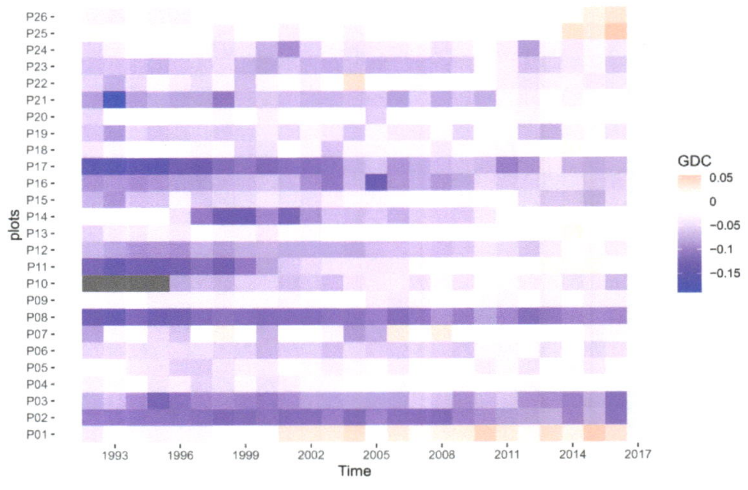

Figure 5. Growth dominance dynamics in each plot.

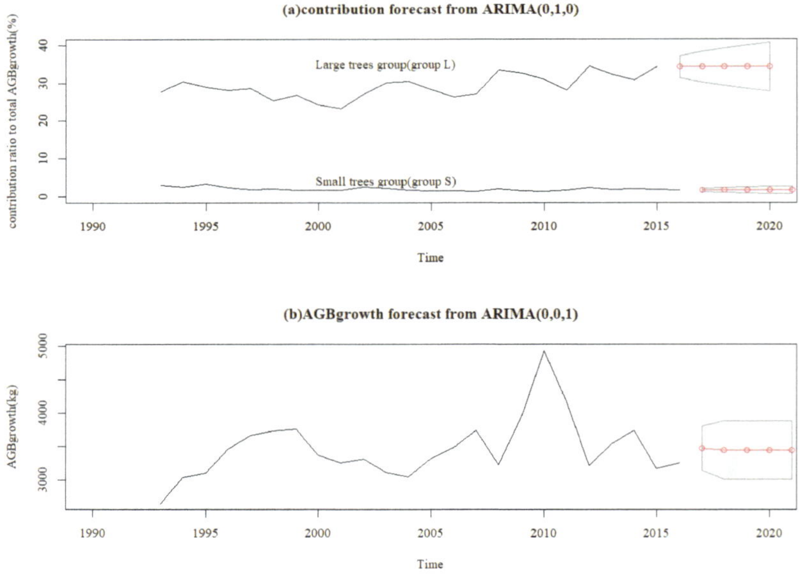

Figure 6. (**a**) Growth contribution of Group S and Group L and total aboveground biomass growth over the next 5 years showed plateauing trend; (**b**) stand AGB growth pattern during study period. The red hollow dots and lines stand for future 5 years changing pattern and the grey area stands for the prediction interval.

4. Discussion

4.1. Response of Trees' Growth in Different Sizes to Drought Events

Drought changed the growth pattern of different-sized *Larix gmelinii* in the Greater Khingan Mountains. The growth of large trees in Region III would increase with better water conditions but decrease in Region I and II (Figure 3). This conclusion is consistent with the previous findings that *Larix gmelinii* shows a decreasing trend in the southern and central parts of the Greater Khingan Mountains [19], but an increasing trend in the northern parts and permafrost areas [17,18]. Our results also showed that larch resistance is inversely related to diameter size in all three regions. Small *Larix gmelinii* individuals exhibited higher drought resilience. After the disturbance event, small trees can all recover to their pre-disturbance growth levels, whereas large-diameter individuals have lower resistance and resilience compared to small trees (Figure 4). This finding contradicts the notion proposed by Lloret [5] that larger individuals exhibit greater recovery. The growth of large trees dominates the upper canopy in the vertical structure and differs from that of smaller trees in the understory [31]. During the drought period, large trees experience growth suppression [36] while small trees can recover to their pre-drought growth levels and even be promoted [35]. Dominant trees rapidly occupy the forest canopy, leading to canopy closure [37] and shading effects [38]. This can alter the microclimate within the plot and mitigate the decline and mortality of nearby small trees during drought conditions. However, the water stress induced by rising temperatures is unfavorable for the growth of dominant individuals occupying the main forest canopy [21]. As a result, larches that had larger diameters exhibited lower resistance and resilience to drought events.

4.2. Tree Resilience under Multiple Drought Events

Low-intensity and continuous drought in the study area reduced the forest productivity [39]. Resilience decreased with the increasing diameter classes in all three regions after the first drought, and trees with larger diameters did not recover to the same growth level as before the drought events (Figure 4c(1st_drought)). After the second drought event, trees did not show further decline in resilience (Figure 4b) and displayed a recovery above 1.0 (Figure 4b(2nd_drought)). Continuous drought might not cause a further growth decline, but would still lead to a stand-level low growth increment. Due to drought legacy, trees might take approximately 16 years to recover [40]. In 2017, severe and extreme drought events occurred. Under these extremely dry conditions, large trees would wilt to reduce solar radiation and water evaporation [41]. This will further weaken the growth of large trees.

4.3. Stand-Level AGB Growth Dynamics under Frequent Droughts

After frequent drought events, the total aboveground biomass increment of *Larix gmelinii* is expected to slow down in the future [42]. Although smaller individuals had a relatively greater contribution during the study period with a GDC of less than 0 in many plots (Figure 5), the growth of large trees still had a greater impact on the total growth of larch, and large trees (Group L) retained a much higher growing level than smaller trees (Group S) (Figure 6). Large trees determine the growth of stands [43], but are more sensitive to droughts. Frequent droughts will result in homogeneous diameter class structures between different regions, with the northern regions exhibiting a lower growth dominance. The extreme drought event in 2018 led to a decrease in tree growth and redistributed the total forest growth to a state that favored small trees [44]. Under climate warming, the mortality rate of large trees has increased [4], and some seedlings might have a greater tendency to distribute to more northern areas [45]. In the future, the study area is likely to have a diameter structure that is dominated by smaller individuals. This, in turn, may lead to a decreased productivity and carbon sequestration capacity. With their more stable contribution to growth instead of canopy-dominating individuals, the roles of smaller trees are relatively greater compared to their proportion of aboveground biomass in forest volume, carbon sequestration, and climate resilience [8]. In the following forest management, we could improve the water availability of large trees in drier stands or maintain a more diverse diameter structure for larch forests to support forests in performing their functions better during drought events.

4.4. Prospects and Outlook

We compared the growth patterns of individuals within different diameter classes, while individuals of the same diameter class but different ages may still exhibit differences in productivity and other aspects. Research on Chinese fir in Guangxi by Huang Xiaorong showed that productivity did not increase with stand age and, in fact, had a negative effect [11]. The total aboveground biomass of larch increased with age [46], but the relationship between growth and age remains uncertain. *Larix gmelinii* growth tends to stabilize after reaching maturity, typically around 100 years of age [47]. Most of the forests in the Greater Khingan Mountains are in the young and middle-aged stages [48]. Although young forests have faster carbon sequestration efficiency compared to older forests, the average volume over large spatial and temporal scales is more critical for carbon flux [37], which means that forests in the study area still have a great potential in carbon sequestration. The data in this study only include results reconstructed from samples of standing trees in plots. Subsequent research could supplement historical biomass change data through records of *L. gmelinii* regeneration and mortality. Also, in this study we did not involve tree heights as a perspective, and comparing the productivity performance of individuals from different vertical layers within the same diameter class may better explain forest stand dynamics under climate change, helping us to make more accurate predictions about the overall changes in the stand level.

5. Conclusions

The study area is located in the south border of boreal forests and faces frequent droughts. In our study, we found that under the influence of two major drought events that occurred in 1999 and 2007, moderate droughts have stimulated the aboveground biomass growth of *Larix gmelinii*. Across all regions, large trees exhibit lower resistance and resilience compared to small trees, resulting in growth suppression. However, with an annual growth of less than 80 kg, small trees cannot fully compensate for the reduced growth of large trees. Large trees still play a dominant role in the growth of pure *Larix gmelinii* forests in the Greater Khingan Mountains. Under the future climate warming and frequent drying, the total aboveground biomass growth of *Larix gmelinii* in the Greater Khingan Mountains is expected to decline, which may lead to the larch forest area in China decreasing. These results provide a reference for future forest management practices and ideas to improve sustainable forest management in the context of frequent drought events. Future forest management could focus on improving water availability for large trees and further prompt small trees' growth.

Author Contributions: Conceptualization, M.L. and L.G.; methodology, M.L.; investigation, M.L. and R.P.; writing—original draft preparation, M.L.; writing—review and editing, Y.Z. and L.G.; supervision, L.G.; funding acquisition, L.G. All authors have read and agreed to the published version of the manuscript.

Funding: This research was funded by the National Natural Science Foundation of China grant number 32371871 and the Key Project of National Key Research and Development Plan, grant number 2022YFD2201001-04.

Data Availability Statement: The data underlying this article cannot be shared publicly for the privacy of lab that participated in this study. The data will be shared on reasonable request to the corresponding author.

Conflicts of Interest: The authors declare no conflicts of interest. The funders had no role in the design of the study; in the collection, analyses, or interpretation of data; in the writing of the manuscript; or in the decision to publish the results.

References

1. Pan, Y.; Birdsey, R.A.; Fang, J.; Houghton, R.; Kauppi, P.E.; Kurz, W.A.; Phillips, O.L.; Shvidenko, A.; Lewis, S.L.; Canadell, J.G.; et al. A Large and Persistent Carbon Sink in the World's Forests. *Science* **2011**, *333*, 988–993. [CrossRef]
2. Seidl, R.; Schelhaas, M.-J.; Rammer, W.; Verkerk, P.J. Increasing Forest Disturbances in Europe and Their Impact on Carbon Storage. *Nat. Clim. Change* **2014**, *4*, 806–810. [CrossRef]
3. Fan, L.; Wigneron, J.-P.; Ciais, P.; Chave, J.; Brandt, M.; Sitch, S.; Yue, C.; Bastos, A.; Li, X.; Qin, Y.; et al. Siberian Carbon Sink Reduced by Forest Disturbances. *Nat. Geosci.* **2023**, *16*, 56–62. [CrossRef]
4. Bennett, A.C.; McDowell, N.G.; Allen, C.D.; Anderson-Teixeira, K.J. Larger Trees Suffer Most during Drought in Forests Worldwide. *Nat. Plants* **2015**, *1*, 15139. [CrossRef] [PubMed]
5. Lloret, F.; Keeling, E.G.; Sala, A. Components of Tree Resilience: Effects of Successive Low-Growth Episodes in Old Ponderosa Pine Forests. *Oikos* **2011**, *120*, 1909–1920. [CrossRef]
6. Wang, Y.-X.; Zhang, S.-G.; Lu, Y.-C.; Meng, J.-H.; Zeng, J.; Bai, S.-B. Instant response of individual size inequality indices to thinning regimes in plantation. *Ying Yong Sheng Tai Xue Bao* **2014**, *25*, 1645–1651. [PubMed]
7. Ovenden, T.S.; Perks, M.P.; Clarke, T.-K.; Mencuccini, M.; Jump, A.S. Life after Recovery: Increased Resolution of Forest Resilience Assessment Sheds New Light on Post-Drought Compensatory Growth and Recovery Dynamics. *J. Ecol.* **2021**, *109*, 3157–3170. [CrossRef]
8. Hubau, W.; De Mil, T.; Van den Bulcke, J.; Phillips, O.L.; Angoboy Ilondea, B.; Van Acker, J.; Sullivan, M.J.P.; Nsenga, L.; Toirambe, B.; Couralet, C.; et al. The Persistence of Carbon in the African Forest Understory. *Nat. Plants* **2019**, *5*, 133–140. [CrossRef]
9. Chen, S.; Zang, S.; Sun, L. Characteristics of Permafrost Degradation in Northeast China and Its Ecological Effects: A Review. *Sci. Cold Arid. Reg.* **2020**, *12*, 1–11.
10. Burrell, A.L.; Sun, Q.; Baxter, R.; Kukavskaya, E.A.; Zhila, S.; Shestakova, T.; Rogers, B.M.; Kaduk, J.; Barrett, K. Climate Change, Fire Return Intervals and the Growing Risk of Permanent Forest Loss in Boreal Eurasia. *Sci. Total Environ.* **2022**, *831*, 154885. [CrossRef] [PubMed]
11. Huang, X. Relationship between Plant Functional Diversity and Productivity of Pinus Massoniana Plantations in Guangxi. *Biodivers. Sci.* **2018**, *26*, 690. [CrossRef]
12. Sun, B.F.; Zhao, H.; Wang, X.K. Spatiotemporal Characteristics of Drought in Northeast China Based on SPEI. *Ecol. Environ. Sci.* **2015**, *24*, 22–28. [CrossRef]
13. Choat, B.; Jansen, S.; Brodribb, T.J.; Cochard, H.; Delzon, S.; Bhaskar, R.; Bucci, S.J.; Feild, T.S.; Gleason, S.M.; Hacke, U.G.; et al. Global Convergence in the Vulnerability of Forests to Drought. *Nature* **2012**, *491*, 752–755. [CrossRef] [PubMed]
14. Gao, Y.; Zhao, H.; Gao, F.; Zhu, H.; Qu, H.; Zhao, F. Climate Change Trend in Future and Its Influence on Wetlands in the Greater Khingan Mountains. *J. Glaciol. Geocryol.* **2016**, *38*, 47–56.
15. Yiliang, Z. *Vegetation Geography in Northeast China*; Science Press: Beijing, China, 1997.
16. Kajimoto, T.; Matsuura, Y.; Osawa, A.; Prokushkin, A.S.; Sofronov, M.A.; Abaimov, A.P. Root System Development of *Larix Gmelinii* Trees Affected by Micro-Scale Conditions of Permafrost Soils in Central Siberia. In *Roots: The Dynamic Interface between Plants and the Earth: The 6th Symposium of the International Society of Root Research, 11–15 November 2001, Nagoya, Japan*; Abe, J., Ed.; Developments in Plant and Soil Sciences; Springer Netherlands: Dordrecht, The Netherlands, 2003; pp. 281–292, ISBN 978-94-017-2923-9.
17. Zhang, P.; Liu, B. Effect of Climate Change on *Larix Gmelinii* Growth in Different Latitudes. *J. Northeast. For. Univ.* **2015**, *43*, 10–13.
18. Chang, Y.; Chen, Z.; Zhang, X.; Bai, X.; Zhao, X.; Li, J.; Lu, X. Responses of Radial Growth to Temperature in *Larix Gmelinii* of the Da Hinggan Ling under Climate Warming. *Chin. J. Plant Ecol.* **2017**, *41*, 279–289.
19. Jiang, Y.; Zhang, J.; Han, S.; Chen, Z.; Setälä, H.; Yu, J.; Zheng, X.; Guo, Y.; Gu, Y. Radial Growth Response of *Larix Gmelinii* to Climate along a Latitudinal Gradient in the Greater Khingan Mountains, Northeastern China. *Forests* **2016**, *7*, 295. [CrossRef]
20. Dapao, Y.; Shunzhong, W. Relationship between tree-ring chronology of *Larix olgensis* in Changbai Mountains and the climae change. *Chin. J. Appl. Ecol.* **2005**, *16*, 14.
21. McDowell, N.G.; Allen, C.D. Darcy's Law Predicts Widespread Forest Mortality under Climate Warming. *Nat. Clim. Change* **2015**, *5*, 669–672. [CrossRef]
22. Yang, L.B.; Sui, X.; Wei, D.; Cui, F.X.; Zhu, D.G.; Ni, H.W. Fungal diversity in the brown coniferous forest soils of Daxing'anling Mountains, Northeast China. *Chin. J. Appl. Ecol.* **2019**, *30*, 3411–3418. [CrossRef]
23. Holmes, R.L. Computer-assisted quality control in tree-ring dating and measurement. *Tree-Ring Bull.* **1983**, *43*, 51–67.
24. Yindelehei, B.A.; Xiang, B.A.O.; Mei, Z.; Pengwu, Z.; Liang, S.H.I.; Liangjie, H.A. Temporal and Spatial Variation Characteristics of Vegetation Cover on the Northern Slope of Daxing'anling Based on MODIS NDVI. *For. Resour. Wanagement* **2018**, *6*, 50. [CrossRef]
25. National Climate Center; China Meteorological Administration Bureau of Forecasting and Network, China Meteorological Bureau Lanzhou Institute of Arid Meteorology Grades of Meteorological Drought 2018. Available online: https://www.nssi.org.cn/nssi/front/107243850.html (accessed on 26 September 2021).
26. Pang, R.; Liu, M.; Gao, L.; Li, S.; Han, X. Dynamic Study on Probability Distribution Zone of *Larix Gmelinii* Forest Based on Species Distribution Model. *J. Northwest For. Univ.* **2023**, *38*, 1–9. [CrossRef]
27. Guerrero-Salazar, P.L.A.; Yevjevich, V.M. Analysis of Drought Characteristics by the Theory of Runs. Ph.D. Thesis, Colorado State University, Fort Collins, CO, USA, 1975.

28. State Forestry Administration of China. Tree biomass models and related parameters to carbon accounting for Larix 2016. Available online: https://www.nssi.org.cn/nssi/front/106651533.html (accessed on 12 October 2021).
29. Qibing, Z.; Ouya, F.; Lixin, L. *Dendroecological Studies on the Tibetan Plateau*; Science Press: Beijing, China, 2019; ISBN 978-7-03-062181-8.
30. Kassambara, A.; Kassambara, M.A. *Package 'Ggpubr'*, R package version 0.1.; R Team: Viena, Austria, 2020; Volume 6.
31. Ali, A.; Wang, L.-Q. Big-Sized Trees and Forest Functioning: Current Knowledge and Future Perspectives. *Ecol. Indic.* **2021**, *127*, 107760. [CrossRef]
32. Baltzer, J.L.; Veness, T.; Chasmer, L.E.; Sniderhan, A.E.; Quinton, W.L. Forests on Thawing Permafrost: Fragmentation, Edge Effects, and Net Forest Loss. *Glob. Change Biol.* **2014**, *20*, 824–834. [CrossRef]
33. R Core Team. *R: A Language and Environment for Statistical Computing*; R Team: Viena, Austria, 2013.
34. Binkley, D. A Hypothesis about the Interaction of Tree Dominance and Stand Production through Stand Development. *For. Ecol. Manag.* **2004**, *190*, 265–271. [CrossRef]
35. Pretzsch, H.; Schütze, G.; Biber, P. Drought Can Favour the Growth of Small in Relation to Tall Trees in Mature Stands of Norway Spruce and European Beech. *For. Ecosyst.* **2018**, *5*, 20. [CrossRef]
36. Bréda, N.; Granier, A.; Aussenac, G. Effects of Thinning on Soil and Tree Water Relations, Transpiration and Growth in an Oak Forest (*Quercus petraea* (Matt.) Liebl.). *Tree Physiol.* **1995**, *15*, 295–306. [CrossRef] [PubMed]
37. Hessburg, P.F.; Charnley, S.; Wendel, K.; White, E.M.; Singleton, P.H.; Peterson, D.W.; Halofsky, J.E.; Gray, A.N.; Spies, T.A.; Flitcrof, R.L. *The 1994 Eastside Screens Large-Tree Harvest Limit: Review of Science Relevant to Forest Planning 25 Years Later*; United States Department of Agriculture, Forest Service Pacific Northwest: Portland, OR, USA, 2020.
38. Pretzsch, H.; Biber, P. Size-Symmetric versus Size-Asymmetric Competition and Growth Partitioning among Trees in Forest Stands along an Ecological Gradient in Central Europe. *Can. J. For. Res.* **2010**, *40*, 370–384. [CrossRef]
39. Lucash, M.S.; Scheller, R.M.; Sturtevant, B.R.; Gustafson, E.J.; Kretchun, A.M.; Foster, J.R. More than the Sum of Its Parts: How Disturbance Interactions Shape Forest Dynamics under Climate Change. *Ecosphere* **2018**, *9*, e02293. [CrossRef]
40. Li, Y.; Zhang, Q.-B.; Fang, O.; Mu, Y.-M.; Jia, H.; Lyu, L. Recovery Time of Juniper Trees Is Longer in Wet than Dry Conditions on the Tibetan Plateau in the Past Two Centuries. *For. Ecol. Manag.* **2021**, *497*, 119514. [CrossRef]
41. An, Y.-Y.; Liang, Z.-S. Staged strategy of plants in response to drought stress. *Chin. J. Appl. Ecol.* **2012**, *23*, 2907–2915. [CrossRef]
42. Sánchez-Pinillos, M.; D'Orangeville, L.; Boulanger, Y.; Comeau, P.; Wang, J.; Taylor, A.R.; Kneeshaw, D. Sequential Droughts: A Silent Trigger of Boreal Forest Mortality. *Glob. Change Biol.* **2022**, *28*, 542–556. [CrossRef] [PubMed]
43. Yuan, Z.; Ali, A.; Sanaei, A.; Ruiz-Benito, P.; Jucker, T.; Fang, L.; Bai, E.; Ye, J.; Lin, F.; Fang, S.; et al. Few Large Trees, Rather than Plant Diversity and Composition, Drive the above-Ground Biomass Stock and Dynamics of Temperate Forests in Northeast China. *For. Ecol. Manag.* **2021**, *481*, 118698. [CrossRef]
44. Bose, A.K.; Rohner, B.; Bottero, A.; Ferretti, M.; Forrester, D.I. Did the 2018 Megadrought Change the Partitioning of Growth between Tree Sizes and Species? A Swiss Case-Study. *Plant Biol.* **2022**, *24*, 1146–1156. [CrossRef]
45. Ni, M.; Vellend, M. Space-for-Time Inferences about Range-Edge Dynamics of Tree Species Can Be Influenced by Sampling Biases. *Glob. Change Biol.* **2021**, *27*, 2102–2112. [CrossRef]
46. Wang, F.; Ye, D.-M.; Zang, C.-F.; Zhang, Q.-L. Tree Biomass Distribution of *Larix Gmelinii* in Daxing'anling Mountains of Inner Mongolia. *J. Northwest For. Univ.* **2017**, *32*, 23–28. [CrossRef]
47. Jin, Y.; Li, J.; Bai, X.; Zhao, Y.; Cui, D.; Chen, Z. High Temperatures Constrain Latewood Formation in *Larix Gmelinii* Xylem in Boreal Forests. *Glob. Ecol. Conserv.* **2021**, *30*, e01767. [CrossRef]
48. Hu, H.Q.; Luo, B.Z.; Wei, S.J.; Wei, S.W.; Wen, Z.M.; Sun, L.; Luo, S.; Wang, L.M.; Ma, H.B. Estimating Biological Carbon Storage of Five Typical Forest Types in the Daxing'anling Mountains, Heilongjiang, China. *Acta Ecol. Sin.* **2015**, *35*, 5745–5760.

forests

MDPI

Article

Forest Adaptation to Climate Change: Altitudinal Response and Wood Variation in Natural-Growth *Cunninghamia lanceolata* in the Context of Climate Change

Gongliang Xie, Sen Liu, Ting Chang and Ninghua Zhu *

College of Forestry, Central South University of Forestry and Technology, Changsha 410004, China; gongliangxie@outlook.com (G.X.)
* Correspondence: zhuninghua@yahoo.com

Abstract: This research delves into the impact of climate change on the wood traits of *Cunninghamia lanceolata* across various altitudinal gradients, aiming to understand the influence of altitude and climatic factors like temperature and precipitation on key wood characteristics. Employing a comprehensive approach, samples were collected from different altitudes for detailed phenotypic analysis. Methods included Pearson correlation, principal component analysis, cluster analysis, and random forest analysis. Results revealed significant variations in wood traits such as heartwood ratio, tracheid length, and width across altitudes. Notably, wood traits in lower- and middle-elevation populations exhibited higher variability compared to higher elevations, indicating greater environmental diversity and genetic adaptability at these altitudes. Climatic factors, particularly temperature and precipitation, were found to increasingly influence wood trait variation with altitude. The research concludes that the adaptation of *Cunninghamia lanceolata* to climate change is significantly influenced by both altitudinal and climatic factors, highlighting their importance in forest genetic breeding and conservation strategies amidst global climate change.

Keywords: *Cunninghamia lanceolata*; wood traits; altitude; climatic adaptation; random forest analysis

Citation: Xie, G.; Liu, S.; Chang, T.; Zhu, N. Forest Adaptation to Climate Change: Altitudinal Response and Wood Variation in Natural-Growth *Cunninghamia lanceolata* in the Context of Climate Change. *Forests* **2024**, *15*, 411. https://doi.org/10.3390/f15030411

Academic Editors: Yassine Messaoud, Jan Světlík and Giorgio Alberti

Received: 26 January 2024
Revised: 18 February 2024
Accepted: 19 February 2024
Published: 21 February 2024

1. Introduction

Cunninghamia lanceolata, endemic to China, is a predominant afforestation species in southern China. It covers one-fourth of the country's artificial arboreal forests in area and one-third in volume, amounting to 9.87 million hectares and 755 million cubic meters, respectively, leading the nation in both aspects [1]. With a long history of cultivation and a wide range of applications, natural occurrences of *Cunninghamia lanceolata* have become exceedingly rare due to prolonged human interference and high levels of gene flow, resulting in minimal genetic differentiation among different sources of *Cunninghamia lanceolata* [2]. This suggests a potential weakening of adaptability against the backdrop of climate change [3]. In 2012, our research team discovered a natural population of *Cunninghamia lanceolata* in the Xiaoxi National Nature Reserve in Hunan Province, a narrowly distributed ecotype. Locally known as "Iron-Heart *Cunninghamia lanceolata*", this variant is distinguished by its dense texture, good rot resistance, high heartwood ratio, and brownish heartwood [4]. Its mechanical properties, such as bending strength, compressive strength, and shrinkage, are significantly superior to other populations of *Cunninghamia lanceolata* [5]. The quality of wood is a crucial indicator for timber forest cultivation [6]. With the shift in China's timber market demand, developing precious native timber species is a key step towards modernizing forestry [7]. The superior wood quality of this regional *Cunninghamia lanceolata* has attracted widespread attention. Extensive research has been conducted on its wood physical properties [4,8], seed traits [9], breeding population construction [10,11], and spatial genetic structure [12]. To further explore the excellent germplasm resources of *Cunninghamia lanceolata* and enhance the sustainable utilization and management of these resources, our research focuses on wood properties of natural populations of *Cunninghamia lanceolata* and their correlation with altitudinal climate change.

Genetic diversity, encompassing the genetic variation within biological populations and individuals [13], forms the foundation of species diversity and represents the core of biodiversity. It reflects the genetic richness of species and determines their potential to adapt to environmental changes [14,15]. Phenotypic diversity, as the external manifestation of genetic diversity, results from the interplay of genes and the environment [16]. The phenotypic variation among individuals within a species is influenced not only by specific habitat conditions but also by historical processes and phylogeny [17]. In different environmental settings, populations of the same plant species may undergo varying selective pressures leading to genetic and phenotypic distinctions among populations [18]. Thus, researching the phenotypic diversity of forest tree populations in diverse ecological environments not only unveils the extent of genetic variation but also aids in understanding the evolutionary level and environmental adaptability of plants [19,20]. Studies have demonstrated that wood traits are significantly influenced by environmental factors such as geographic location (latitude and longitude) [21], altitude [22], light [23], soil [24], and climate [25,26].

According to relevant statistics, by 2020, the global temperature had risen by approximately 0.99 °C compared to the pre-industrial era (1850–1900), leading to an increase in extreme weather events worldwide [27]. Global climate change is accelerating its impact on forest disturbance mechanisms and adaptability [28], with effects on forests being dual-natured; they can be either positive or negative [29]. The adaptive capacity of tree populations to climate change depends on intraspecific genetic variation and phenotypic plasticity [30]. Altitude, a pivotal ecological factor, influences plant growth through gradient changes in temperature, precipitation, light, soil, and other factors [31]. This contributes to adaptive variation in plant phenotypic traits, making altitude an ideal subject for researching genetic variation [32,33]. While previous research has explored wood trait variation across altitudinal gradients from ecological [34,35], physiological [36–38], and dendrochronological perspectives [39,40], there remains a gap in understanding the genetic variation of these traits and their impact on the environmental adaptability of trees particularly from a genetic standpoint. The upper limit of the vertical distribution of *Cunninghamia lanceolata* often varies with different topographical and climatic conditions In the high-latitude Dabie Mountains, it grows below an altitude of 700 m, while in the lower latitude region of Dali, Yunnan, it can be found at elevations as high as 2500 m [41] Investigating the variation of *Cunninghamia lanceolata* across different altitudinal gradients and the influence of altitudinal climatic factors on wood traits can provide insights into the species' response to climate change and the role of the environment in shaping traits [42]

Wood traits encompass a range of characteristics, primarily including anatomical traits such as tracheids and microfibril angles, physical traits like wood density, shrinkage and swelling properties, mechanical properties, and chemical traits involving the content of cellulose, hemicellulose, lignin, and various metabolites [43]. Wood density is a pivotal factor affecting plant ecological strategies, influenced by vascular traits and tree growth rates [44,45]. Denser wood is not only harder and more decay-resistant but also exhibits greater wind resistance [46]. In this research, seven representative wood traits were meticulously selected for comprehensive analysis. The annual ring width, indicative of the wood layer formed during a tree's one-year growth cycle, serves not only as a crucial indicator of growth rate [47] but also mirrors historical climate changes, ecosystem dynamics, and trees' adaptability to environmental shifts [48]. The heartwood ratio, wood density, and water absorption rate are pivotal in assessing wood quality. The heartwood ratio often signifies wood quality, while wood density and water absorption directly influence wood's processing performance, strength, and durability [49]. Wood with high water absorption is susceptible to swelling, deformation, and biodegradation [50]. Moreover, variations in wood density, a vital attribute in trees' carbon storage strategy, emerge as significant indicators for climate change assessment [51,52]. Lastly, the dimensions, shape, and arrangement of tracheids, crucial xylem cells in gymnosperms responsible for water and nutrient transport, not only impact the wood's physical and mechanical properties [53] but also reflect the plant's adaptability to its growth environment [54].

In this research, wood core samples were meticulously collected from four distinct altitudinal populations of *Cunninghamia lanceolata* within its natural habitat. We conducted a thorough analysis of phenotypic variation in seven key wood traits. The objective was to unravel the patterns of phenotypic differentiation and variation in *Cunninghamia lanceolata* wood traits along altitudinal gradients and to investigate the correlations between these traits and corresponding climatic factors at different altitudes. The findings of this research significantly enhance our understanding of the growth characteristics and adaptive mechanisms of *Cunninghamia lanceolata* across various altitudes. This is crucial for the discovery and conservation of high-quality germplasm resources of *Cunninghamia lanceolata*, as well as for bolstering its conservation and sustainable utilization. Furthermore, in the context of global climate change, these insights are pivotal in forecasting future shifts in forest ecosystems and elucidating the mechanisms through which trees adapt to environmental changes.

2. Materials and Methods

2.1. Overview of the Research Area

Xiaoxi National Nature Reserve (hereinafter referred to as Xiaoxi) is located in Xiangxi Tujia and Miao Autonomous Prefecture of Hunan Province (Figure 1), east of the Yunnan–Guizhou Plateau and the central part of the Wuling Mountain Range, which is the intersection of China's second and third ladders, with the highest altitude of 1327.1 m, and latitude and longitude of 110°6′50″–110°21′35″ E, 28°42′15″–28°53′55″ N. The climate is a subtropical humid monsoon climate, a warm and humid climate with abundant rainfall and an annual average temperature of 13~15 °C, annual precipitation of 1300~1400 mm, and relative humidity of 79%. The soil in Xiaoxi shows distinct zonal characteristics: below 400 m in altitude, there are mountain red soils; between 400 m and 500 m, mountain yellow–red soils; from 500 m to 800 m, mountain yellow soils; and above 800 m, mountain yellow–brown soils. Due to the diversity of habitats in Xiaoxi, the area boasts an exceptionally rich variety of plant species. According to surveys and analyses, there are a total of 2252 species of seed plants in Xiaoxi, including 20 species of gymnosperms and 2232 species of angiosperms. The reserve is home to 43 species of nationally protected plants of Classes I and II, such as *Manglietia decidua*, *Davidia involucrata Baill*, *Taxus wallichiana var. mairei*, and *Bretschneidera sinensis* [55]. *Cunninghamia lanceolata* predominantly exists as scattered individuals in primary and secondary forests, mainly distributed at altitudes between 500 m and 1200 m.

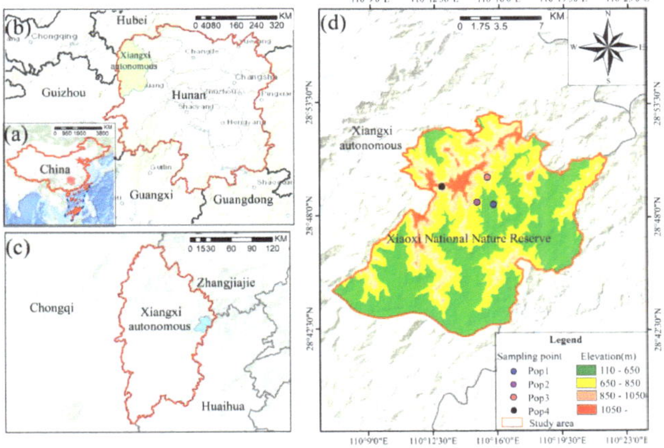

Figure 1. Schematic map of the sampling research area of *Cunninghamia lanceolata* at different altitudes. (**a**) Administrative map of China; (**b**) Administrative map of Hunan; (**c**) Tujia–Miao autonomous continent of Xiangxi Prefecture; (**d**) Xiaoxi National Nature Reserve.

2.2. Sample Collection

In November 2022, a comprehensive field survey was conducted on the naturally grown *Cunninghamia lanceolata* in the Xiaoxi National Nature Reserve. The species was categorized into four altitudinal populations, each separated by an interval of 200 m population 1 (Pop1) at 550 m, population 2 (Pop2) at 750 m, population 3 (Pop3) at 950 m, and population 4 (Pop4) at 1150 m. For each population, 30 adult individual plants exhibiting normal growth and free from apparent pests and diseases were selected, with a consistent diameter at breast height (DBH, 1.3 m). Due to prior human interference and the species' altitudinal distribution limits, only 8 sample trees were collected for Pop4. A minimum distance of 100 m was maintained between each plant. Using a 5 mm growth cone, a core sample was extracted through the pith from south to north at a height of 1.3 m from the target tree. The core was then placed into a PVC pipe and numbered for subsequent wood trait analysis. Concurrently, the DBH and height of each sample tree were measured, and GPS coordinates were recorded to locate the sample sites. Detailed information about the sample sites and associated meteorological factors is presented in Table S1.

2.3. Measurement of Traits

Wood properties assessed in this research encompass seven key attributes: average annual ring width (Rb), heartwood ratio (P), basic wood density (WHD), water absorption (Hy), tracheid length (L), tracheid width (D), and the tracheid length-to-width ratio (L/D).

As per the technical guidelines, the wood core was first smoothed with 600# sandpaper A line was drawn perpendicular to the annual rings in the radial direction, and the total width of the complete annual rings was measured along this line. The total width was precisely measured to 0.01 mm using a ruler, and the number of annual rings within the measurement range was counted. The average width of the annual rings was then calculated by dividing the total width by the number of rings [56]. The heartwood ratio (P) was determined by dividing the disc area, calculated using the radius from the edge of the heartwood to the pith, by the total disc area. This was carried out by measuring the length of the heartwood and the cores, with P calculated as $r^2/R^2 \times 100\%$, where r and R represent the lengths of the heartwood and the core, respectively (Figure 2a). Basic wood density (WHD) was measured using the saturated drainage method. The cores were immersed in water, with water changes every two days, and weighed periodically until a constant saturated water content weight (W1) was achieved. The cores were then dried in an oven at 103 + 2 °C to a constant weight (W2); WHD was calculated as $1(W1/W2 - 0.346)$, and Hy as $(W1 - W2)/W2$. The cores were divided radially into three sections (near the pith middle section, near the bark) and dissociated using a mixture of glacial acetic acid and 30% hydrogen peroxide solution. The dissociated samples were then observed under a bio-digital microscope (model OLYMPUS-BX51) to analyze the tracheids' morphological characteristics (Figure 2b). The lengths and widths of the tracheids were measured 20 times for each sample using Photoshop 2022 software, with the average of these measurements from the three samples taken as the final value. Finally, the tracheid length-to-width ratio was calculated [57].

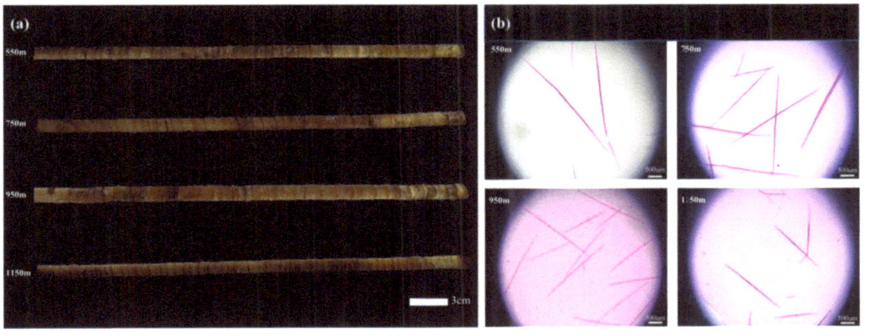

Figure 2. Schematic diagrams of *Cunninghamia lanceolata* cores and tracheids at different altitudes; 550 m, 750 m, 950 m, and 1150 m from Pop1, Pop2, Pop3, and Pop4 populations, respectively, representing the average altitude sampled, (**a**) *Cunninghamia lanceolata* cores collected from different altitudes; (**b**) tracheids under the bio-digital microscope.

2.4. Data Processing and Analysis

Preliminary processing of the measured data was carried out using Excel, and the population mean, standard deviation, intra- and interpopulation coefficients of variation, and Nestle's analysis of variance (ANOVA) were calculated for each trait using SPSS 27.

(1) Inter- and intrapopulation variation was analyzed based on Nestle's ANOVA model (significance tests were performed using the Duncan method of multiple comparisons), with a linear model of:

$$Y_{ijk} = \mu + \tau_i + \delta_{j(i)} + \varepsilon_{ijk} \quad (1)$$

where: Y_{ijk} is the kth observation of the jth monoculture in the ith population, μ is the overall mean, τ_i is the between-population effect value, $\delta_{j(i)}$ is the within-population monoculture random effect value, and ε_{ijk} is the random error.

(2) The coefficient of variation (CV) between populations was calculated using the following formula:

$$CV = \delta / \overline{x} \quad (2)$$

where: δ and $\overline{x}$ represent the standard deviation and mean of the trait, respectively.

The climate data for this research were sourced from the World Climate Database (https://worldclim.org/data/worldclim21.html, accessed on 26 October 2023). This database primarily compiles climate observation records from weather stations globally, spanning from 1950 to 2000. The data were generated through interpolation, utilizing a global climate database with a spatial resolution of 30′ [58]. Utilizing ArcGIS 10.2 software, we obtained estimated meteorological data for each altitudinal sampling point. Subsequently, six representative meteorological factors were extracted from a total of nineteen climatic factors through principal component analysis. These factors included the annual mean temperature (BIO1), the average temperature of the wettest season (BIO8), the average temperature of the driest season (BIO9), the annual precipitation (BIO12), the precipitation of the wettest season (BIO16), and the precipitation of the driest season (BIO17). The specifics of these factors are detailed in Table S1.

Data processing and analysis in this research were conducted within the R language environment. Principal component analysis (PCA) was executed using the FactoMineR package, complemented by graphical presentations created with the factoextra and ggplot2 packages. Cluster analysis employed the Euclidean distance matrix, computed via the ape package, and hierarchical clustering was performed using the Ward.D method, with graphical outputs generated by the RColorBrewer package. The linkET package facilitated correlation analysis, calculating Pearson correlation coefficients and significance p-values, which were further validated through the Mantel test. Data processing and transformation were adeptly handled by the dplyr and magrittr packages, with all graphs being plotted using ggplot2.

To uncover the primary drivers of wood trait variation, four traits—heartwood ratio, tracheid length, tracheid width, and tracheid aspect ratio—showing significant correlations with altitudinal climatic factors, were selected based on correlation analysis results. These traits were then subjected to random forest (RF) analysis [59], utilizing the randomForest package to calculate feature importance. The contribution rate of altitudinal climatic factors to the variation of these four wood traits was determined based on this feature importance The contributions were then visualized using Origin 2022 software.

3. Results

3.1. Analysis of Variance of Wood Traits of *Cunninghamia lanceolata* at Different Altitudes

Nested ANOVA conducted on the wood traits of *Cunninghamia lanceolata* across different altitudes revealed significant variations (refer to Table 1). There were highly significant differences observed among *Cunninghamia lanceolata* populations in terms of heartwood ratio, tracheid length, and tracheid length-to-width ratio ($p < 0.001$). Notable differences were also evident in the average annual ring width and tracheid width ($p < 0.05$). However no significant differences were found in basic wood density and water absorption rate The within-population ANOVA indicated a lack of significant differences across all seven wood traits.

Table 1. Nested ANOVA of wood property traits of natural populations of *Cunninghamia lanceolata* at different altitudes.

Wood Property Traits	Mean Square			F	
	Among Populations	Within Populations	Random Error	Among Populations	Within Populations
R_b, mm	207.6424	53.2672	67.8218	3.0616 *	0.7854
P	1031.1240	102.7945	63.9987	16.1116 ***	1.0662
WBD, gm^3	0.0023	0.0014	0.0024	0.9622	0.5977
Hy, %	763.9594	492.2978	820.6657	0.9309	0.5999
L, um	1,076,974.1572	68,727.2615	64,557.9490	16.6823 ***	1.0646
D, um	59.0698	21.0064	20.0651	2.9439 *	1.0469
L/D	385.5733	44.5341	78.5751	4.9071 ***	0.5668

Note: * $p < 0.05$; *** $p > 0.001$. Rb, Average Annual Ring Width; P, Heartwood Ratio; WBD, Basic Wood Density Hy, Water Absorption Rate; L, Tracheid Length; D, Tracheid Width; L/D, Tracheid Length-to-Width Ratio.

Multiple comparisons and comparing the mean values of wood traits at different altitudes (Table 2) showed that the basic wood density, tracheid length, tracheid width, and tracheid length-to-width ratio were the largest in Pop1, reaching 0.42 g/cm^3, 2888.06 mm 45.22 mm, and 67.45, respectively; and they were the smallest in Pop4, with 0.39 g/cm^3 2283.09 mm, 40.99 mm, and 58.20. The mean heartwood ratio and water absorption were the largest in Pop3, reaching 30.51 mm, and the smallest in Pop1, with only 24.35 mm The average annual ring width was the largest in Pop3, reaching 30.51 mm, and the smallest in Pop1, only 24.35 mm. Heartwood ratio and water absorption were the largest in Pop3, 77.51% and 191.9%, respectively, and the smallest in Pop1, only 53.95% and 174.97%, respectively.

The variation of each wood trait showed (see Table 2) that the coefficient of variation of wood traits was 13.55% on average, and the level of variation of different wood traits varied greatly, with the average variation of each trait ranging from 9.40% to 24.31%, with the smallest variation being in tracheid length (9.40%) and the largest being in the average annual ring width (24.31%). The magnitude of variation of wood traits within populations at different altitudes was also significantly different, with the largest magnitude of variation in Pop3 (9.4%~36.38%) and the smallest magnitude of variation in Pop4 (7.81%~13.38%) In terms of the average coefficient of variation of wood traits within the populations the largest and smallest coefficients of variation were found in Pop3 (15.14%) and Pop4 (10.27%), respectively.

Table 2. Variations of wood property traits of natural populations of *Cunninghamia lanceolata* at different altitudes.

Wood Property Traits	Population I		Population Pop2		Population Pop3		Population Pop4		Among Populations	
	Mean ± SD	CV (%)	Mean ± SD	CV (%)	Mean ± SD	CV (%)	Mean ± SD	CV (%)	Mean ± SD	CV (%)
R_b, mm	24.35 ± 6.66b	27.35	28.01 ± 5.88ab	20.99	30.51 ± 11.1a	36.38	29.21 ± 3.61ab	12.36	28.02 ± 6.81	24.31
P	53.95 ± 7.58c	14.04	57.963 ± 8.99c	15.51	63.58 ± 9.64b	15.15	77.51 ± 7.95a	10.25	63.05 ± 8.54	13.50
WBD, g/cm^3	0.42 ± 0.05a	11.90	0.42 ± 0.05a	11.90	0.41 ± 0.04a	9.76	0.39 ± 0.04a	10.26	0.41 ± 0.05	10.98
Hy, %	174.97 ± 27.86a	15.92	177.64 ± 28.43a	16.00	180.61 ± 24.25a	13.43	191.90 ± 25.67a	13.38	181.28 ± 26.55	14.65
L, um	2888.06 ± 266.18a	9.22	2589.22 ± 215.29b	8.31	2503.64 ± 296.24b	11.83	2283.09 ± 187.22c	8.20	2629.51 ± 241.23	9.40
D, um	45.22 ± 4.54a	10.04	44.65 ± 4.92a	11.02	43.21 ± 4.32ab	10.00	40.99 ± 3.2b	7.81	43.52 ± 4.25	9.75
L/D	67.45 ± 10.06a	14.91	61.08 ± 9b	14.73	59.14 ± 5.56b	9.40	58.20 ± 5.60b	9.62	61.47 ± 7.56	12.29
average value	-	14.77	-	14.07	-	15.14	-	10.27	-	13.55

Note: SD, standard deviation; CV, coefficient of variation. The underline represents the maximum value and the letters a, b, and c represent the results of multiple comparisons. Rb, Average Annual Ring Width; P, Heartwood Ratio; WBD, Basic Wood Density; Hy, Water Absorption Rate; L, Tracheid Length; D, Tracheid Width; L/D, Tracheid Length-to-Width Ratio.

3.2. *Principal Component Analysis and Cluster Analysis of Wood Traits of Cunninghamia lanceolata at Different Altitudes*

3.2.1. Principal Component Analysis of *Cunninghamia lanceolata* Wood Traits at Different Altitudes

The principal component analysis (PCA) conducted on the *Cunninghamia lanceolata* populations at different altitudes identified three principal components with eigenvalues exceeding 1. The cumulative contribution rate of these first three principal components to the variance in the seven wood traits was 79.79% (as shown in Figure 3, Tables S2 and S3), effectively encapsulating most of the information pertinent to the characteristic variables of the wood traits. The first principal component, accounting for 37.98% of the total variance, included key variables such as water absorption rate, basic wood density, average annual ring width, and the aspect ratio of *Cunninghamia lanceolata*. The second principal component, encompassing tracheid length and heartwood ratio, explained 23.14% of the total variance. The third principal component, comprising tracheid width and tracheid aspect ratio, accounted for 10.82% of the total variance. The cumulative contributions of these principal components, in descending order of significance, were for tracheid width, basic wood density, water absorption, tracheid aspect ratio, tracheid length, mean annual ring width, and heartwood ratio.

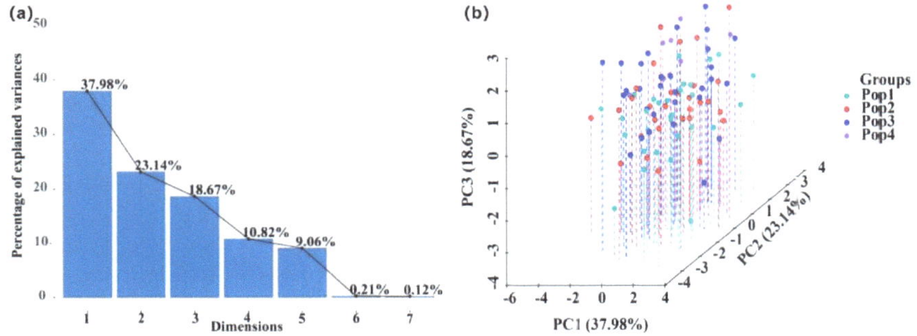

Figure 3. (**a**) Fragmentation plot; (**b**) Principal component analysis 3D plot. Pop1, Pop2, Pop2, Pop4 represent four *Cunninghamia lanceolata* populations at different elevations.

3.2.2. Cluster Analysis of Wood Traits of *Cunninghamia lanceolata* at Different Altitudes

The clustering analysis of 12 wood traits from four *Cunninghamia lanceolata* populations at different altitudes revealed distinct clusters, as illustrated in Figure 4. The analysis categorized the 98 samples into four clusters. Cluster I comprised 27 samples, representing 27.55% of the total, predominantly including 16 samples from Pop1 and 7 from Pop2. Cluster II, with the fewest samples, contained 10 samples, accounting for 12.20% of the total, mainly from Pop3. Cluster III encompassed 20 samples, constituting 20.41% of the total, primarily including eight samples from Pop3 and six from Pop4. Cluster IV, the largest population, contained 41 samples, making up 41.84% of the total, with 16 samples from Pop2, 13 from Pop1, and 12 from Pop3.

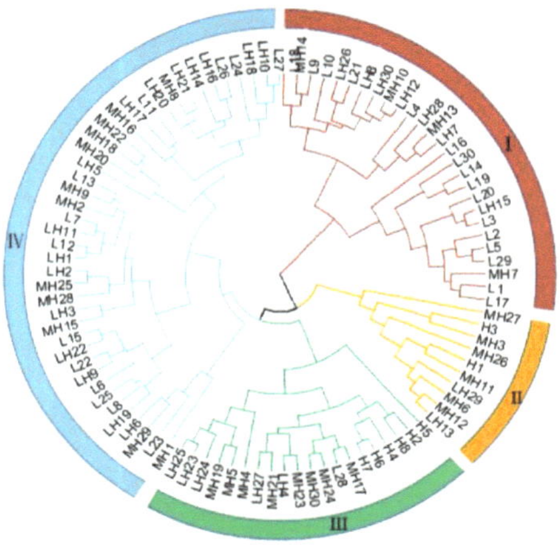

Figure 4. Cluster analysis diagram. L1–L30 represent 30 samples from Pop1; LH1–LH30 represent 30 samples from Pop2; MH1–MH30 represent 30 samples from Pop1; H1–H8 represent 8 samples from Pop4.

3.3. *Correlation between Cunninghamia lanceolata Wood Traits and Altitude Climate Factors at Different Altitudes*

3.3.1. Correlation Analysis between Wood Traits of *Cunninghamia lanceolata* at Different Altitudes

Pearson's correlation analysis was conducted among various wood traits of *Cunninghamia lanceolata* (as depicted in Figure 5 and detailed in the accompanying Table S4). The analysis revealed that the mean annual ring width was significantly negatively correlated with the tracheid aspect ratio (−0.2449 *) and tracheid length (−0.2363 *). It also showed a significant negative correlation with wood density (r = −0.4641 ***) and a significant positive correlation with water absorption rate (r = 0.4668 ***). The heartwood ratio exhibited a significant negative correlation with the tracheid aspect ratio (−0.3011 **) and a highly significant negative correlation with tracheid length (−0.3525 ***). Wood density was found to have a significant positive correlation with the tracheid aspect ratio (0.2182 *) and a highly significant negative correlation with water absorption (−0.9908 ***). The tracheid aspect ratio demonstrated a highly significant negative correlation with water absorption (−0.2280 *) and tracheid width (−0.4466 ***), indicating significant and highly significant negative correlations, respectively. Furthermore, tracheid length showed a highly significant positive correlation with tracheid width (0.3570 ***) and the tracheid length-to-width ratio (0.6573 ***).

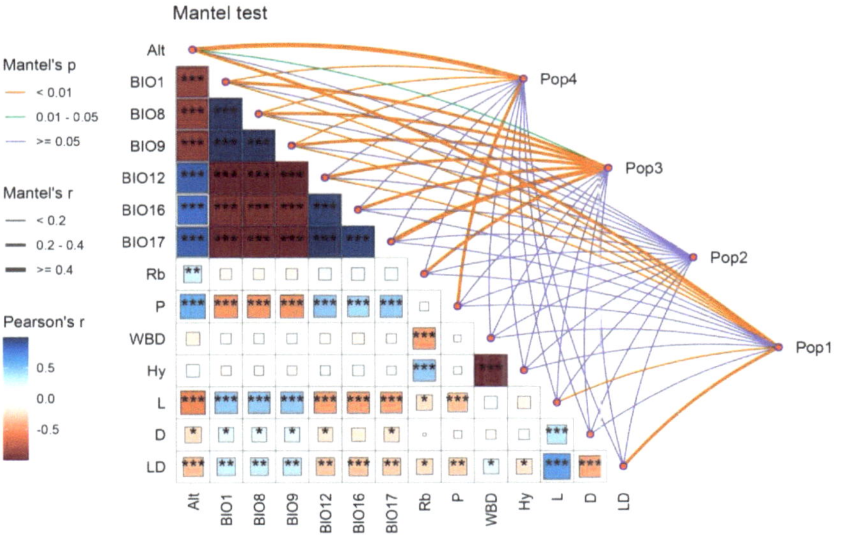

Figure 5. Correlation analysis and Mantel test between altitude climatic factors and wood traits. * $p < 0.05$; ** $p < 0.01$; *** $p < 0.001$. BIO1, Mean Annual Temperature; BIO8, Mean Wettest Season Temperature; BIO9, Mean Driest Season Temperature; BIO12, Annual Precipitation; BIO16, Wettest Season Precipitation; BIO17, Driest Season Precipitation. Rb, Average Annual Ring Width; P, Heartwood Ratio; WBD, Basic Wood Density; Hy, Water Absorption Rate; L, Tracheid Length; D, Tracheid Width; L/D, Tracheid Length-to-Width Ratio.

3.3.2. Correlation Analysis of Wood Traits of *Cunninghamia lanceolata* at Different Altitudes with Altitude Climatic Factors

The correlation analysis between wood traits and altitudinal–climatic factors revealed highly significant correlations for three traits: heartwood ratio, tracheid length, and tracheid length–width ratio (Figure 5 and Table S4). The results of detailed analyses are as follows: there was a significant positive correlation between the average annual ring width and altitude (0.2852 *). A highly significant positive correlation was observed between the heartwood ratio and altitude (0.5793 ***), annual precipitation (0.4752 ***), precipitation of the wettest season (0.4153 ***), and precipitation of the driest season (0.4687 ***), along with a highly significant negative correlation with annual average temperature (−0.5019 ***), average temperature of the wettest season (−0.5004 ***), and average temperature of the driest season (−0.5061 ***). Tracheid length demonstrated a highly significant negative correlation with altitude (−0.5525 ***), annual precipitation (−0.4516 ***), precipitation of the wettest season (−0.4497), and precipitation of the driest season (−0.4369 ***) and a highly significant positive correlation with annual average temperature (0.4634 ***), average temperature of the wettest season (0.4626 ***), and average temperature of the driest season (0.4630 ***). Tracheid width showed a significant negative correlation with altitude (−0.2488 *), annual precipitation (−0.2030 *), and precipitation of the driest season (−0.2123 *) and a significant positive correlation with annual average temperature (0.2111 *), average temperature of the wettest season (0.2103 *), and average temperature of the driest season (0.2135 *). The tracheid length–width ratio exhibited a highly significant negative correlation with altitude (−0.3608 ***) and precipitation of the wettest season (−0.3303), a significant negative correlation with annual precipitation (−0.3100 *) and precipitation of the driest season (−0.2960 *), and a significant positive correlation with annual average temperature (0.3095 **), average temperature of the wettest season (0.3095 **), and average temperature of the driest season (0.3080 **). Notably, wood density and water absorption rate showed no significant correlation with altitudinal–climatic factors.

3.3.3. Mantel Test of *Cunninghamia lanceolata* Populations at Different Altitudes with Altitudinal Climatic Factors

The Mantel test conducted on the wood traits of *Cunninghamia lanceolata* at different altitudes in relation to altitudinal climatic factors revealed (as shown in Figure 5, Table S5) a highly significant correlation between the wood traits of the Pop3 population and the driest seasonal precipitation (BIO17). Additionally, a highly significant correlation was observed between the wood traits of the Pop4 population and altitude. The wood traits of the Pop3 population exhibited significant or highly significant correlations with altitudinal climate factors. In contrast, the wood traits of Pop1 and Pop4 populations showed partial significant correlations with these factors, whereas the wood traits of the Pop2 population did not display any significant correlation with altitudinal climate factors.

3.4. Random Forest Analysis of Wood Traits of Cunninghamia lanceolata at Different Altitudes

Random forest analysis results (illustrated in Figure 6a) suggested that the altitude factor predominantly influenced the variation of four wood traits: heartwood ratio, tracheid length, tracheid width, and tracheid length-to-width ratio, with contribution rates of 48.81%, 45.71%, 61.94%, and 60.92%, respectively. The variances of these four wood traits were differentially impacted by climatic factors. For instance, the average temperature during the wettest season was the most influential climatic factor for the heartwood ratio (10.18%), tracheid length (9.97%), and tracheid length-to-width ratio (7.13%). In contrast, the precipitation during the driest season had the most significant impact on tracheid width (7.55%). Additionally, the cumulative contributions of temperature factors (BIO1, BIO8, BIO9) to the heartwood ratio, tracheid length, tracheid width, and tracheid length-to-width ratio were 29.72%, 28.22%, 19.53%, and 20.70%, respectively. These were higher than the cumulative contributions of precipitation factors (BIO12, BIO16, and BIO17) to the four wood traits, which were 21.46%, 26.07%, 18.53%, and 18.38%, respectively. This indicates that among the climatic factors, temperature had a greater contribution to the variation of wood traits compared to precipitation.

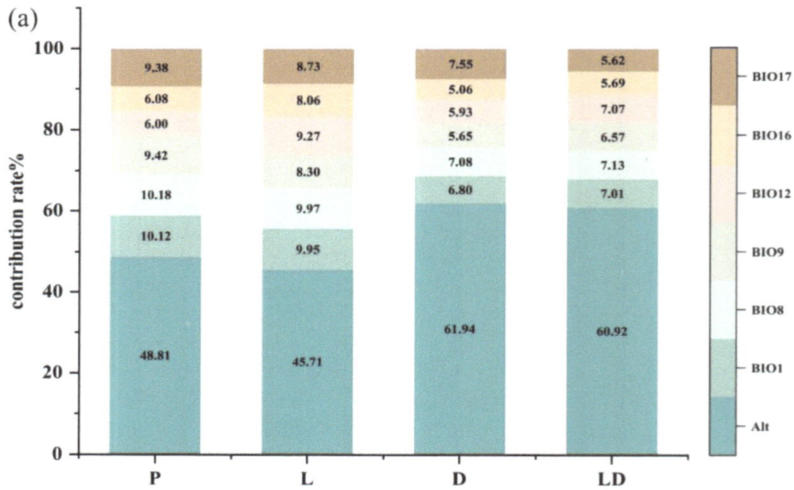

Figure 6. *Cont.*

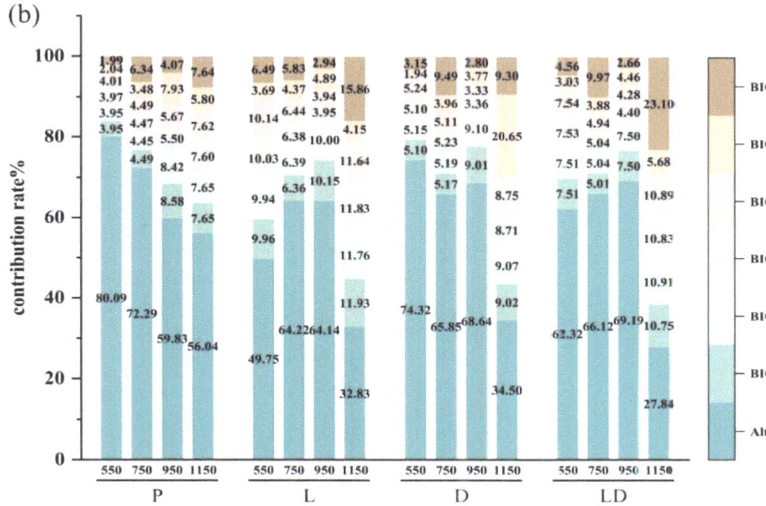

Figure 6. (**a**) Stacked diagram of the sources of variation in wood traits. (**b**) Stacked diagram of the sources of variation in wood traits at different altitudes. Random forest analysis. Values represent the contribution of altitudinal climatic factors to variation in wood traits. BIO1, Mean Annual Temperature; BIO8, Mean Wettest Season Temperature; BIO9, Mean Driest Season Temperature; BIO12, Annual Precipitation; BIO16, Wetter Season Precipitation; BIO17, Driest Season Precipitation; P, Heartwood Ratio; L, Tracheid Length; D, Tracheid Width; L/D, Tracheid Length-to-Width Ratio; 550, 750, 950, 1150 for different altitudes.

Further analysis revealed significant differences in the contribution of altitudinal climatic factors to the variability of four wood traits across the altitudinal gradient (as shown in Figure 6b). In the middle- and low-altitude gradients (populations Pop1, Pop2, and Pop3), the altitude factor was the predominant influence on the variation of the four wood traits. However, with increasing altitude, the influence of the altitude factor on the variance of heartwood ratio decreased, while its effects on tracheid length and tracheid aspect ratio initially increased and then decreased. The impact on tracheid width showed an overall decreasing trend. This decrease in the influence of the altitude factor was accompanied by an increased impact of the climatic factor on the four wood traits. Notably, except for the heartwood ratio, the cumulative contribution of climatic factors to tracheid length, tracheid width, and tracheid aspect ratio was greater than that of the altitude factor at higher altitudes (population Pop4), reaching 67.17%, 65%, and 72.16%, respectively. These factors became the primary influences on the wood traits at higher altitudes.

4. Discussion

4.1. Analysis of Variation in Wood Traits of *Cunninghamia lanceolata* at Different Altitudes

Plant phenotypic traits, resulting from the interplay between genetics and environment, not only mirror genetic diversity but are also pivotal in identifying superior germplasm resources [12]. This research delved into the variation of wood traits in *Cunninghamia lanceolata* across different altitudes, examining the influence of altitudinal climatic factors. Nested ANOVA results indicated highly significant differences ($p < 0.001$) in heartwood ratio, tracheid length, and tracheid length-to-width ratio among various altitudes. Mean annual ring width and tracheid width also showed significant differences ($p < 0.05$). These findings align with Elif Topaloğlu et al.'s research [60] on the anatomical properties of oriental beech wood, underscoring the variability of wood traits across altitudes. However, basic wood density and water absorption rate did not exhibit significant differences across altitudes, suggesting a stronger genetic influence on these traits, with a lesser role for environmental factors [61]. The within-population ANOVA revealed no significant differences in any of the seven wood traits, contrasting with Hongjing Duan et al.'s findings [57], which showed significant within-population variation in *Cunninghamia lanceolate* wood traits from different seed sources. This discrepancy might stem from factors like similar selection pressures [62], inbreeding [63], gene flow [64], and human interference [65] in *Cunninghamia lanceolata* populations in the same altitudinal gradient. The variability of *Cunninghamia lanceolata* wood traits across altitudes ranged from 9.40% to 24.31%, relatively modest compared to the variability (6.43% to 45.55%) in *Cunninghamia lanceolate* wood traits from different seed sources [57]. The coefficients of variation for middle- and low-altitude populations (Pop1, Pop2, Pop3) were comparable, yet significantly higher than the 10.27% for the high-altitude population Pop4. This could reflect the more complex and diverse habitats at middle and low altitudes, while the high-altitude region might offer more severe and uniform environmental conditions, constraining phenotypic variation [66].

To delve deeper into the contribution of the seven wood traits to overall variance, we conducted a principal component analysis (PCA). This analysis identified the key variables of *Cunninghamia lanceolata* wood traits at various altitudes and quantified their contributions to total variance. This approach offers valuable insights into the adaptation and growth characteristics of *Cunninghamia lanceolata* across different altitudinal ranges. The PCA results indicated that the first three principal components accounted for a cumulative contribution of 79.79% to the total variance. The order of contribution of the seven wood traits, from highest to lowest, was tracheid width, basic wood density, water absorption, tracheid length-to-width ratio, tracheid length, average annual ring width, and heartwood ratio. This analysis underscored the principal factors influencing wood traits [67], providing a deeper understanding of *Cunninghamia lanceolata*'s adaptation and growth at different altitudes [68] and aiding in the selection of superior germplasm resources [69].

Cluster analysis of *Cunninghamia lanceolata* wood traits across different altitudes revealed that the 98 samples were categorized into four taxa. Interestingly, these clusters did not align strictly with altitude, suggesting that while traits like heartwood ratio, tracheid length, and tracheid length-to-width ratio varied significantly among altitudes, these variations were not solely altitude-dependent. This indicates the potential influence of genetic factors or other environmental elements beyond altitude on wood trait formation. For instance, studies on wood traits in species such as the spruce *Picea asperata* [70], Japanese larch [71], and poplar [72] have demonstrated that traits like tracheid size and wood density are strongly genetically controlled. Additionally, environmental factors like temperature [73], precipitation [74], soil [75], and their interactions [76] also significantly impact wood traits. Therefore, the variation in *Cunninghamia lanceolata* wood traits could be attributed to a combination of factors, including but not limited to altitude.

4.2. Variation Patterns of Wood Traits and Ecological Adaptations of Cunninghamia lanceolata *at Different Altitudes*

Mountains, as unique ecosystems, exhibit high habitat heterogeneity and climatic diversity. Environmental factors such as temperature, light, moisture, and soil exhibit gradient variations with altitude [77]. This gradient provides a distinctive setting for investigating species' genetic variation and their responses to climate change [78]. As altitude increases, plant growth and development are constrained by temperature and moisture, leading to notable changes in adaptive traits, which in turn influence the wood traits of trees [79,80].

In our Pearson correlation analysis of *Cunninghamia lanceolata* wood traits at different altitudes, we observed intricate relationships among wood traits and between wood traits and climatic factors at various altitudes. These relationships highlight the significance of quantitative traits in trees, where correlations between two traits might be due to strong linkage disequilibrium or pleiotropic effects of related genes [81]. Our findings revealed that the mean annual ring width was significantly negatively correlated with tracheid length and aspect ratio and highly significantly negatively correlated with wood density. Conversely, it showed highly significant positive correlations with water absorption. These results align with previous studies that underscore the interplay between wood traits and environmental influences [57,82]. Notably, the significant negative correlation between heartwood ratio and tracheid aspect ratio and the highly significant negative correlation with tracheid length suggest a robust connection between heartwood formation and the morphological characteristics of tracheids. Moreover, wood density exhibited a significant positive correlation with tracheid aspect ratio and a highly significant negative correlation with water absorption, underscoring the pivotal role of wood density in determining wood traits.

Further examination of the correlation between wood traits and climatic factors at different altitudes revealed that the heartwood ratio exhibited highly significant positive correlations with altitude, annual precipitation, and precipitation during the wettest and driest seasons and highly significant negative correlations with mean annual air temperature and mean air temperature during the wettest and driest seasons. This pattern suggests that heartwood formation is smaller in environments with higher temperatures and lower precipitation, reflecting the adaptive survival strategies of trees under varying hydrothermal conditions. These observations align with Almeida et al.'s findings [83] regarding the impact of climate on eucalyptus heartwood. Additionally, a significant positive correlation was observed between tracheid length, width, and aspect ratio and altitude and air temperature, indicating an increase in tracheid structural dimensions with rising altitude and air temperature. This trend might be attributed to the influence of climatic conditions on plant growth at higher altitudes [84], as suggested by Fonti et al. [85], where a warming climate is conducive to tracheid growth. Conversely, tracheid structure showed a significant negative correlation with annual precipitation and precipitation during the wettest and driest seasons, implying that excessive moisture may inhibit tracheid growth, highlighting the regulatory role of water conditions on plant cell structure. These insights corroborate previous studies [86,87], underscoring the critical role of environmental factors in shaping plant cell structure. Nonetheless, it is essential to consider the variability across different tree species, geographical locations, and environmental conditions [88,89].

4.3. Analysis of the Contribution Rate of Altitude Climatic Factors to the Variation of Wood Traits of *Cunninghamia lanceolata*

Random forest analysis results underscored the significant impact of altitude on the variability of wood traits like heartwood ratio and tracheid length, width, and aspect ratio, with respective contribution rates of 48.81%, 45.71%, 61.94%, and 60.92%. This evidence points to the pivotal role of altitude in shaping wood traits, particularly in middle and lower altitudinal gradients where it emerges as the dominant environmental influence. Additionally, the analysis revealed that temperature's contribution to wood trait variability surpassed that of precipitation, aligning with Castagneri et al. [90] and Dang, Haishan et al. [34], who highlighted temperature's limiting role in tree growth at higher altitudes.

As altitude increases, the influence of altitude factors on wood trait variation diminishes, while climatic factors become more pronounced. At higher altitudes, climatic factors' cumulative contribution to tracheid length, width, and aspect ratio surpasses that of altitude, becoming the primary determinants of wood traits. This shift could be attributed to low temperatures and altered precipitation patterns at higher altitudes significantly impacting tree growth and wood traits. For instance, low temperatures may restrict cell division and elongation, influencing tracheid formation and size [91]. Changes in precipitation, on the other hand, can affect trees' water use efficiency and growth rates [92], thereby altering wood quality and structure. These physiological responses indicate trees' adaptive adjustments to their environment. At higher altitudes, temperature and precipitation's limiting effects on tree growth become more evident, making wood traits more sensitive to these climatic factors [93]. Research has shown that at the upper limits of a species' altitudinal distribution, low temperatures during the growing season can limit metabolic processes like photosynthesis fixation, subsequently affecting cell differentiation and xylem formation [22,90].

5. Conclusions

In this research, the phenotypic variation in wood traits of *Cunninghamia lanceolata* across diverse altitudinal gradients was meticulously analyzed. Our findings revealed significant differences in traits like heartwood ratio, tracheid length, and tracheid length-to-width ratio among various elevations, with mean annual ring width and tracheid width also exhibiting notable variations. This underscores the substantial variability in *Cunninghamia lanceolata* wood traits across different elevation zones. Notably, the coefficients of variation for these wood traits were markedly higher in populations at middle and lower elevations compared to those at higher elevations, suggesting a greater influence of diverse environmental conditions and enhanced genetic adaptability at these altitudes.

Principal component analysis identified key variables influencing wood trait variation, with tracheid width, basic wood density, and water absorption emerging as significant contributors to overall variability. These insights are crucial for comprehending the adaptation and growth dynamics of *Cunninghamia lanceolata* at varying altitudes and are instrumental in identifying superior germplasm resources. Cluster analysis further revealed that wood trait variation is influenced not only by altitude but also by an interplay of genetic and other environmental factors.

Pearson correlation analysis delineated intricate relationships among wood traits and between these traits and climatic factors at different altitudes. These relationships highlight the significance of quantitative traits in trees, where correlations between traits could be attributed to strong linkage disequilibrium or pleiotropic effects of associated genes. Random forest analysis underscored the pivotal role of elevation in influencing wood traits such as heartwood ratio, tracheid length, width, and aspect ratio. As elevation increases, the influence of elevation factors on wood trait variation diminishes, while climatic factors, particularly temperature, play an increasingly limiting role in tree growth at higher elevations.

These findings offer novel perspectives on tree adaptation to climate change and bear significant implications for forest genetic breeding and resource conservation. The pronounced variability observed in populations at lower and middle elevations presents a wealth of genetic resources for breeding programs. Concurrently, preserving high-elevation populations is essential for maintaining their unique adaptive traits. In the context of global climate change, these insights are vital for anticipating future shifts in forest ecosystems.

Future research should delve into the specific mechanisms through which different environmental factors, such as soil and light, impact wood traits in *Cunninghamia lanceolata*. Understanding how these variations influence the ecological functions and ecosystem services of these trees will enrich our knowledge of forest ecology and genetics, guiding future forest management and conservation strategies.

Supplementary Materials: The following supporting information can be downloaded at: https://www.mdpi.com/article/10.3390/f15030411/s1, Table S1: Basic information of sampling population of different altitudes of Cunninghamia lanceolata; Table S2: Principal component analysis of wood traits of *Cunninghamia lanceolata* at different altitudes; Table S3: Contribution of variation in wood traits of *Cunninghamia lanceolata* at different elevations; Table S4: Correlation coefficients between wood traits of *Cunninghamia lanceolata* at different elevations and climatic factors at elevation; Table S5: Mantel test for wood traits of *Cunninghamia lanceolata* at different elevations in relation to climatic factors at elevation.

Author Contributions: Conceptualization, N.Z.; Methodology, G.X.; software, G.X.; investigation, G.X., S.L., T.C.; Data curation, G.X.; Writing—original draft, G.X.; Writing—review and editing, N.Z.; Funding acquisition, N.Z. All authors have read and agreed to the published version of the manuscript.

Funding: This research was supported by a sub-project of the National Key Research and Development Program of China (2021YFD2201303) and a project of Science and Technology Innovation Fund for Postgraduate Students of Central South Forestry University of Science and Technology (2023CX02052).

Data Availability Statement: Data will be made available on request.

Acknowledgments: We thank Yamei Gong for her help and support in the field sampling.

Conflicts of Interest: The authors declare no conflicts of interest.

References

1. Cui, C.W.; Peng, L.H.; Ma, D.X.; Wang, J.Q.; Jiang, X.Q.; Jiang, X.G.; Ma, X.Q.; Lin, K. Effects of Thinning on Soil Microbial Necromass Carbon in *Cunninghamia lanceolata* Plantation. *Sci. Silvae Sin.* **2023**, *59*, 41–52. [CrossRef]
2. Li, Y.X. Genetic Diversity and Genetic Differentiation of Geographic Seed Sources of *Cunninghamia lanceolata*. Master's Thesis, Chinese Academy of Forestry Sciences, Beijing, China, 2015.
3. Li, Y.; Li, M.; Li, C.; Liu, Z. Optimized Maxent Model Predictions of Climate Change Impacts on the Suitable Distribution of *Cunninghamia lanceolata* in China. *Forests* **2020**, *11*, 302. [CrossRef]
4. Zhang, X.; Peng, J.Y.; Shi, J.T.; Xu, Q.Q.; Xu, Z.K.; Huang, F. Wood Anatomical Characteristics and Physical—Mechanical Properties of Dark—Brown Heart *Cunningham lanceolate* from Hunan. *J. Southwest For. Univ. (Nat. Sci.)* **2021**, *41*, 155–160.
5. You, R.; Zhu, N.; Deng, X.; Wang, J.; Liu, F. Variation in Wood Physical Properties and Effects of Climate for Different Geographic Sources of *Chinese Fir* in Subtropical Area of China. *Sci. Rep.* **2021**, *11*, 4664. [CrossRef] [PubMed]
6. Zhou, J.; Yu, C.; Cheng, S.D.; Xu, Q.L.; Ma, X.Q.; Li, M. Comparison of Early Growth and Lateral Branch Development in Characteristics of *Cunninghamia lanceolata* Clones. *J. For. Environ.* **2023**, *43*, 561–568. [CrossRef]
7. Jiang, Y.; Wu, J.Y.; Yao, X.Q.; Lin, Q.; Long, Y.Z. Research and Development Countermeasures on Precious Indigenous Timber Tree Species in Hunan Province. *Hunan For. Sci. Technol.* **2012**, *40*, 1–4. [CrossRef]
8. Wang, J. Genetic and Variations of Black Heartwood of *Cunninghamia lanceolata*. Master's Thesis, Central South Forestry University of Science and Technology, Changsha, China, 2018. [CrossRef]
9. Hunag, R.C.; Zhu, N.H.; Yang, J.; Yang, X.W. Research on Cone and Seed Qualities among Different Families of Black-heart Wood *Cunninghamia lanceolata*. *Hunan For. Sci. Technol.* **2021**, *48*, 40–44. [CrossRef]
10. Yang, X.W. Construction of Parent Population for Breeding Programme in the Black-Heart Wood *Cunninghamia lanceolata*. Master's Thesis, Central South Forestry University of Science and Technology, Changsha, China, 2021. [CrossRef]
11. Yang, X.W.; Zhu, N.H.; Han, Z.Q.; Liu, T.Y.; Li, H.; Yang, J.; Zhu, X.C. Identification of Paternal Parents of Half-sib Progeny Superior Plants of the Black-heart Wood *Cunninghamia lanceolata*. *J. Cent. South Univ. For. Technol.* **2021**, *41*, 26–35. [CrossRef]
12. Zhu, N.; Yang, X.; Han, Z.; Can, X. Research Progress on *Cunninghamia lanceolata*. In *Conifers—Recent Advances*; Cristina Gonçalves, A., Fonseca, T., Eds.; IntechOpen: Chongqing, China, 2022. [CrossRef]
13. Hu, S.R.; Xia, M.; Guo, C.Y.; Lu, X.C. Overview of Research Methods on Genetic Diversity of Forest Trees. *J. Northeast For. Univ.* **2001**, *29*, 72–75.
14. Wen, Y.F.; Han, W.J.; Wu, S. Plant Genetic Diversity and its Influencing Factors. *J. Cent. South Univ. For. Technol.* **2010**, *30*, 80–87.
15. Hughes, A.R.; Inouye, B.D.; Johnson, M.T.J.; Underwood, N.; Vellend, M. Ecological Consequences of Genetic Diversity. *Ecol. Lett.* **2008**, *11*, 609–623. [CrossRef]
16. Ma, D.C.; Wang, L.H.; Liang, J. Research Progress on Application of Morphological Markers in Plants. *Jiangsu Agric. Sci.* **2022**, *50*, 55–62. [CrossRef]
17. Alcántara-Ayala, O.; Oyama, K.; Ríos-Muñoz, C.A.; Rivas, G.; Ramirez-Barahona, S.; Luna-Vega, I. Morphological Variation of Leaf Traits in the *Ternstroemia lineata* Species Complex (Ericales: Penthaphylacaceae) in Response to Geographic and Climatic Variation. *PeerJ* **2020**, *8*, e8307. [CrossRef]
18. Albarrán-Lara, A.L.; Petit, R.J.; Kremer, A.; Caron, H.; Peñaloza-Ramírez, J.M.; Gugger, P.F.; Dávila-Aranda, P.D.; Oyama, K. Low Genetic Differentiation between Two Morphologically and Ecologically Distinct Giant-Leaved Mexican Oaks. *Plant Syst. Evol.* **2019**, *305*, 89–101. [CrossRef]
19. Bai, T.D.; Yu, C.L.; Gan, Z.C.; Lai, H.R.; Yang, Y.C.; Huang, H.C.; Jiang, W.X. Association of Cone and Seed Traits of *Pinus yunnanensis* var. *tenuifolia* with Geo-meteorological Factors. *Chin. J. Plant Ecol.* **2020**, *44*, 1224–1235. [CrossRef]
20. Zhang, T.J.; Chen, X.H.; Kang, X.K.; Liu, J. Phenotypic Diversity of Leaf Morphologic Traits of *Davidia involucrata* Natural Populations in Sichuan Province. *Chin. J. Ecol.* **2019**, *38*, 35–43. [CrossRef]
21. Yan, Y.; Li, B.J.; He, Z.M.; Liu, Y.H.; Li, M.; Yu, X.T.; Ma, X.Q. Comparative Research of Wood Properties of *Cunninghamia lanceolate* from Different Seed Sources. *For. Res.* **2021**, *34*, 49–57. [CrossRef]
22. Petit, G.; Anfodillo, T.; Carraro, V.; Grani, F.; Carrer, M. Hydraulic Constraints Limit Height Growth in Trees at High Altitude. *New Phytol.* **2011**, *189*, 241–252. [CrossRef]
23. Gebauer, R.; Volařík, D.; Urban, J.; Børja, I.; Nagy, N.E.; Eldhuset, T.D.; Krokene, P. Effects of Different Light Conditions on the Xylem Structure of Norway Spruce Needles. *Trees* **2012**, *26*, 1079–1089. [CrossRef]
24. D'Orangeville, L.; Côté, B.; Houle, D.; Morin, H.; Duchesne, L. A Three-Year Increase in Soil Temperature and Atmospheric N Deposition Has Minor Effects on the Xylogenesis of Mature Balsam Fir. *Trees* **2013**, *27*, 1525–1536. [CrossRef]
25. Dufour, B.; Morin, H. Climatic Control of Tracheid Production of Black Spruce in Dense Mesic Stands of Eastern Canada. *Tree Physiol.* **2013**, *33*, 175–186. [CrossRef] [PubMed]
26. Arenas-Navarro, M.; Oyama, K.; García-Oliva, F.; Torres-Miranda, A.; De La Riva, E.G.; Terrazas, T. The Role of Wood Anatomical Traits in the Coexistence of Oak Species along an Environmental Gradient. *AoB Plants* **2021**, *13*, plab066. [CrossRef] [PubMed]

27. You, Q.; Jiang, Z.; Yue, X.; Guo, W.; Liu, Y.; Cao, J.; Li, W.; Wu, F.; Cai, Z.; Zhu, H.; et al. Recent Frontiers of Climate Changes in East Asia at Global Warming of 1.5 °C and 2 °C. *npj Clim. Atmos. Sci.* **2022**, *5*, 80. [CrossRef]
28. Mary Petritan, A.; Beloiu Schwenke, M. Forest Functioning under Climate Warming and Future Perspectives on Forest Disturbances. *Forests* **2023**, *14*, 2302. [CrossRef]
29. Harrington, C.A.; Gould, P.J.; Cronn, R. Site and Provenance Interact to Influence Seasonal Diameter Growth of *Pseudotsuga menziesii*. *Front. For. Glob. Chang.* **2023**, *6*, 1173707. [CrossRef]
30. Harrington, C.A.; Dang, Q.-L.; Man, R.; Inoue, S.; Tedla, B. Editorial: Changing Seasons: How Is Global Warming Affecting Forest Phenology? *Front. For. Glob. Chang.* **2023**, *6*, 1257096. [CrossRef]
31. Jiang, A.P.; Jiang, J.M.; Liu, J. Responses of Leaf Traits of *Sassafras tsumu* (Hemsl.) Hemsl along an Altitudinal Gradient. *J. Ecol.* **2016**, *35*, 1467–1474. [CrossRef]
32. Shen, D.; Bo, W.; Xu, F.; Wu, R. Genetic Diversity and Population Structure of the Tibetan Poplar (*Populus szechuanica* Var. *tibetica*) along an Altitude Gradient. *BMC Genet.* **2014**, *15*, S11. [CrossRef]
33. Reisch, C.; Rosbakh, S. Patterns of Genetic Variation in European Plant Species Depend on Altitude. *Divers. Distrib.* **2021**, *27*, 157–163. [CrossRef]
34. Dang, H.; Zhang, Y.; Zhang, K.; Jiang, M.; Zhang, Q. Climate-Growth Relationships of Subalpine Fir (*Abies fargesii*) across the Altitudinal Range in the Shennongjia Mountains, Central China. *Clim. Chang.* **2013**, *117*, 903–917. [CrossRef]
35. Zhang, L.; Jiang, Y.; Zhao, S.; Kang, X.; Zhang, W.; Liu, T. Lingering Response of Radial Growth of Picea Crassifolia to Climate at Different Altitudes in the Qilian Mountains, Northwest China. *Trees* **2017**, *31*, 455–465. [CrossRef]
36. Carrer, M.; Castagneri, D.; Prendin, A.L.; Petit, G.; Von Arx, G. Retrospective Analysis of Wood Anatomical Traits Reveals a Recent Extension in Tree Cambial Activity in Two High-Elevation Conifers. Front. *Plant Sci.* **2017**, *8*, 737. [CrossRef]
37. Keleş, S.Ö. The Effect of Altitude on the Growth and Development of Trojan fir (*Abies nordmanniana* subsp. *equi-trojani* [Asch. & Sint. ex Boiss] Coode & Cullen) Saplings. *Cerne* **2020**, *26*, 381–392. [CrossRef]
38. Cocozza, C.; Palombo, C.; Tognetti, R.; La Porta, N.; Anichini, M.; Giovannelli, A.; Emiliani, G. Monitoring Intra-Annual Dynamics of Wood Formation with Microcores and Dendrometers in Picea Abies at Two Different Altitudes. *Tree Physiol.* **2016**, *36*, 832–846. [CrossRef] [PubMed]
39. Cai, Q.; Liu, Y.; Qian, H.; Liu, R. Inverse Effects of Recent Warming on Trees Growing at the Low and High Altitudes of the Dabie Mountains, Subtropical China. *Dendrochronologia* **2020**, *59*, 125649. [CrossRef]
40. Zhuang, L.; Axmacher, J.C.; Sang, W. Different Radial Growth Responses to Climate Warming by Two Dominant Tree Species at Their Upper Altitudinal Limit on Changbai Mountain. *J. For. Res.* **2017**, *28*, 795–804. [CrossRef]
41. Plant Intelligence. Available online: https://www.iplant.cn/info/%E6%9D%89%E6%9C%A8 (accessed on 25 December 2023).
42. Gong, H.; Yang, M.; Wang, C.; Tian, C. Leaf Phenotypic Variation and Its Response to Environmental Factors in Natural Populations of Eucommia Ulmoides. *BMC Plant Biol.* **2023**, *23*, 562. [CrossRef]
43. Xu, M.; Ren, H.Q.; Guo, W. Research Progress of Variation for Wood Properties *Cunninghamia lanceolata*. *J. Northwest Coll. For* **2008**, 185–189.
44. Wimmer, R. Wood Anatomical Features in Tree-Rings as Indicators of Environmental Change. *Dendrochronologia* **2002**, *20*, 21–36 [CrossRef]
45. Nagavciuc, V.; Mursa, A.; Ionita, M.; Sfeclă, V.; Popa, I.; Roibu, C.-C. An Overview of Extreme Years in Quercus Sp. Tree Ring Records from the Northern Moldavian Plateau. *Forests* **2023**, *14*, 894. [CrossRef]
46. Iglesias, C.; Santos, A.; Martínez, J.; Pereira, H.; Anjos, O. Influence of Heartwood on Wood Density and Pulp Properties Explained by Machine Learning Techniques. *Forests* **2017**, *8*, 20. [CrossRef]
47. Thybring, E.E.; Fredriksson, M. Wood Modification as a Tool to Understand Moisture in Wood. *Forests* **2021**, *12*, 372. [CrossRef]
48. Arsić, J.; Stojanović, M.; Petrovičová, L.; Noyer, E.; Milanović, S.; Světlík, J.; Horáček, P.; Krejza, J. Increased Wood Biomass Growth Is Associated with Lower Wood Density in *Quercus petraea* (Matt.) Liebl. Saplings Growing under Elevated CO_2. *PLoS ONE* **2021**, *16*, e0259054. [CrossRef]
49. McDowell, N.G.; Allen, C.D.; Anderson-Teixeira, K.; Aukema, B.H.; Bond-Lamberty, B.; Chini, L.; Clark, J.S.; Dietze, M. Grossiord, C.; Hanbury-Brown, A.; et al. Pervasive Shifts in Forest Dynamics in a Changing World. *Science* **2020**, *368*, eaaz9463 [CrossRef] [PubMed]
50. Ziemińska, K.; Butler, D.W.; Gleason, S.M.; Wright, I.J.; Westoby, M. Fibre Wall and Lumen Fractions Drive Wood Density Variation across 24 Australian Angiosperms. *AoB Plants* **2013**, *5*, plt046. [CrossRef]
51. Yeboah, D.; Burton, A.J.; Storer, A.J.; Opuni-Frimpong, E. Variation in Wood Density and Carbon Content of Tropical Plantation Tree Species from Ghana. *New For.* **2014**, *45*, 35–52. [CrossRef]
52. Bouriaud, O.; Leban, J.-M.; Bert, D.; Deleuze, C. Intra-Annual Variations in Climate Influence Growth and Wood Density of Norway Spruce. *Tree Physiol.* **2005**, *25*, 651–660. [CrossRef] [PubMed]
53. Dahlen, J.; Nabavi, M.; Auty, D.; Schimleck, L.; Eberhardt, T.L. Models for Predicting the Within-Tree and Regional Variation of Tracheid Length and Width for Plantation Loblolly Pine. *For. Int. J. For. Res.* **2021**, *94*, 127–140. [CrossRef]
54. Palla, B.; Ladányi, M.; Cseke, K.; Buczkó, K.; Höhn, M. Wood Anatomical Traits Reveal Different Structure of Peat Bog and Lowland Populations of *Pinus sylvestris* L. in the Carpathian Region. *Forests* **2021**, *12*, 494. [CrossRef]
55. Xu, L.; Chen, G.X.; Zhang, D.G.; Li, X.T.; Zhang, Y. Flora of Seed Plants in Xiaoxi National Nature Reserve, Hunan, China. *Acta Bot. Boreali-Occident. Sin.* **2010**, *30*, 2307–2316.

56. *GB/T 1927314-2022*; State Administration of Market Supervision, National Standardization Administration. Test Method for Physical and Mechanical Properties of Wood in Small Specimens without Defects Part 3: Determination of Growth Wheel Width and Latewood Rate. China Standard Press: Beijing, China, 2022.
57. Duan, H.J.; Cao, S.; Zheng, H.; Hu, D.; Lin, J.; Lin, H.; Hu, R.; Sun, Y.; Li, Y. Variation in the Growth Traits and Wood Properties of *Cunninghamia lanceolata* from Six Provinces of Southern China. *Forests* **2016**, *7*, 192. [CrossRef]
58. Xu, Y.L. Genetic Variation of Natural Population in *Pinus yunnanensis* Franch. Ph.D. Thesis, Beijing Forestry University, Beijing, China, 2015.
59. Jiao, S.; Chen, W.; Wang, J.; Du, N.; Li, Q.; Wei, G. Soil Microbiomes with Distinct Assemblies through Vertical Soil Profiles Drive the Cycling of Multiple Nutrients in Reforested Ecosystems. *Microbiome* **2018**, *6*, 146. [CrossRef] [PubMed]
60. Topaloğlu, E.; Ay, N.; Altun, L.; Serdar, B. Effect of Altitude and Aspect on Various Wood Properties of Oriental Beech (*Fagus orientalis* Lipsky) Wood. *Turk. J. Agric. For.* **2016**, *40*, 397–406. [CrossRef]
61. Rocha, S.M.G.; Vidaurre, G.B.; Pezzopane, J.E.M.; Almeida, M.N.F.; Carneiro, R.L.; Campoe, O.C.; Scolforo, H.F.; Alvares, C.A.; Neves, J.C.L.; Xavier, A.C.; et al. Influence of Climatic Variations on Production, Biomass and Density of Wood in Eucalyptus Clones of Different Species. *For. Ecol. Manag.* **2020**, *473*, 118290. [CrossRef]
62. Audigeos, D.; Buonamici, A.; Belkadi, L.; Rymer, P.; Boshier, D.; Scotti-Saintagne, C.; Vendramin, G.G.; Scotti, I. Aesqeaurchaaprtioclerins in the Wild: Natural Genetic Diversity and Selective Pressure in the PIP Gene Family in Five Neotropical Tree Species. *BMC Evol. Biol.* **2010**, *10*, 202. [CrossRef]
63. Huang, R.; Zeng, W.; Deng, H.; Hu, D.; Wang, R.; Zheng, H. Inbreeding in *Cunninghamia lanceolata*: Insight into the Rare Self-Fertilizing Event from a Genetic View. *Genes* **2022**, *13*, 2105. [CrossRef]
64. Silva-Montellano, A.; Eguiarte, L.E. Geographic Patterns in the Reproductive Ecology of Agave Lechuguilla (Agavaceae) in the Chihuahuan Desert. II. Genetic Variation, Differentiation, and Inbreeding Estimates. *Am. J. Bot.* **2003**, *90*, 700–706. [CrossRef]
65. Jin, Z.X.; Li, J.M. Analysis of the Genetic Diversity of *Hepatacodium miconioides* at Different Altitude and in Tiantai Mountain in Zhejiang Province and its Relationship with Environmental Factors. *J. Zhejiang Univ. (Sci. Ed.)* **2005**, 452–458. [CrossRef]
66. Schellenberger Costa, D.; Gerschlauer, F.; Kiese, R.; Fischer, M.; Kleyer, M.; Hemp, A. Plant Niche Breadths along Environmental Gradients and Their Relationship to Plant Functional Traits. *Divers. Distrib.* **2018**, *24*, 1869–1882. [CrossRef]
67. Zheng, G.B.; Ma, L.; Zhang, Z.M.; Li, M.L.; Liu, S.Y.; Zang, Y.N.; Sun, Y.; Ma, X.Y. Principal Component Analysis and Comprehensive Evaluation of Ningxia Wolfberry Fruits under Water Stress Conditions at Different Fertility Periods. *Water Sav. Irrig.* **2022**, 47–57.
68. Liang, K.L.; Jiang, W.Q.; Zhou, Z.Y.; Guo, X.; Li, X.Z.; Dai, W.A.; Wang, R.; Liu, X.Y. Variation in Main Morphological Characteristics of *Amorpha fruticosa* Plants in the Qinghai-Tibet Plateau. *Acta Ecol. Sin.* **2012**, *32*, 311–318. [CrossRef]
69. Jia, Q.B.; Zhang, H.G.; Zhang, L. Variation in Carbon Conten of Hybrid larch Families and Superior Families Selection. *J. Northeast For. Univ.* **2016**, *44*, 1–6.
70. Luo, J.X.; Li, X.Q.; Sun, P.; Wang, L.H.; Hu, G.R.; Wang, F.L.; Zheng, W. Variation Pattern of Tracheid and Wood Density in Natural Population of *Picea asperata*. *J. Beijing For. Univ.* **2004**, 80–85.
71. Ma, S.X.; Wang, J.H.; Zhang, S.T.; Sun, X.M.; Liang, B.S.; Zhou, D.Y.; Liu, S.M. Genetic Variation of Wood Properties in Japanese larch Clones. *For. Res.* **2008**, 69–73.
72. Yu, H.Y.; Pang, Z.Y.; Yin, C.H.; Ding, C.J.; Wang, F.S.; Zhang, J.W.; Wang, F.L.; Jin, P.L.; Fu, Z.X.; Qu, G.Z.; et al. Variation Analysis on the Growth and Wood Properties Variation of 100 Poplar Clones. *J. Northeast For. Univ.* **2023**.
73. Ramananantoandro, T.; Ramanakoto, M.F.; Rajoelison, G.L.; Randriamboavonjy, J.C.; Rafidimanantsoa, H.P. Influence of Tree Species, Tree Diameter and Soil Types on Wood Density and Its Radial Variation in a Mid-Altitude Rainforest in Madagascar. *Ann. For. Sci.* **2016**, *73*, 1113–1124. [CrossRef]
74. Camarero, J.J.; Fernández-Pérez, L.; Kirdyanov, A.V.; Shestakova, T.A.; Knorre, A.A.; Kukarskih, V.V.; Voltas, J. Minimum Wood Density of Conifers Portrays Changes in Early Season Precipitation at Dry and Cold Eurasian Regions. *Trees* **2017**, *31*, 1423–1437. [CrossRef]
75. Fang, S.; Sun, D.; Shang, X.; Fu, X.; Yang, W. Variation in Radial Growth and Wood Density of Cyclocarya Paliurus across Its Natural Distribution. *New For.* **2020**, *51*, 453–467. [CrossRef]
76. Zhang, H.; Zhang, S.; Chen, S.; Xia, D.; Yang, C.; Zhao, X. Genetic Variation and Superior Provenances Selection for Wood Properties of Larix Olgensis at Four Trials. *J. For. Res.* **2022**, *33*, 1867–1879. [CrossRef]
77. Jiang, A.P.; Jiang, J.M.; Liu, J. Genetic Diversity and Genetic Structure of Sassafras tsumu Populations along Altutudinal Gradients in Tianmushan Mountain, China. *Chin. J. Appl. Ecol.* **2016**, *27*, 1829–1836. [CrossRef]
78. Zu, K.; Wang, Z. Research progress on the elevational distribution of mountain species in response to climate change. *Biodivers. Sci.* **2022**, *30*, 123–137. [CrossRef]
79. Pan, H.L.; Li, M.H.; Cai, S.H.; Wu, J.; Du, Z.; Liu, X.L. Responses of Growth and Ecological of Plant to Altitude. *Ecol. Environ. Sci.* **2009**, *18*, 722–730. [CrossRef]
80. Fajardo, A.; Piper, F.I.; García-Cervigón, A.I. The Intraspecific Relationship between Wood Density, Vessel Diameter and Other Traits across Environmental Gradients. *Funct. Ecol.* **2022**, *36*, 1585–1598. [CrossRef]
81. Du, Q.; Gong, C.; Wang, Q.; Zhou, D.; Yang, H.; Pan, W.; Li, B.; Zhang, D. Genetic Architecture of Growth Traits in Populus Revealed by Integrated Quantitative Trait Locus (QTL) Analysis and Association Studies. *New Phytol.* **2016**, *209*, 1067–1082. [CrossRef] [PubMed]

82. Bouslimi, B.; Koubaa, A.; Bergeron, Y. Intra-Ring Variations and Interrelationships for Selected Wood Anatomical and Physical Properties of *Thuja occidentalis* L. *Forests* **2019**, *10*, 339. [CrossRef]
83. Almeida, M.N.F.D.; Vidaurre, G.B.; Pezzopane, J.E.M.; Lousada, J.L.P.C.; Silva, M.E.C.M.; Câmara, A.P.; Rocha, S.M.G.; Oliveira, J.C.L.D.; Campoe, O.C.; Carneiro, R.L.; et al. Heartwood Variation of Eucalyptus Urophylla Is Influenced by Climatic Conditions. *For. Ecol. Manag.* **2020**, *458*, 117743. [CrossRef]
84. Olano, J.M.; Almería, I.; Eugenio, M.; Von Arx, G. Under Pressure: How a M Editerranean High-mountain Forb Coordinates Growth and Hydraulic Xylem Anatomy in Response to Temperature and Water Constraints. *Funct. Ecol.* **2013**, *27*, 1295–1303. [CrossRef]
85. Fonti, P.; Bryukhanova, M.V.; Myglan, V.S.; Kirdyanov, A.V.; Naumova, O.V.; Vaganov, E.A. Temperature-induced Responses of Xylem Structure of *Larix sibirica* (Pinaceae) from the Russian Altay. *Am. J. Bot.* **2013**, *100*, 1332–1343. [CrossRef] [PubMed]
86. Zhao, X.P.; Guo, M.H. Response of Wood Tracheid Morphological Features of *Pinus koraiensis* Plantation to Climate Change. *J Northeast For. Univ.* **2009**, *37*, 1–4.
87. Zhu, L.J.; Li, Z.S.; Wang, X.C. Anatomical Characteristics of Xylem in Tree Rings and its Relationship with Environmental. *Chin. J. Plant Ecol.* **2017**, *41*, 238–251. [CrossRef]
88. Xu, J.M.; Zhang, R.; Lv, J.X.; Robert, E. Climatic Response in of Picea crassifolia along Elevation Gradient in Qilian Mountains, Northwestern China. *J. Beijing For. Univ.* **2015**, *37*, 102–108. [CrossRef]
89. Xu, J.M.; Lu, J.X.; Bao, F.C.; Huang, R.F.; Zhao, Y.K. Influence of Climatic Factors on Wood Cell Structure. *For. Sci.* **2011**, *47*, 151–158.
90. Castagneri, D.; Petit, G.; Carrer, M. Divergent Climate Response on Hydraulic-Related Xylem Anatomical Traits of *Picea abies* along a 900-m Altitudinal Gradient. *Tree Physiol.* **2015**, *35*, 1378–1387. [CrossRef]
91. Nagelmüller, S.; Hiltbrunner, E.; Körner, C. Low Temperature Limits for Root Growth in Alpine Species Are Set by Cell Differentiation. *AoB Plants* **2017**, *9*, plx054. [CrossRef] [PubMed]
92. Heilman, K.A.; Trouet, V.M.; Belmecheri, S.; Pederson, N.; Berke, M.A.; McLachlan, J.S. Increased Water Use Efficiency Leads to Decreased Precipitation Sensitivity of Tree Growth, but Is Offset by High Temperatures. *Oecologia* **2021**, *197*, 1095–1110. [CrossRef] [PubMed]
93. Maxime, C.; Hendrik, D. Effects of Climate on Diameter Growth of Co-Occurring *Fagus sylvatica* and *Abies alba* along an Altitudinal Gradient. *Trees* **2011**, *25*, 265–276. [CrossRef]

forests

MDPI

Article

Tracheids vs. Tree Rings as Proxies for Dendroclimatic Reconstruction at High Altitude: The Case of *Pinus sibirica* Du Tour

Mikhail S. Zharkov [1,*], Bao Yang [2], Elena A. Babushkina [1,3], Dina F. Zhirnova [1,3], Eugene A. Vaganov [1,4] and Vladimir V. Shishov [1,*]

[1] Mathematical Methods and IT Department, Siberian Federal University, Krasnoyarsk 660041, Russia; babushkina70@mail.ru (E.A.B.); dina-zhirnova@mail.ru (D.F.Z.); eavaganov@hotmail.com (E.A.V.)
[2] School of Geography and Ocean Science, Nanjing University, Nanjing 210023, China; yangbao@nju.edu.cn
[3] Khakass Technical Institute, Siberian Federal University, Abakan 655000, Russia
[4] Sukachev Institute of Forest, Siberian Branch of the Russian Academy of Science, Krasnoyarsk 660036, Russia
* Correspondence: mzharkov@sfu-kras.ru (M.S.Z.); vlad.shishov@gmail.com (V.V.S.)

Abstract: Siberian pine (*Pinus sibirica* Du Tour) is a widespread and long-lived species in the northern hemisphere, which makes it a good potential proxy for climatic data. However, the tree-ring growth of this species weakly correlates with climatic conditions, which prevents its use in dendroclimatic reconstruction. It was proposed to use the measurements of tracheid characteristics as model predictors to reconstruct the smoothed temperature of the key periods in tree growth. In this study, algorithms for preprocessing tracheids and temperature data, as well as for model cross-validation, were developed to produce reliable high-resolution (weekly-based) temperature reconstructions. Due to the developed algorithms, the key time periods of Siberian pine growth were identified during the growing season—early June (most active cell development) and mid-July (setting new buds for the next growing season). For these time periods, reliable long-term temperature reconstructions ($R^2 > 0.6$, $p < 10^{-8}$) were obtained over 1653–2018. The temperature reconstructions significantly correlated ($p < 10^{-8}$) with independent reanalysis data for the 19th century. The developed approach, based on preprocessing tracheid and temperature data, shows new potential for Siberian pine in high-resolution climate reconstructions and can be applied to other tree species that weakly respond to climate forcing.

Keywords: wood anatomy; cell measurements; radial cell diameter; cell wall thickness; tracheidograms; tree-ring response; poorly sensitive to climate; temperature reconstruction

Citation: Zharkov, M.S.; Yang, B.; Babushkina, E.A.; Zhirnova, D.F.; Vaganov, E.A.; Shishov, V.V. Tracheids vs. Tree Rings as Proxies for Dendroclimatic Reconstruction at High Altitude: The Case of *Pinus sibirica* Du Tour. *Forests* **2024**, *15*, 167. https://doi.org/10.3390/f15010167

Academic Editor: Jesús Julio Camarero

Received: 13 November 2023
Revised: 19 December 2023
Accepted: 10 January 2024
Published: 12 January 2024

1. Introduction

Siberian pine forests are the most complex ecosystems in the Siberian taiga, characterized by regenerative-age dynamics, stability, spatial and temporal structure, and biodiversity [1]. The distribution range of Siberian pine (*Pinus sibirica* Du Tour), a forest-forming species of the "cedar" forest formation, extends from the northeast of European Russia to the south of East Siberia, reaching Mongolia in the southern part (http://agroatlas.ru/ru/content/related/Pinus_sibirica/map/, accessed on 27 June 2023).

The observed global warming of recent decades has contributed to a shift in the timespan of the active growing season and its heat availability [2–4]. The consequences of these changes are most acute at the edges of the growing range, where trees often grow at the limits of physiological endurance [5,6]. Moreover, the responses of tree plants to climate change are mixed, from range expansion to suppression or death [7–11]. Under current climate changes, it is therefore critical not only to understand the response of tree vegetation as an important part of the planetary carbon cycle [12–15] but also to obtain a quantitative assessment through robust reconstructions of long-term climatic fluctuations.

The use of long-term chronologies of the anatomical parameters of tree rings allows one to obtain a more accurate and detailed understanding of past and recent variability in principal climatic factors compared to simply analyzing tree-ring width (TRW) [16–19]. The seasonal development of xylem anatomical structure significantly determines tree productivity and survival; hence, forests' vulnerability to climate change and their ability to fix carbon dioxide [20,21]. This approach also enhances opportunities to study the physiological mechanisms of plant adaptation to changing environmental conditions, due to changes in both hydraulic and mechanical functions of woody tissue [22–24]. Therefore, understanding how and at which intervals of the growing season the principal climatic factors (temperature or precipitation) modify the tree-ring structure turns out to be important both for reconstructing past climate-tree ring relationships and developing adequate prediction models of climatic factors influencing the anatomical structure of tree rings [24–29]. The application of the quantitative wood anatomy approach is particularly relevant for the species (i.e., *Pinus sibirica* Du Tour) for which the study of climate response is extremely difficult due to the limited climate sensitivity of their radial growth [30,31].

A unique 495-year chronology of wood anatomical characteristics for Siberian pine has recently been obtained for the timberline in the Western Sayan, which has made it possible to estimate the climatic response of tree-ring widths and integral anatomical characteristics (namely, TRW indices, mean and maximum radial cell diameters, and cell wall thickness) [19].

However, the dataset of detailed anatomical measurements used in this study, namely the radial cell diameter and cell wall thickness of *Pinus Sibirica* tracheidograms, allows us to estimate the century-old records of climate factors with much higher time resolution. In this work, we exploit the full potential of the anatomical structure of tree rings in a detailed analysis of their climatic response, with a focus on dendroclimatic reconstructions, using previously developed [32] and new approaches.

In this study, we employ tracheid measurements with the objective of revealing the potential of *Pinus sibirica* Du Tour for high-resolution temperature reconstructions Algorithms for tracheid and temperature preprocessing and accurate model evaluation were developed. Additionally, two hypotheses were tested: (1) key periods of tree growth can be determined by the quality of the smoothed daily temperature reconstruction models and (2) the use of TRW for modeling would not achieve the same reconstruction reliability as using tracheid data.

2. Materials and Methods

2.1. Climate Data

Temperature and precipitation data were obtained from the Tashtyp meteorological station (WMO 29956), 52.8 N, 89.9 E, 455 m a.s.l., and they included mean daily temperature and cumulative daily precipitation records from 1929 to 2016. Mean daily NOAA/CIRES/DOE 20th Century Reanalysis V3 temperatures were obtained from the Climate Explorer web platform (http://climexp.knmi.nl/select.cgi?field=c3t2m_daily, accessed on 29 August 2023), using the grid box region: longitude from 89.5 to 90.5 E, latitude from 52.5 to 53.5 N (Figure 1).

To create a dependent variable for the reconstruction from the raw temperature data a two-way sequential smoothing algorithm was applied (Figure 2).

To suppress high-frequency fluctuations in the mean daily temperature, intra-annual smoothing with varying lengths w of the sliding window (from 1 up to 14 days) was applied (Figures 2 and S1). The optimal length of the sliding window was determined by optimization of the temperature reconstruction models as described below.

In addition to the intra-annual smoothing, an inter-annual smoothing procedure for temperature was applied to develop reconstruction models of the smoothed temperature based on anatomical characteristics. The second smoothing was done to reduce annual variance in temperature. We used a moving average with a sliding window of W years (from 1 to 11 years) (Figures 2 and S2), using the same criteria to choose the length of W.

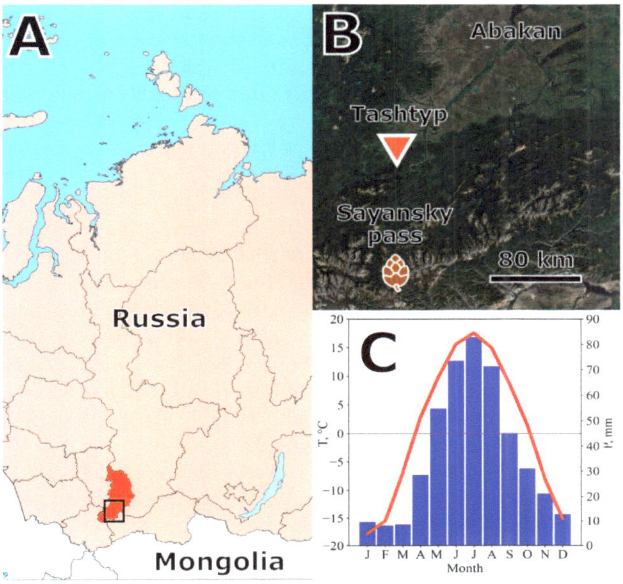

Figure 1. The locations of the study plot (brown pinecone for *Pinus sibirica* Du Tour) and the climate station Tashtyp (red triangle) (**A**,**B**), and the average monthly mean temperature (T, °C) and monthly total precipitation (P, mm) in Tashtyp from 1929 to 2016 (**C**).

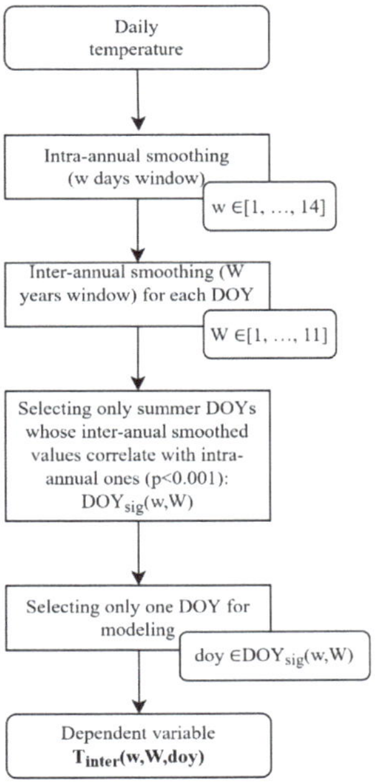

Figure 2. The two-way sequential smoothing algorithm for obtaining the dependent temperature variable T_{inter} (w, W, doy), where w is the length of the intra-annual sliding window; W is the length of the inter-annual sliding window; *doy* is the day of the year for which the reconstruction model is developed.

For each pair $(w,\ W)$ of intra- and inter-annual sliding windows, the set $DOY_{sig}(w,\ W) \subseteq \{152,\ \ldots,\ 243\}$ was created as a subset of tree-ring growing days (DOYs, growing season). $DOY_{sig}(w,\ W)$ is the set of DOYs for which the Pearson correlation between inter-annual and intra-annual smoothed temperatures was highly significant ($p < 0.001$).

As a result of the two-way sequential smoothing algorithm, we obtained dependent variables $T_{inter}(w,\ W,\ doy)$ of inter-annual smoothed temperatures for the model.

2.2. Tree Data Collection and Processing

Wood samples were collected from seven trees of *Pinus sibirica* Du Tour located on the border of the Republic of Khakassia and the Republic of Tyva (51°42.8′ N 89°51.9′ E, 1970–2020 m a.s.l.) (Figure 1). Tracheidogram measurements of tree rings over 1653–2018, obtained earlier (see [19] for more details), were used in this work (Figure S3).

For each tree ring, lumens (Ls) and double cell wall thickness (DCWT) were measured for 5 rows of cells (Figure S3). Then, the radial cell diameter (Cell Diameter, D) was calculated as the sum of lumen and double cell wall, and cell wall thickness (CWT) was determined as half of DCWT [33]. The obtained measurement were verified by an independent cell measuring tool [34].

The resulting tracheidograms of averaged D and CWT from the five rows of the measured series (Figure S4A,C) were then standardized to 15 cells (mean seasonal cell production over 1653–2018), resulting in standardized sD and sCWT series for each year of each tree [35]. Absolute tree ring width (TRW) values were obtained by summarizing DMeans of the corresponding rings.

The TRW values of all trees in a year were averaged to obtain the site tree-ring chronology. A standardized chronology was developed by bi-weight robust averaging individual tree indices, which were obtained by removing the age-related trends using cubic smoothing splines with a 50% frequency response at 67% of the series length [36].

The standardized tracheidograms of individual trees were year-to-year averaged (Figure S5). As a result, a single 30-dimensional object was obtained for each year (growing season), consisting of the values of cell diameters (D1–D15) and cell wall thicknesses (CWT1–CWT15) of the corresponding standardized curves (Figure S6).

Finally, 15 radial-cell-diameter and 15 cell-wall-thickness chronologies were obtained over 1653–2018 (Figure S7). Figure S8 shows examples of smoothed ("inter-seasonal smoothing") chronologies of radial cell size and cell wall thickness. All the smoothed time series had significant Pearson correlations (p-value $< 10^{-16}$) with D1–D15, CWT1–CWT15 series for all the sliding windows.

Principal component analysis (PCA) was applied to the radial-cell-size and cell-wall-thickness chronologies to reduce the dimensionality of the resulting 30-dimensional objects. It was shown that 4 principal components explained 95% of the variance of the series, while 9 principal components explained 99% of the variance (Figure S10). In addition, the application of PCA allows one to avoid multicollinearity in the obtained series, as noted above [37].

The first P principal components, smoothed with the inter-annual window W (see the example for $P = 5$, Figure S9):

$$PC_i(W) = \left\{ pc_i^W(y) \middle| y \in \{1653 \ldots 2018\} \right\},\ i \in [1,\ \ldots,\ P] \quad (1)$$

were used as predictors (independent variables) in the model development process ($P \in [4, \ldots, 9]$). The principal component transformation matrix is presented in Table S1.

2.3. Reconstruction Models Development

For the best model fit, a triplet of hyperparameters $(w,\ W,\ P)$ was varied as follows: $w \in \{1, \ldots, 14\},\ W \in \{1,\ \ldots, 11\},\ P \in \{4,\ \ldots, 9\}$.

For each triplet (w, W, P), we obtained the set of independent variables $\{PC_1(W), \ldots, PC_P(W)\}$ (Figure 3) and the set of dependent variables $T_{inter}(w, W, doy)$, $doy \in DOY_{sig}(w, W)$ (Figure 2).

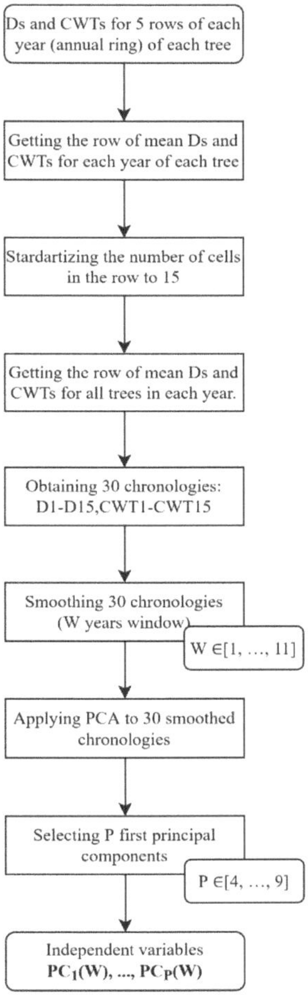

Figure 3. The reduction of tracheidograms to smoothed principal components algorithm for obtaining independent variables from the raw tracheid data (Diameters (Ds) and Cell Wall Thicknesses (CWTs)).

For each $doy \in DOY_{sig}(w, W)$, a separate multiple linear regression (MLR) model $MLR_{w,\,W,\,P,doy}\,(year)$ was developed.

The final MLR models were considered as ensembles of individual MLR models obtained in a new rolling leave-one-out cross-validation (RLCO CV) procedure:

$$MLR_{w,\,W,\,P,doy}(year) = k_0 + \sum\nolimits_{l=1}^{P} k_l \cdot pc_i^{W}(y), \tag{2}$$

where $k_l = k_l(w, W, P, doy)$ is the lth coefficient of the final MLR model ($l = \overline{0, P}$), estimated as $k_l = \frac{\sum_{\theta=1}^{N(W)} k_l^{\theta}}{N}$.

We note that $k_l^{\theta} = k_l^{\theta}(w,\ W,\ P, doy)$ is the lth coefficient of the θth individual MLR model, and $N = 2016 - 1929 + 1 = 88$ is the total number of individual MLR models obtained in the RLOO CV procedure.

Each θth individual MLR model can be described by this formula:

$$MLR^{\theta}_{w,\ W,\ P,doy}(year) = k^{\theta}_0 + \sum_{l=1}^{P} k^{\theta}_l \cdot pc^{W}_i(y), \quad \theta = \overline{1929, 2016} \tag{3}$$

To obtain the θth individual MLR model, $T_{inter}(w,\ W,\ doy)$ was split into calibration ($T^{cal}_{inter}(w,\ W,\ doy)$) and verification ($T^{ver}_{inter}(w,\ W,\ doy)$) sets by the rules of the RLOO CV procedure:

1. Each θth element (year) of $T_{inter}(w,\ W,\ doy)$ is considered as a verification set.
2. The elements from $\left[\theta - \left|\frac{W}{2}\right|, \theta\right) \cup \left(\theta, \theta + \left|\frac{W}{2}\right|\right]$ are omitted ($\left|\frac{W}{2}\right|$ is the floored division). This is done to prevent the data from the θth element from getting into the calibration set due to smoothing with the W inter-annual sliding window and affecting the elements from $\left[\theta - \left|\frac{W}{2}\right|, \theta\right) \cup \left(\theta, \theta + \left|\frac{W}{2}\right|\right]$. All indices from $\theta - \left|\frac{W}{2}\right| < 1929$ or $\theta + \left|\frac{W}{2}\right| > 2016$ are ignored.
3. All other elements are considered as a calibration set.

In this study, the RLOO CV procedure was developed as an extension of the LOO CV procedure [38] for smoothed data.

After obtaining the calibration and verification sets, the coefficients k^{θ}_l of the θth individual MLR model are obtained by training the model on the calibration set.

To evaluate the individual models on the calibration sets, the coefficient of determination ($R^2_{cal,\theta}$) and the Root Mean Squared Error ($RMSE_{cal,\theta}$) were calculated between $T^{cal}_{inter}(w,\ W,\ doy)$ and $\left\{MLR^{\theta}_{w,\ W,\ P,doy}(year) \middle| year \in T^{cal}_{inter}(w,\ W,\ doy)\right\}$.

After training the $N = 88$ models, one for each year, the chronology of the verification values was obtained as:

$$CRN^{ver}_{w,\ W,\ P,doy} = \left\{MLR^{1929}_{w,\ W,\ P,doy}(1929), \ldots, MLR^{2016}_{w,\ W,\ P,doy}(2016)\right\} \tag{4}$$

and the mean metrics $R^2_{cal} = \frac{\sum_{\theta=1929}^{2016} R^2_{cal,\theta}}{N}$, $RMSE_{cal} = \frac{\sum_{\theta=1929}^{2016} RMSE_{cal,\theta}}{N}$ were calculated to evaluate the total quality of the individual models on the calibration set.

To evaluate the individual models on the verification set, R^2_{ver} and $RMSE_{ver}$ were calculated between $CRN^{ver}_{w,\ W,\ P,doy}$ and $T_{inter}(w,\ W,\ doy)$.

After individual evaluation, the final MLR model $MLR_{w,\ W,\ P,doy}(year)$ was developed by averaging the coefficients of the individual models.

To evaluate the final model, R^2_{sim} and $RMSE_{sim}$ (sim—simulated) were calculated between $T_{inter}(w, W, doy)$ and $\left\{MLR_{w,W,P,doy}(year) \middle| year \in T_{inter}(w, W, doy)\right\}$.

All data processing algorithms were implemented in Python and can be downloaded from: https://github.com/mikewellmeansme/dendroclimatic-reconstructor/ (accessed on 9 January 2024).

3. Results

3.1. Reconstruction of Temperature Dynamics

We chose $R^2 > 0.5$ as a threshold for the selection of qualitatively reconstructed periods on the calibration and verification sets to be sure that the model explained most of the variance. After varying the triplet of hyperparameters (w, W, P), the next heatmap was obtained (Figure 4). Obviously, reliable reconstructions appeared when the annual smoothing windows W was not less than 8 years.

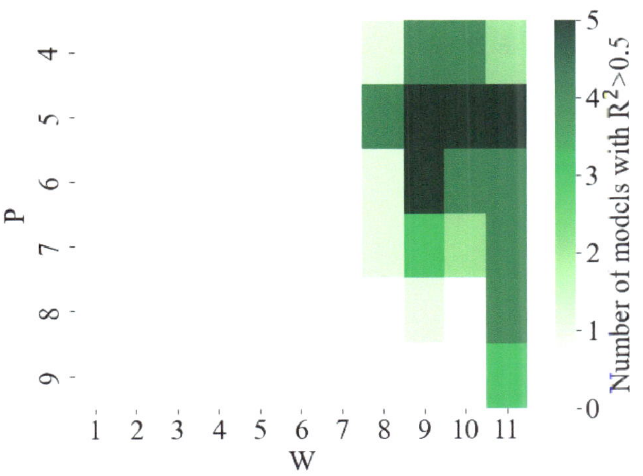

Figure 4. Heatmap showing the number of models for which R^2 on the calibration and verification sets was greater than 0.5 for different pair of hyperparameters W (inter-annual sliding window) and P (number of first principal components), and $w = 7$ (7-day intra-annual sliding window of temperatures).

The approach of choosing the optimal triplet $(w,\ W,\ P)$ of hyperparameters is debatable, and different combinations of the triplet values sometimes result in the reconstruction of different days of the growth season. Mainly, temperatures at the beginning of June (weeks with DOY 152–154 at the center) and the middle of July (weeks with DOY 195-198 at the center) are adequately simulated by the tracheidograms.

There are also a couple of models for late June ($P = 4$, DOY 175) and August (DOY 220 and 232), but these results are not sustainable because they do not appear with other triplets of hyperparameters.

As an example, we chose the triplet $w = 7$, $W = 9$, $P = 5$ to demonstre the applicability of our approach to dendroclimatic reconstruction of summer temperatures (Figure 5).

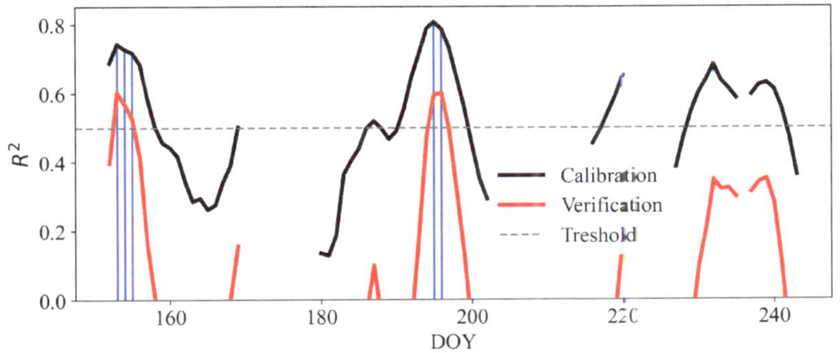

Figure 5. The values of R^2 metric per day of the year (DOY) on the calibration (black line) and verification (red line) sets for the models of temperature reconstruction with $w = 7$, $W = 9$, $P = 5$. The dotted gray line shows the threshold for selecting acceptable reconstructed periods ($R^2 > 0.5$ on both calibration and verification sets). The blue lines show DOYs with acceptable metrics. The gaps in the curves show DOYs where the intra-annual temperatures did not correlate with their smoothed curves (see Figure S1).

We compared the new reconstructions with the observed and reanalyzed temperatures for two main sustainable periods: A—DOY 154 (the mean daily temperature from 30 May to 5 June) and B—DOY 197 (the mean temperature of the week from 12 July to 18 July) (Figure 6).

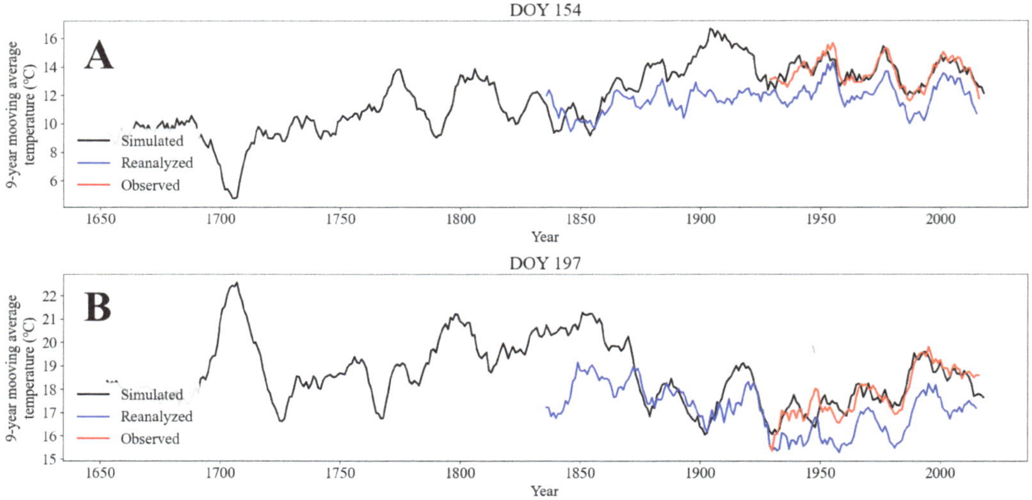

Figure 6. Smoothed (W = 9, 1 year step) mean Tashtyp temperature of two periods: DOY 154 (**A**) and DOY 197 (**B**). Observed (red line), reconstructed (black line), reanalysis V3 temperatures (blue line).

3.2. Reconstruction Model Evaluation

For each model, we obtained R^2 and RMSE metrics for the calibration set—performance of the models on the data on which they were trained; and for the verification set—performance of the models on the data that were not used in the training process. We also calculated simulated metrics—final performance of the models on the whole dataset. The obtained metrics are shown in Table 1.

Table 1. Mean R^2 and RMSE obtained for the models during their calibration and verification.

Period	DOY	R^2 Calibration	R^2 Verification	R^2 Simulated	RMSE Calibration	RMSE Verification	RMSE Simulated
A	154	0.74 ± 0.03	0.60	0.74	0.48 ± 0.15	0.60	0.48
B	197	0.79 ± 0.03	0.60	0.78	0.46 ± 0.15	0.63	0.46

Since different R^2 and RMSE were obtained for each iteration of the rolling cross-validation for the calibration set, Table 1 summarizes the mean values of the statistics on the calibration sample (±s.d.).

The obtained metrics show a reasonably high quality of the models, with more than 74% of the explained variance with RMSE less than 0.49 °C on the calibration set, and 60% of the explained variance with RMSE less than 0.63 °C on the verification set.

Figure 6 shows that, over the period of instrumental observations (1929–2016), the modeled (black line) and observed (red line) temperatures for both periods (A and B) have an extremely high degree of synchrony (R > 0.8, $p < 10^{-26}$, Table 2), as do the observed and reanalyzed temperatures (blue line) (R > 0.96, $p < 10^{-45}$, Table 2), but the reanalyzed data are significantly underestimated relative to the observed temperature (>1 °C) (see Figure 7 for more detail).

Table 2. The Pearson correlations (Rs) and their significance (p) between the observed, reanalyzed, and simulated temperatures.

Years	Period	Observed\Reanalyzed		Observed\Simulated		Simulated\Reanalyzed	
		Pearson R	p-Value	Pearson R	p-Value	Pearson R	p-Value
1836–1928	A	-	-	-	-	0.67	$<10^{-12}$
	B	-	-	-	-	0.57	$<10^{-8}$
1929–2016	A	0.97	$<10^{-51}$	0.86	$<10^{-26}$	0.79	$<10^{-18}$
	B	0.96	$<10^{-45}$	0.89	$<10^{-29}$	0.81	$<10^{-20}$

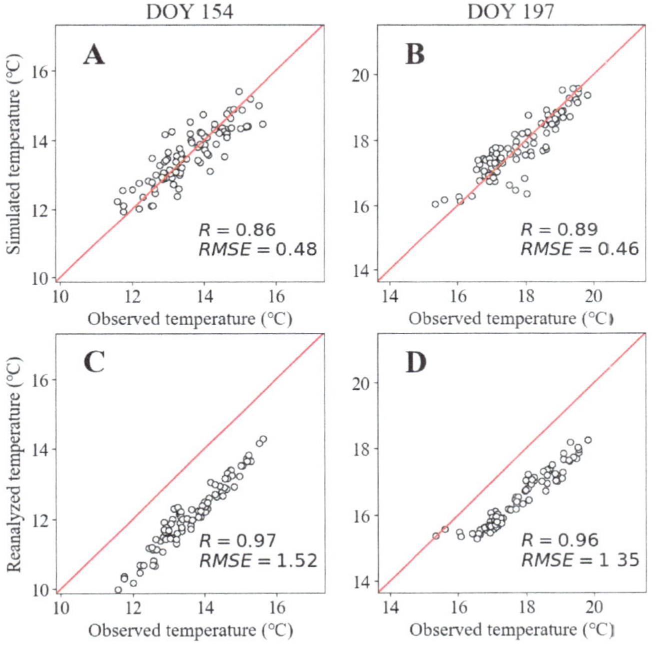

Figure 7. Scatterplots for the simulated/reanalyzed temperatures versus the observed temperatures obtained for two periods: DOY 154 (**A**,**C**) and DOY 197 (**B**,**D**) (1929 to 2016).

Due to their strong correlation with the observed temperature, despite the underestimation, the reanalyzed data can be taken as a reliable source of information on temperature trends. In this regard, the significant correlation (R > 0.56, $p < 10^{-8}$, Table 2) between simulated and reanalyzed temperatures for the years with no direct climate observations (1836–1928) indicates that the models are sufficiently correct in indicating temperature trends for the modeled periods (A and B). In this paper, reanalyzed data are used for additional assessment of model quality beyond the metrics from Table 1.

The reanalyzed data are consistently underestimated (from 1.35 to 1.78 °C) relative to the observed data. This can be explained by the altitude difference between Tashtyp and the stations on the basis of which the reanalyzed data were constructed. As the altitude increases by 100 m, the temperature may proportionally decrease by a value close to 0.5 °C [39], so it can be assumed that the reanalyzed data are constructed on the basis of the data from the stations located 300–400 m above Tashtyp.

We noted that the obtained tree-ring width chronology did not correlate ($p > 0.05$) with the smoothed observed temperature of the Tashtyp weather station for intra-annual intervals up to 14-day smoothing.

Moreover, the enumeration of inter-annual sliding windows (Ws) for the standard TRW chronology from the range of 1 to 11 years did not allow one to obtain any models with $R^2 > 0.5$.

4. Discussion

It is known that both the width of tree rings and their anatomical structure are controlled by internal and external factors [35,40–43]. However, when comparing the variance and sensitivity coefficient of tree-ring width chronologies for different species under the same growing conditions, Siberian pine is shown to have rather low values of growth variability indices, even under climatically limited growing conditions [44,45]. There is no doubt that the low variability in Siberian pine growth is the result of a greater contribution of internal factors to its seasonal and perennial growth. These factors include adaptations of its physiology to the slow accumulation and utilization of nutrients [46,47], adaptations to wetter growing conditions (average variability of growth in other similar species of Siberian taiga, Siberian spruce, and Siberian fir [48,49]), and other ecophysiological features [50]. It is possible that the minor response of Siberian pine growth to low temperatures at high elevations is related to its genome size, which is larger than that in most tree species [51,52]

However, the long lifespan (up to 800–900 years [50,53]), wide distribution, and high economic and ecological value of Siberian pine have determined the ongoing attempts of dendrochronologists to obtain centuries-long annals of its tree rings and decode the contribution of the dynamics of environmental conditions, including climate forcing, on the formed rings. One of the approaches, due to a greater number of low-frequency fluctuations compared to annual fluctuations in tree-ring growth, is the use of smoothed time series with a window width of several years. The effectiveness of this approach was demonstrated in this work, and corresponds well with previously published results [30,54].

Another approach is to search more sensitive indicators of climate variability among the parameters of annual rings, including the use of quantitative wood anatomy. The use of long-term chronologies of tree-ring anatomical characteristics to reconstruct long-term changing climates over centuries is a relatively new and rapidly developing research direction [27,28,55,56].

We hypothesized that the ability to reconstruct seasonal, weekly-based temperature changes could be enhanced by using cell chronologies of Siberian pine anatomical characteristics, which was confirmed by both previously published results [19] and the new temperature reconstructions developed in the current research. Since the effects of internal factors are fairly uniform from year to year, by special treatment and combining the available cell measurements, we can accumulate the effects of external climate forcing and potentially reveal "hidden" correlations with climatic variables that cannot be observed based on traditional dendroclimatology techniques. To realize this approach, we developed several procedures to treat initial climate and anatomical data (Figures 2 and 3). We also developed a new reconstruction procedure of seasonal temperatures (Figure 4) based on the principal component decomposition of cell measurements, specifically radial cell sizes and cell wall thickness.

The search for optimal windows for temperature generalization within a season and on a long-term scale allowed us not only to obtain reconstruction models that had high convergence with instrumental and reanalysis data but also established physiologically based regularities. The high first-order autocorrelation within a decade, accounting for the increased contribution of low-frequency oscillations and the possibility of preferential reconstruction of smoothed climatic series, can, at least partly, be attributed to the perennial needles of the species. It is known that needles of different species can persist for up to 14 years (e.g., spruce at the limit of distribution in the mountains; [57]). Siberian pine has a needle life span of 4–6 years at lower elevations and up to 9–10 years at higher elevations [58,59]. The age of needles of a related species, European cedar *Pinus cembra*, can reach 9–12 years [60,61].

The temperature intervals during the season, for which the reconstruction models showed the highest and most reliable statistical estimates, are important indicators of xylogenesis phenology. Thus, the first period in the first week of early June indicates the time interval of the most active growth of Siberian pine under certain local conditions, with sufficient humidity and increasing daylight but low heat availability. The second period (the week in the middle of July) corresponds to the interval of the end of primary growth and the budding for the development of shoots and needles for the next season [62–64].

The resulting models had quite high metrics. The metrics of the models based on tracheid data ($R > 0.86$) are much higher than those of models based on tree ring widths ($R \in [0.2, 0.6]$), both in the current work and compared to standard metric values of models based on climate-insensitive trees [19,65]. The obtained metric values are comparable to those from recent works on temperature reconstruction based on quantitative wooden anatomy of climate-insensitive species ($R^2 > 0.74$ on the calibration set) [65], which suggests that the cellular indices of these species are robust in explaining 60%–75% of the variance in temperature reconstruction.

In the current work, to test the quality of the models, in addition to the verification sample, whose data did not participate in the model training process and therefore can be considered independent, we also used V3 reanalysis grid data, with the modeled temperature also significantly correlating with these data, both during the years with instrumental measurements ($R > 0.79$) and outside them ($R > 0.57$).

Nevertheless, the potential of the proposed approach to modeling short-term climatic fluctuations requires further research. Various transformations or standardization of anatomical measurements may, for example, suppress a part of the non-climatic signal (including size-age) or the climatic signal inherited from previous stages of xylogenesis [66]. Different approaches to selecting relevant predictors among PCs and identifying promising intervals for intra- and inter-annual averaging, such as wavelet analysis, are worth exploring. The proposed approach can also be widely tested on new data from different forest stands with the same or other tree species that have low sensitivity to climate, as well as in different climatic zones.

5. Conclusions

The use of tracheid measurements has made it possible to realize the potential of *Pinus sibirica* Du Tour for high-resolution temperature reconstruction.

Using the developed algorithms, reliable models for temperature reconstruction were obtained. The seasonal, weekly-based intervals for which the models were obtained correlated with the key growth periods of *Pinus sibirica* Du Tour—the period of the highest cellular activity (early June) and the period of budding for the next season (mid-July), which confirmed our first hypothesis. Also, the smoothing windows, for which the first adequate reconstructions appeared, correlated with the lifespan of *Pinus sibirica* needles.

The use of tree ring widths as predictors did not allow one to obtain reliable reconstruction models for any seasonal time interval or with a single smoothing window due to their poor correlation with the temperature data. At the same time, the results of reconstructions derived from tracheid data did not only show good metric values for the verification set ($R^2 > 0.6$) but also had a significant correlation with the independent V3 reanalysis data for the 19th century ($p < 10^{-8}$), which also confirmed the second hypothesis of this work.

Further modifying the developed algorithms can improve the quality of temperature reconstruction, but it is already evident at this stage that, despite the greater complexity of obtaining tracheidogram measurements compared to annual rings, quantitative wooden anatomy allows one to reconstruct temperature with much higher accuracy and resolution than classical approaches.

Supplementary Materials: The following supporting information can be downloaded at: https://www.mdpi.com/article/10.3390/f15010167/s1, Figure S1. Example of daily (or intra-annual) smoothing. The raw (red dashed) and smoothed ($w = 7$, 1 day step) daily Tashtyp temperature; Figure S2. Example of unsmoothed and 9-year (or inter-annual) smoothing of temperature characteristics:

mean values of the 1st week of May (red dotted curve), June (green dotted curve), and July (blue dotted curve) and their smoothed analogs (solid thick lines), respectively; Figure S3. Example of cell measurements for the year 1653 of Tree №2; Figure S4. Dmean (A) and CWTmean (C) tracheidograms for the year 1653 of Tree №1, and their standardization to 15 cells (B,D); Figure S5. Example of the mean standardized tracheidograms (thick black curves) for the 1653 year: radial cell diameter (A) and cell wall thickness (B); Figure S6. Obtained tracheidogram objects; Figure S7. The obtained 30 cell chronologies: 15 mean standardized cell diameters (A) and corresponding 15 cell wall thicknesses (B); Figure S8. Example of 9-year smoothed cell chronologies: mean standardized cell diameters (A) and corresponding cell wall thicknesses; Figure S9. First five principal components (PCs) of the smoothed (9-year sliding window) tracheid chronologies; Figure S10. Cumulative explained variance of the tracheidogram objects; Figure S11. Visualization of the Rolling Leave-One-Out Cross Validation procedure for the data with the sliding windows $w = 7$, $W = 9$, for the $doy = 152$. The red cells are considered as a verification set for the corresponding model, the gray cells are omitted, and the white cells are considered as a calibration set for the corresponding model. Table S1. Examples of the temperature time series T_{inter} (w, W, doy) for $w = 7$, $W = 9$, $|DOY_{sig}(7{,}9)| = 63$; Table S2. Examples of thirty ($2n = 30$) mean tracheidogram chronologies; Table S3. Example of inter-annual smoothed tracheid chronologies for $W = 9$; Table S4. Example of PC chronologies for $W = 9$; Table S5. PCA transformation matrix for first five principal components; Table S6. Example of the calibration and verification sets for w = 7, $W = 9$, $doy = 152$, $\theta = 2000$; Table S7. Model coefficients.

Author Contributions: Conceptualization, M.S.Z., E.A.V. and V.V.S.; methodology, M.S.Z. and V.V.S.; software, M.S.Z.; validation, B.Y., E.A.V. and V.V.S.; formal analysis, M.S.Z., B.Y., E.A.V. and V.V.S.; investigation, M.S.Z., B.Y., E.A.V. and V.V.S.; resources, E.A.B.; data curation, M.S.Z. and V.V.S.; writing—original draft preparation, M.S.Z., B.Y., D.F.Z., E.A.V. and V.V.S.; writing—review and editing, M.S.Z., B.Y. and V.V.S.; visualization, M.S.Z.; supervision, E.A.V. and V.V.S.; project administration, V.V.S.; funding acquisition, E.A.V. and V.V.S. All authors have read and agreed to the published version of the manuscript.

Funding: This research was funded by the Russian Science Foundation, grant no. 23-44-0006? (data analysis, reconstruction models, software development), and the National Natural Science Foundation of China, grant no. 42261134537, in the framework of a joint Russian–Chinese project (state-of-the-art analysis and international collaboration expenses), and by the Russian Ministry of Science and Higher Education, grant number FSRZ-2023-0007 (climate and wooden data collection and measurements).

Data Availability Statement: The data presented in this study are available upon reasonable request from the authors. All data processing algorithms were implemented in Python and can be found at https://github.com/mikewellmeansme/dendroclimatic-reconstructor/ (accessed on 9 January 2024).

Acknowledgments: The authors express their deep gratitude to Yulia Olegovna Zharkova, Krasnoyarsk Science Center, Siberian Branch of the Russian Academy of Sciences, for linguistic and stylistic corrections of the text.

Conflicts of Interest: The authors declare no conflicts of interest. The funders had no role in the design of this study; in the collection, analyses, or interpretation of data; in the writing of the manuscript; or in the decision to publish the results.

References

1. Danchenko, A.M.; Beh, I.A. *Cedar Forests of Western Siberia*; Tomsk State University: Tomsk, Russia, 2010.
2. White, A.; Cannell, M.G.R.; Friend, A.D. CO_2 Stabilization, Climate Change and the Terrestrial Carbon Sink. *Glob. Chang. Bio* **2000**, *6*, 817–833. [CrossRef]
3. Menzel, A.; Sparks, T.H.; Estrella, N.; Koch, E.; Aaasa, A.; Ahas, R.; Alm-Kübler, K.; Bissolli, P.; Braslavská, O.; Briede, A.; et a European Phenological Response to Climate Change Matches the Warming Pattern. *Glob. Chang. Biol.* **2006**, *12*, 1969–197 [CrossRef]
4. Huang, J.G.; Zhang, Y.; Wang, M.; Yu, X.; Deslauriers, A.; Fonti, P.; Liang, E.; Mäkinen, H.; Oberhuber, W.; Rathgeber, C.B.K.; et a A Critical Thermal Transition Driving Spring Phenology of Northern Hemisphere Conifers. *Glob. Chang. Biol.* **2023**, *29*, 1606–161 [CrossRef]
5. Körner, C. *Alpine Plant Life*; Springer: Berlin/Heidelberg, Germany, 2003.
6. Holtmeier, F.-K. *Mountain Timberlines*; Springer: Dordrecht, The Netherlands, 2009.

7. Harsch, M.A.; Hulme, P.E.; McGlone, M.S.; Duncan, R.P. Are Treelines Advancing? A Global Meta-Analysis of Treeline Response to Climate Warming. *Ecol. Lett.* **2009**, *12*, 1040–1049. [CrossRef] [PubMed]
8. Allen, C.D.; Macalady, A.K.; Chenchouni, H.; Bachelet, D.; McDowell, N.; Vennetier, M.; Kitzberger, T.; Rigling, A.; Breshears, D.D.; Hogg, E.H.; et al. A Global Overview of Drought and Heat-Induced Tree Mortality Reveals Emerging Climate Change Risks for Forests. *For. Ecol. Manag.* **2010**, *259*, 660–684. [CrossRef]
9. Jochner, M.; Bugmann, H.; Nötzli, M.; Bigler, C. Tree Growth Responses to Changing Temperatures across Space and Time: A Fine-Scale Analysis at the Treeline in the Swiss Alps. *Trees-Struct. Funct.* **2018**, *32*, 645–660. [CrossRef]
10. Lett, S.; Dorrepaal, E. Global Drivers of Tree Seedling Establishment at Alpine Treelines in a Changing Climate. *Funct. Ecol.* **2018**, *32*, 1666–1680. [CrossRef]
11. McDowell, N.G.; Allen, C.D.; Anderson-Teixeira, K.; Aukema, B.H.; Bond-Lamberty, B.; Chini, L.; Clark, J.S.; Dietze, M.; Grossiord, C.; Hanbury-Brown, A.; et al. Pervasive Shifts in Forest Dynamics in a Changing World. *Science* **2020**, *368*, eaaz9463. [CrossRef]
12. Lindner, M.; Maroschek, M.; Netherer, S.; Kremer, A.; Barbati, A.; Garcia-Gonzalo, J.; Seidl, R.; Delzon, S.; Corona, P.; Kolström, M.; et al. Climate Change Impacts, Adaptive Capacity, and Vulnerability of European Forest Ecosystems. *For. Ecol. Manag.* **2010**, *259*, 698–709. [CrossRef]
13. Parks, C.G.; Bernier, P. Adaptation of Forests and Forest Management to Changing Climate with Emphasis on Forest Health: A Review of Science, Policies and Practices. *For. Ecol. Manag.* **2010**, *259*, 657–659. [CrossRef]
14. Morin, X.; Fahse, L.; Jactel, H.; Scherer-Lorenzen, M.; García-Valdés, R.; Bugmann, H. Long-Term Response of Forest Productivity to Climate Change Is Mostly Driven by Change in Tree Species Composition. *Sci. Rep.* **2018**, *8*, 5627. [CrossRef]
15. Pugh, T.A.M.; Lindeskog, M.; Smith, B.; Poulter, B.; Arneth, A.; Haverd, V.; Calle, L. Role of Forest Regrowth in Global Carbon Sink Dynamics. *Proc. Natl. Acad. Sci. USA* **2019**, *116*, 4382–4387. [CrossRef] [PubMed]
16. Fonti, P.; Von Arx, G.; García-González, I.; Eilmann, B.; Sass-Klaassen, U.; Gärtner, H.; Eckstein, D. Studying Global Change through Investigation of the Plastic Responses of Xylem Anatomy in Tree Rings. *New Phytol.* **2010**, *185*, 42–53. [CrossRef] [PubMed]
17. Wang, H.; Shao, X.; Fang, X.; Yin, Z.; Chen, L.; Zhao, D.; Wu, S. Responses of Pinus Koraiensis Tree Ring Cell Scale Parameters to Climate Elements in Changbai Mountains. *Ying Yong Sheng Tai Xue Bao* **2011**, *22*, 2643–2652.
18. Wang, H.; Shao, X.; Fang, X.; Jiang, Y.; Liu, C.; Qiao, Q. Relationships between Tree-Ring Cell Features of Pinus Koraiensis and Climate Factors in the Changbai Mountains, Northeastern China. *J. For. Res.* **2017**, *28*, 105–114. [CrossRef]
19. Zhirnova, D.F.; Belokopytova, L.V.; Upadhyay, K.K.; Tripathi, S.K.; Babushkina, E.A.; Vaganov, E.A. 495-Year Wood Anatomical Record of Siberian Stone Pine (Pinus Sibirica Du Tour) as Climatic Proxy on the Timberline. *Forests* **2022**, *13*, 247. [CrossRef]
20. Anderegg, W.R.L. Spatial and Temporal Variation in Plant Hydraulic Traits and Their Relevance for Climate Change Impacts on Vegetation. *New Phytol.* **2015**, *205*, 1008–1014. [CrossRef]
21. Sperry, J.S.; Love, D.M. What Plant Hydraulics Can Tell Us about Responses to Climate-Change Droughts. *New Phytol.* **2015**, *207*, 14–27. [CrossRef]
22. Chave, J.; Coomes, D.; Jansen, S.; Lewis, S.L.; Swenson, N.G.; Zanne, A.E. Towards a Worldwide Wood Economics Spectrum. *Ecol. Lett.* **2009**, *12*, 351–366. [CrossRef]
23. Von Arx, G.; Archer, S.R.; Hughes, M.K. Long-Term Functional Plasticity in Plant Hydraulic Architecture in Response to Supplemental Moisture. *Ann. Bot.* **2012**, *109*, 1091–1100. [CrossRef]
24. Castagneri, D.; Fonti, P.; Von Arx, G.; Carrer, M. How Does Climate Influence Xylem Morphogenesis over the Growing Season? Insights from Long-Term Intra-Ring Anatomy in Picea Abies. *Ann. Bot.* **2017**, *119*, 1011–1020. [CrossRef] [PubMed]
25. Tumajer, J.; Shishov, V.V.; Ilyin, V.A.; Camarero, J.J. Intra-Annual Growth Dynamics of Mediterranean Pines and Junipers Determines Their Climatic Adaptability. *Agric. For. Meteorol.* **2021**, *311*, 108685. [CrossRef]
26. Tumajer, J.; Kašpar, J.; Kuželová, H.; Shishov, V.V.; Tychkov, I.I.; Popkova, M.I.; Vaganov, E.A.; Treml, V. Forward Modeling Reveals Multidecadal Trends in Cambial Kinetics and Phenology at Treeline. *Front. Plant Sci.* **2021**, *12*, 613643. [CrossRef] [PubMed]
27. Pritzkow, C.; Wazny, T.; Heußner, K.U.; Słowiński, M.; Bieber, A.; Liñán, I.D.; Helle, G.; Heinrich, I. Minimum Winter Temperature Reconstruction from Average Earlywood Vessel Area of European Oak (Quercus Robur) in N-Poland. *Palaeogeogr. Palaeoclimatol. Palaeoecol.* **2016**, *449*, 520–530. [CrossRef]
28. Ziaco, E.; Biondi, F.; Heinrich, I. Wood Cellular Dendroclimatology: Testing New Proxies in Great Basin Bristlecone Pine. *Front. Plant Sci.* **2016**, *7*, 223658. [CrossRef]
29. Valeriano, C.; Gutiérrez, E.; Colangelo, M.; Gazol, A.; Sánchez-Salguero, R.; Tumajer, J.; Shishov, V.; Bonet, J.A.; Martínez de Aragón, J.; Ibáñez, R.; et al. Seasonal Precipitation and Continentality Drive Bimodal Growth in Mediterranean Forests. *Dendrochronologia* **2023**, *78*, 126057. [CrossRef]
30. D'Arrigo, R.; Jacoby, G.; Frank, D.; Pederson, N.; Cook, E.; Buckley, B.; Nachin, B.; Mijiddorj, R.; Dugarjav, C. 1738 Years of Mongolian Temperature Variability Inferred from a Tree-Ring Width Chronology of Siberian Pine. *Geophys. Res. Lett.* **2001**, *28*, 543–546. [CrossRef]
31. Cerrato, R.; Salvatore, M.C.; Gunnarson, B.E.; Linderholm, H.W.; Carturan, L.; Brunetti, M.; De Blasi, F.; Baroni, C. A *Pinus cembra* L. Tree-Ring Record for Late Spring to Late Summer Temperature in the Rhaetian Alps, Italy. *Dendrochronologia* **2019**, *53*, 22–31. [CrossRef]

32. Zharkov, M.S.; Huang, J.G.; Yang, B.; Babushkina, E.A.; Belokopytova, L.V.; Vaganov, E.A.; Zhirnova, D.F.; Ilyin, V.A.; Popkova, M.I.; Shishov, V.V. Tracheidogram's Classification as a New Potential Proxy in High-Resolution Dendroclimatic Reconstructions. *Forests* **2022**, *13*, 970. [CrossRef]
33. Silkin, P.P. *Methods of Multiparameter Analysis of Conifers Tree-Ring Structure*; Siberian Federal University: Krasnoyarsk, Russia, 2010
34. Dyachuk, P.; Arzac, A.; Peresunko, P.; Videnin, S.; Ilyin, V.; Assaulianov, R.; Babushkina, E.A.; Zhirnova, D.; Belokopytova, L.; Vaganov, E.A.; et al. AutoCellRow (ACR)–A New Tool for the Automatic Quantification of Cell Radial Files in Conifer Images. *Dendrochronologia* **2020**, *60*, 125687. [CrossRef]
35. Vaganov, E.A.; Hughes, M.K.; Shashkin, A.V. *Growth Dynamics of Conifer Tree Rings*; Springer: Berlin/Heidelberg, Germany, 2006
36. Cook, E.; Peters, K. The Smoothing Spline: A New Approach to Standardizing Forest Interior Tree-Ring Width Series for Dendroclimatic Studies. *Tree-Ring Bull.* **1981**, *41*, 45–53.
37. Perez, L.V. *Principal Component Analysis to Address Multicollinearity*; Whitman College: Whitman, MA, USA, 2017.
38. Hastie, T.; Tibshirani, R.; Friedman, J. *The Elements of Statistical Learning*; Springer Series in Statistics; Springer: New York, NY, USA, 2009; ISBN 978-0-387-84857-0.
39. Chen, B.X.; Sun, Y.F.; Zhang, H.B.; Han, Z.H.; Wang, J.S.; Li, Y.K.; Yang, X.L. Temperature Change along Elevation and Its Effect on the Alpine Timberline Tree Growth in the Southeast of the Tibetan Plateau. *Adv. Clim. Chang. Res.* **2018**, *9*, 185–191. [CrossRef]
40. Baas, P.; Wheeler, E.A. Wood Anatomy and Climate Change. *Clim. Chang. Ecol. Syst.* **2011**, *78*, 141–155. [CrossRef]
41. Arzac, A.; Fonti, M.V.; Vaganov, E.A. An Overview on Dendrochronology and Quantitative Wood Anatomy Studies of Conifers in Southern Siberia (Russia). *Prog. Bot.* **2021**, *83*, 161–181. [CrossRef]
42. Pandey, S. Climatic Influence on Tree Wood Anatomy: A Review. *J. Wood Sci.* **2021**, *67*, 24. [CrossRef]
43. Anderson-Teixeira, K.J.; Herrmann, V.; Rollinson, C.R.; Gonzalez, B.; Gonzalez-Akre, E.B.; Pederson, N.; Alexander, M.R.; Allen, C.D.; Alfaro-Sánchez, R.; Awada, T.; et al. Joint Effects of Climate, Tree Size, and Year on Annual Tree Growth Derived from Tree-Ring Records of Ten Globally Distributed Forests. *Glob. Chang. Biol.* **2022**, *28*, 245–266. [CrossRef] [PubMed]
44. Nazarov, A.N.; Myglan, V.S. Application of Siberian Cedar for Reconstruction of Climate and Geomorphological Events in Altai *Izv. Ross. Akad. Nauk. Seriya Geogr.* **2015**, *2*, 43–51. [CrossRef]
45. Shah, S.; Liu, Q.; Ahmad, A.; Mannan, A. Climate Growth Response of Pinus Sibirica (SIBERIAN PINE) in the Altai Mountains, Northwestern China. *Pak. J. Bot.* **2020**, *52*, 593–600. [CrossRef]
46. Reich, P.B. The World-Wide 'Fast–Slow' Plant Economics Spectrum: A Traits Manifesto. *J. Ecol.* **2014**, *102*, 275–301. [CrossRef]
47. Ning, Q.R.; Gong, X.W.; Li, M.Y.; Hao, G.Y. Differences in Growth Pattern and Response to Climate Warming between Larix Olgensis and Pinus Koraiensis in Northeast China Are Related to Their Distinctions in Xylem Hydraulics. *Agric. For. Meteorol* **2022**, *312*, 108724. [CrossRef]
48. Jiao, L.; Jiang, Y.; Wang, M.; Kang, X.; Zhang, W.; Zhang, L.; Zhao, S. Responses to Climate Change in Radial Growth of Picea Schrenkiana along Elevations of the Eastern Tianshan Mountains, Northwest China. *Dendrochronologia* **2016**, *40*, 117–127 [CrossRef]
49. Lei, J.; Feng, X.; Shi, Z.; Bai, D.; Xiao, W. Climate–Growth Relationship Stability of Picea Crassifolia on an Elevation Gradient Qilian Mountain, Northwest China. *J. Mt. Sci.* **2016**, *13*, 734–743. [CrossRef]
50. Ignatenko, M.M. *Siberian Cedar: (Biology, Introduction, Culture)*; Nauka: Moscow, Russia, 1988.
51. MacGillivray, C.W.; Grime, J.P. Genome Size Predicts Frost Resistance in British Herbaceous Plants: Implications for Rates of Vegetation Response to Global Warming. *Funct. Ecol.* **1995**, *9*, 320. [CrossRef]
52. Pellicer, J.; Leitch, I.J. The Plant DNA C-Values Database (Release 7.1): An Updated Online Repository of Plant Genome Size Data for Comparative Studies. *New Phytol.* **2020**, *226*, 301–305. [CrossRef] [PubMed]
53. Nazarov, A.N.; Myglan, V.S. The Possibility of Construction of the 6000-Year Chronology for Siberian Pine in the Central Altai. *J Sib. Fed. Univ. Biol.* **2012**, *5*, 70–88.
54. Kharuk, V.I.; Ranson, K.J.; Im, S.T.; Dvinskaya, M.L. Response of Pinus Sibirica and Larix Sibirica to Climate Change in Southern Siberian Alpine Forest–Tundra Ecotone. *Scand. J. For. Res.* **2009**, *24*, 130–139. [CrossRef]
55. González-Cásares, M.; Camarero, J.J.; Colangelo, M.; Rita, A.; Pompa-García, M. High Responsiveness of Wood Anatomy to Water Availability and Drought near the Equatorial Rear Edge of Douglas-Fir. *Can. J. For. Res.* **2019**, *49*, 1114–1123. [CrossRef]
56. Seftigen, K.; Fuentes, M.; Ljungqvist, F.C.; Björklund, J. Using Blue Intensity from Drought-Sensitive Pinus Sylvestris in Fennoscandia to Improve Reconstruction of Past Hydroclimate Variability. *Clim. Dyn.* **2020**, *55*, 579–594. [CrossRef]
57. Reich, P.B.; Oleksyn, J.; Modrzynski, J.; Tjoelker, M.G. Evidence That Longer Needle Retention of Spruce and Pine Populations at High Elevations and High Latitudes Is Largely a Phenotypic Response. *Tree Physiol.* **1996**, *16*, 643–647. [CrossRef]
58. Sobchak, R.O.; Zotikova, A.P. Influence of High Altitude Conditions on Anatomo-Physiological Indices of Siberian Pine Needles *Bull. Tomsk. State Univ.* **2009**, *326*, 200–202.
59. Bender, O.G.; Zotikova, A.P.; Bender, A.G. The State of Photosynthetic Apparatus of Different-Aged Needles of Siberian Cedar at the Southern Limit of Growth in the Altai Mountains. In Proceedings of the Forest Biogeocenoses of the Boreal Zone: Geography Structure, Functions, Dynamics, Krasnoyarsk, Russia, 16 September 2014; Sukachev Institute of Forest SB RAS, Federal Research Center "Krasnoyarsk Science Center SB RAS": Krasnoyarsk, Russia, 2014; pp. 380–383.
60. Günthardt, M.S.; Wanner, H. Die Menge des Cuticulären Wachses auf Nadeln von *Pinus cembra* L. und *Picea abies* (L.) Karsten in Abhängigkeit von Nadelalter und Standort. *Flora* **1982**, *172*, 125–137. [CrossRef]
61. Nebel, B.; Matile, P. Longevity and Senescence of Needles in *Pinus cembra* L. *Trees* **1992**, *6*, 156–161. [CrossRef]

62. Cooke, J.E.K.; Eriksson, M.E.; Junttila, O. The Dynamic Nature of Bud Dormancy in Trees: Environmental Control and Molecular Mechanisms. *Plant Cell Environ.* **2012**, *35*, 1707–1728. [CrossRef] [PubMed]
63. Khare, S.; Drolet, G.; Sylvain, J.D.; Paré, M.C.; Rossi, S. Assessment of Spatio-Temporal Patterns of Black Spruce Bud Phenology across Quebec Based on MODIS-NDVI Time Series and Field Observations. *Remote Sens.* **2019**, *11*, 2745. [CrossRef]
64. Silvestro, R.; Brasseur, S.; Klisz, M.; Mencuccini, M.; Rossi, S. Bioclimatic Distance and Performance of Apical Shoot Extension: Disentangling the Role of Growth Rate and Duration in Ecotypic Differentiation. *For. Ecol. Manag.* **2020**, *477*, 118483. [CrossRef]
65. Lopez-Saez, J.; Corona, C.; von Arx, G.; Fonti, P.; Slamova, L.; Stoffel, M. Tree-Ring Anatomy of Pinus Cembra Trees Opens New Avenues for Climate Reconstructions in the European Alps. *Sci. Total Environ.* **2023**, *855*, 158605. [CrossRef]
66. Meko, D.M.; Baisan, C.H. Pilot Study of Latewood-Width of Conifers as an Indicator of Variability of Summer Rainfall in the North American Monsoon Region. *Int. J. Climatol.* **2001**, *21*, 697–708. [CrossRef]

forests

MDPI

Article

Growth Responses to Climate and Drought in Relict Cork Oak Populations as a Benchmark of the Species Tolerance

J. Julio Camarero [1,*], Antonio Gazol [1], Cristina Valeriano [1] Michele Colangelo [2] and Álvaro Rubio-Cuadrado [1,3]

[1] Instituto Pirenaico de Ecología (IPE-CSIC), Avda. Montaña 1005, 50192 Zaragoza, Spain; agazol@ipe.csic.es (A.G.); cvaleriano@ipe.csic.es (C.V.); alvaro.rubio.cuadrado@upm.es (Á.R.-C.)
[2] Scuola di Scienze Agrarie, Forestali, Alimentari e Ambientali, Università della Basilicata, Viale dell'Ateneo Lucano 10, 85100 Potenza, Italy; michele.colangelo@unibas.it
[3] Departamento de Sistemas y Recursos Naturales, Escuela Técnica Superior de Ingeniería de Montes, Forestal y del Medio Natural, Universidad Politécnica de Madrid, Ciudad Universitaria s/n, 28040 Madrid, Spain
* Correspondence: jjcamarero@ipe.csic.es

Abstract: We still lack information on the long-term growth responses to climate of relict tree populations, which often persist in topoclimatic refugia. To fill that research gap, we studied three relict cork oak (*Quercus suber*) populations located in northern Spain using dendrochronology. The sites were subjected to humid (Zarautz), continental (Bozoó) and xeric (Sestrica) climate conditions. Cool–wet conditions during the current spring enhanced growth in Bozoó and Sestrica, whereas wet conditions in the previous October enhanced growth in Zarautz. In this site, growth also increased in response to dry conditions in the prior winter linked to high North Atlantic Oscillation indices. Correlations between the precipitation summed from the previous September to the current May peaked at the driest site (Sestrica). The strongest growth responses to drought severity were also found at this site, where growth negatively responded to 9-month early-summer droughts, followed by the continental Bozoó site, where growth was constrained by 1-month July droughts. Growth declined in response to 6-month January droughts in the wettest site (Zarautz), where cork oak was vulnerable to previous late-summer to autumn drought stress. Despite warmer and drier spring conditions that would negatively impact cork oak at the Bozoó and Sestrica sites, trees from these populations could tolerate further aridity.

Keywords: dendroecology; drought; Mediterranean climate; North Atlantic oscillation; *Quercus suber*; SPE

Citation: Camarero, J.J.; Gazol, A.; Valeriano, C.; Colangelo, M.; Rubio-Cuadrado, Á. Growth Responses to Climate and Drought in Relict Cork Oak Populations as a Benchmark of the Species Tolerance. *Forests* **2024**, *15*, 72. https://doi.org/10.3390/f15010072

Academic Editors: Yassine Messaoud, Jan Světlík and Giorgio Alberti

Received: 14 December 2023
Revised: 27 December 2023
Accepted: 27 December 2023
Published: 29 December 2023

1. Introduction

The Mediterranean Basin is a climate change hotspot subjected to increasingly warming conditions and ongoing aridification and an area where further warming and drying are forecasted [1,2]. In this region, forests have been used and managed by people for millennia to extract timber, charcoal, fodder, fruits and also cork from the cork oak (*Quercus suber* L.) bark [3]. However, climate warming may menace the persistence of some cork oak populations because hotter droughts lead to growth decline and increase mortality due to severe water deficits or greater pathogen incidence [4–10]. Furthermore, in addition to dieback and growth decline, climate change may result in a 20% decrease in cork production by the late 21st century [11]. Therefore, a better long-term assessment of climate impacts and growth responses to drought is needed for cork oak, given its wide distribution (ca. 2.5 million ha) and socio-economic and ecological relevance in the western Mediterranean basin.

The cork oak is an evergreen tree species with diffuse-porous wood widely distributed in areas with mild climate conditions characterized by acidic, often nutrient-poor soils, where it is usually managed for cork production [12]. Cork is stripped from the trunks of mature trees at intervals ranging from 8 to 14 years [12]. Cork oak provides multiple ecological services, such as enhancing biodiversity, promoting carbon sequestration, protecting soils from erosion and supplying economic (cork, wood, timber) and cultural services [3]. This oak responds to the typical Mediterranean summer drought through changes in several traits (leaf osmotic adjustments, rapid stomata closure, isohydric behavior, uptake of deep water sources, reduction in xylem vessel diameter), which make it a drought-avoiding species [13–18]. The drought-avoiding strategy of cork oak should help it to alleviate the negative impacts of severe water shortages on radial growth and reduce the risk of hydraulic failure through controlling stomatal conductance, tissue hydration, xylem embolism and root access to deep water sources [13–19]. Nevertheless, it may also succumb in response to severe, lasting droughts [5,20]. Declining trees have shown tissue dehydration and low water-use efficiency, suggesting poor resistance to severe droughts [21].

As in other Mediterranean oaks, growth decline in cork oak associated with canopy dieback and drought stress is reflected not only by lower radial growth rates (narrow tree rings) but also by reduced cork production (narrow cork rings) [22]. In cork oak, intra-annual radial growth peaks in late spring (June), shows minimum values in summer and winter (dormancy) and is constrained by dry conditions in the previous winter [22]. Therefore, dry conditions in the winter prior to tree-ring growth should induce lower growth rates, leading to higher adult mortality rates as drought stress intensifies [21,23]. Adult mortality is high in cork oak populations from xeric sites subjected to increasing aridity, albeit survival and growth also depend on stand density and pathogen incidence (e.g., *Phytophthora cinnamomi* Rands.) [24]. Consequently, a lower growth rate and higher growth responsiveness or sensitivity to climate stressors, including drought, could be expected in the most arid sites. According to studies based on cork oak seedlings, it was concluded that greater aboveground (root) development is expected in humid (xeric) sites [25,26]. Therefore, higher growth rates are expected in humid sites subjected to mild conditions (e.g., coastal sites), while sites subjected to continental conditions (e.g., inland sites) could be negatively impacted by low temperatures in winter and early spring.

Here, we compared the growth responses to climate variables (maximum and minimum temperatures, precipitation) and a drought index of three relict cork oak populations located in northern Spain using dendrochronology. Isolated, relict tree populations, often located in harsh sites, constitute ideal settings to assess growth sensitivity to climate and drought stress because they may have adapted to marginal site conditions through phenotypic plasticity and/or genotypic variation [27,28]. We compared three cork oak populations situated in a humid coastal site (Zarautz), a humid continental site (Bozoó) and a xeric continental site (Sestrica). We expected that the trees growing in the driest (wettest) sites would display the lowest (highest) growth rates and the highest (lowest) responsiveness (change in growth rate, correlation coefficient) to precipitation and drought, whereas trees growing in the coldest site would be more responsive to cold conditions in the early growing season (spring) when cambium resumption occurs [22].

2. Materials and Methods

2.1. Study Sites and Field Sampling

We selected three relict cork oak populations subjected to contrasting climatic conditions in northern Spain (Figures 1 and 2, Table 1). In the three sites, the last debarking of trees to extract cork occurred at least 15–20 years before sampling. Therefore, the study sites were considered not to be managed for the past two decades. We measured the location and elevation of trees using a GPS with a 5 m resolution. We also recorded the mean slope and exposition of each site. The stand composition (main accompanying tree species in terms of relative basal area) was also characterized. The main tree species, in addition to cork oak, were holm oak (*Quercus ilex* L.) and maritime pine (*Pinus pinaster* Ait.).

According to 0.1°-gridded climate data [29], Zarautz and Sestrica were the wettest and driest study sites, whereas Bozoó was the most continental site, showing the widest thermal range (Table 1). In all sites, soils were acidic and sandy, but soil rockiness was particularly notable in Sestrica. The Zarautz and Bozoó forests are public, but the Sestrica forest is private. The summer drought was evident in Bozoó and Sestrica.

Table 1. Main characteristics of the three study sites, including the mean annual temperature (MAT) and the total annual precipitation (TAP). Dbh is the diameter at breast height. Different letters indicate significant ($p < 0.05$) differences between sites according to Mann–Whitney tests.

Site	Latitude N	Longitude W	Elevation (m a.s.l.)	Exposition	Slope (%)	MAT, Range (°C)	TAP (mm)	KOI [1]	Dbh (cm)	Height (m)	Other Tree Species
Zarautz	43°17′20″	2°10′52″	141	E	15–20	12.4, 3.9–22.8	1417	26.3	34.8 ± 1.9 c	7.5 ± 1.2 b	*Arbutus unedo, Laurus nobilis*
Bozoó	42°44′15″	3°06′31″	759	NE	5–10	10.1, −2.5–23.4	789	21.5	27.3 ± 1.6 a	5.9 ± 0.3 a	*Quercus ilex, Pinus pinaster*
Sestrica	41°30′02″	1°38′11″	865	E, SE	10–45	11.8, −1.6–27.8	455	15.3	31.4 ± 1.2 b	6.6 ± 0.9 ab	*Quercus ilex, Quercus faginea*

[1] KOI is the Kerner Oceanity Index, which is the ratio of the mean monthly air temperature difference between October and April and the difference between mean monthly temperatures of the warmest and coldest months. Small and large KOI values indicate oceanic—high continentality (11 ≤ KOI < 20)—and hyperoceanic—low continentality (21 ≤ KOI < 50)—conditions, respectively.

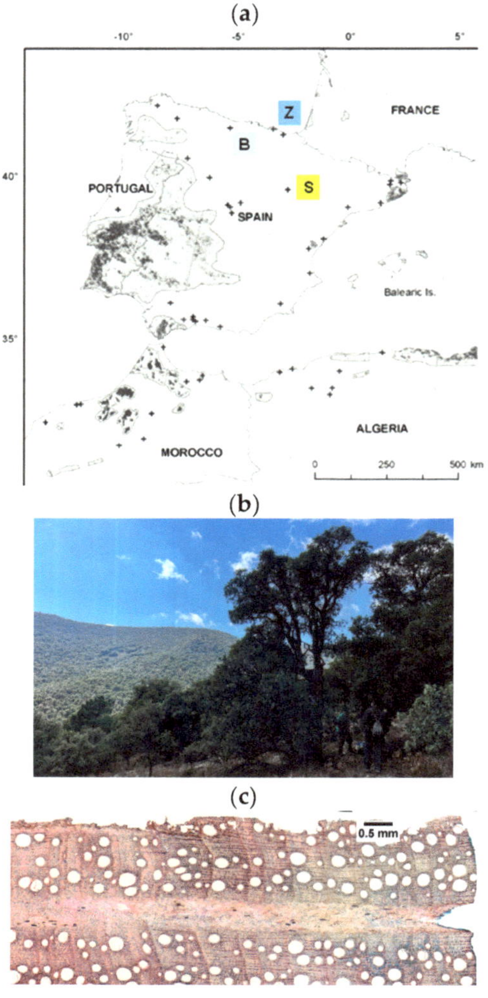

Figure 1. (**a**) Distribution of cork oak (*Q. suber*) in the western Mediterranean Basin and locations of the three relict populations sampled in northern Spain (Z, Zarautz, dark blue square; B, Bozoó, light blue square; S, Sestrica, yellow square). Light gray is the species distribution; dark gray is the data from forest inventories; and crosses are isolated or relict populations. Modified from [30]. (**b**) View of sampled site in Sestrica. (**c**) Cross-section of a cork oak core sampled at the same site and showing annual rings.

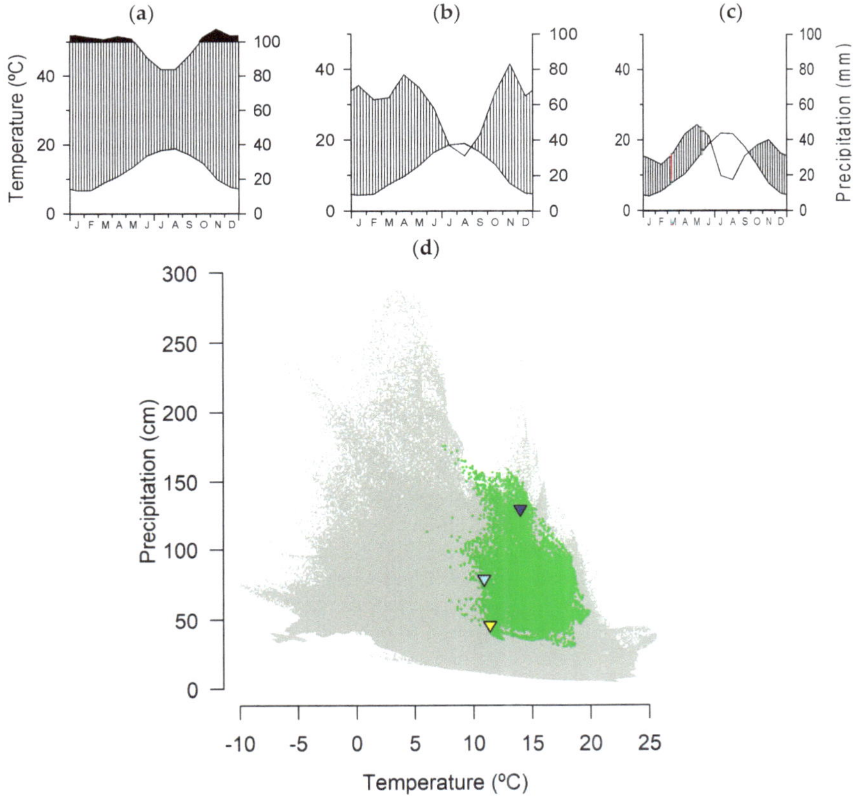

Figure 2. Climate diagrams of the three study sites ((**a**) Zarautz, (**b**) Bozoó and (**c**) Sestrica) and (**d**) locations in the climatic space. In the lower plots, the grey and green dots show annual temperature and precipitation data for Europe (12° W–60° E, 32–72 °N) and for those dominated by cork oak, respectively. The study sites are represented using dark blue (Zarautz), light blue (Bozoó) and yellow (Sestrica) triangles.

Sampling was carried out in spring 2022. In each site, we selected 10 to 14 mature, apparently healthy cork oak trees for sampling across an area of ca. 2 ha We measured their diameter at breast height (dbh) and total height using tapes and a laser rangefinder, respectively (Nikon Forestry Pro II, Nikon, Japan). Then, two cores per tree were taken at 1.3 m and perpendicular to the maximum slope using 5-mm Pressler increment borers (Haglöf, Långsele, Sweden). The mean basal area of cork oak was similar among sites, with values ranging from 12.2 (Sestrica) to 15.7 (Zarautz) m^2 ha^{-1}.

2.2. Climate Data and Drought Index

To avoid problems with heterogeneity or missing values, we obtained 0.1°-gridded monthly climate (maximum and minimum temperatures, total precipitation) data from the European E-OBS v. 28.0e dataset [30]. To measure drought severity, we used the standardized precipitation-evapotranspiration index (SPEI) for the period 1970–2021. The SPEI is a standardized multi-scalar drought index based on the accumulated water deficit, calculated as differences between precipitation and potential evapotranspiration, with negative (positive) values indicating dry (wet) conditions [31]. We used weekly SPEI data and considered 1- (SPEI 1), 3- (SPEI 3), 6- (SPEI 6) and 9-month (SPEI 9) SPEI temporal resolutions. These data were downloaded at a 1.1 km^2 spatial resolution from the Spanish SPEI dataset [32]. For the analyses performed with daily climate, the 5 km grid calculated by the Spanish Meteorological Agency (AEMET) from more than 3200 and 1800 precipitation and temperature local stations, respectively, located in Spain was used (https://www.aemet.es/es/serviciosclimaticos/cambio_climat/datos_diarios?w=2 accessed on 11 June 2023).

To characterize the climatic marginality of the study relict populations, we first obtained distribution data of the species in Europe, considering the area encompassed by coordinates 12° W–60° E and 32–72° N [33]. Then, two relevant climatic variables (mean annual temperature, total precipitation) were downloaded at 1-km^2 spatial resolution from the Worldclim database, considering the period 1970–2000 [34]. Finally, the raster R package was used to select the 1-km^2 grids where cork oak was present [35]. The three sites occupied marginal locations within the cork oak climatic distribution space in Europe, with the Zarautz and Sestrica sites corresponding to the wettest and driest locations, respectively (Figure 2d).

2.3. Processing Wood Samples and Tree-Ring Width Data

Cores were air-dried in the laboratory. Then, they were glued onto wooden mounts and sanded until tree rings were clearly visible [36]. All samples were visually cross-dated, and tree-ring width was measured with a 0.001 mm resolution using scanned images (EPSON 10,000 XL, Nagano, Japan) obtained at 2400 dpi resolution. Ring widths were measured and cross-dated using the CooRecorder and CDendro software v. 9.3.1 [37]. The quality of cross-dating was checked using the COFECHA software, which calculates moving correlations between individual series of ring width values and the mean site chronologies [38]. In some trees, particularly those most difficult to cross-date, thin (15–25 µm) wood cross-sections were prepared to help with their accurate cross-dating (Figure 1c). Wood cross-sections were obtained by using a sledge microtome, and the resulting samples were stained using safranin [39]. The estimation of tree age was based on cores taken at a height of 1.3 m and reaching the pith.

Individual tree-ring width series were detrended using negative linear or exponential functions. The resulting ring width indices were pre-whitened by fitting an autoregressive model and removing the first-order autocorrelation. A bi-weight robust mean was used to calculate the mean site residual (pre-whitened) series or chronologies of ring width indices. Then, to describe ring width data and chronologies, several dendrochronological statistics were calculated [40]: the mean tree-ring width and its first-order autocorrelation, a measure of year-to-year growth persistence, the mean sensitivity, which measures relative changes in standardized (not pre-whitened) ring width indices, and the mean correlation among the series of pre-whitened ring width indices (rbar). These analyses were carried out using the dplR package [41]. We also calculated the Expressed Population Signal (EPS) of indexed ring width site series to assess their internal coherence and replication [42].

2.4. Statistical Analyses

Comparisons between sites of tree size (dbh, height) and tree-ring statistics (ring width, first-order autocorrelation, mean sensitivity) were performed using Mann–Whitney tests. Pearson correlations were used to assess relationships between growth indices, climate variables and the SPEI drought index. Correlations with monthly climate variables were calculated from September of the previous year up to September of the growth year. Correlations were also calculated with the summed precipitation from the previous September until the current May, which accounts for most of the precipitation of the hydrological year. In this case, slopes were used as a measure of responsiveness to climate and they were compared among sites using ANCOVAs. A similar analysis was carried out by relating estimated tree age and mean tree-ring width.

Correlations between weekly SPEI data and ring width indices were calculated from January to November, when most of the annual ring was already formed. Relationships between pre-whitened series of ring width and climate variables or drought indices were assessed using the R package treeclim [43].

Since the winter climate conditions over northern Spain are tightly coupled with the North Atlantic Oscillation (NAO), moving correlations between the NAO index of the previous November and ring width indices were calculated using 20-year intervals shifted year by year. The NAO variability is related to shifts in the position of the Iceland low-pressure and Azores high-pressure systems linked to changes in the direction and strength of the westerlies [44]. High (low) NAO values in late autumn and winter are related to low (high) precipitation over most of the northern Iberian Peninsula [45]. Therefore, if relationships between winter precipitation and ring width indices are negative (positive), we would expect correlations with NAO indices to be positive (negative).

Radial growth is a continuous process that is not limited by monthly boundaries. Furthermore, the study of growth–climate relationships can give very different results depending on whether continuous periods of time are used or whether the climate of each month is analyzed separately [46]. To address this issue, we calculated, for each climatic variable, the time period in which the growth–climate relationship is maximal (best climate window) using the R package climwin [47,48], which offers important advantages in the field of dendroecology [46]. To select the best climatic window based on daily climate data, all possible linear and quadratic models relating a series of ring width indices and the climate variable are first fitted. In each model, a different climate window is tested. Then, the model that minimizes the corrected Akaike Information Criterion (AICc) is chosen [49]. The mean and sum of temperatures and precipitation, respectively, in each time window considered were used as the aggregate statistics. The large numbers of models that are fitted to test all possible climatic windows increase the possibility of obtaining models with low AICc by chance. To solve this problem, randomization tests were performed with 1000 replications to obtain a probability value (p AICc), which determines the likelihood that the AICc value of the selected model has occurred by chance [47,48]. All statistical analyses were performed within the R software [50].

3. Results

3.1. Growth Patterns and Variability

The smallest trees were sampled in the continental Bozoó site, whereas the largest trees were sampled in the humid Zarautz site (Tables 1 and 2). The estimated age ranged between 54 and 65 years. The mean ring width was the highest in Zarautz (2.27 mm) and lowest in Bozoó (1.14 mm) (Figure 3), while mean sensitivity was highest in Sestrica (0.45) as well as rbar (0.70) (Table 2). The highest first-order autocorrelation (0.74) corresponded to Bozoó. The common, best-replicated period was 1967–2021, albeit EPS values were below the 0.85 threshold, which is usually considered to define well-replicated and coherent chronologies. There were negative relationships between tree age and mean ring width, which were significant ($p < 0.05$) in all sites except Bozoó (Figure S1).

Table 2. Variables and statistics related to tree-ring data. Values are means ± SD. Different letters indicate significant ($p < 0.05$) differences between sites according to Mann–Whitney tests. Abbreviations: rbar, mean correlation among all series; EPS, expressed population signal.

Site	No. Trees (No. Radii)	Period	Estimated Tree Age (yrs.)	Tree-Ring Width (mm)	First-Order Autocorrelation	Mean Sensitivity	rbar	EPS
Zarautz	10 (14)	1949–2021	54 ± 11	2.27 ± 0.22 [c]	0.50 ± 0.09 [a]	0.27 ± 0.04 [a]	0.39	0.61
Bozoó	12 (21)	1940–2021	64 ± 10	1.14 ± 0.16 [a]	0.74 ± 0.11 [b]	0.30 ± 0.05 [a]	0.57	0.72
Sestrica	14 (23)	1937–2021	65 ± 14	1.71 ± 0.18 [b]	0.51 ± 0.06 [a]	0.45 ± 0.09 [b]	0.70	0.81

Both the Bozoó and Sestrica series of tree-ring widths showed significant ($p < 0.001$) negative trends ($r = -0.92$ and $r = -0.78$, respectively; Figure 3a). This growth decline is a common ontogenetic pattern that does not correspond to a loss in tree vigor (dieback). No significant correlation was found between the site series of ring width indices, indicating they did not share common regional climate signals (Figure 3b).

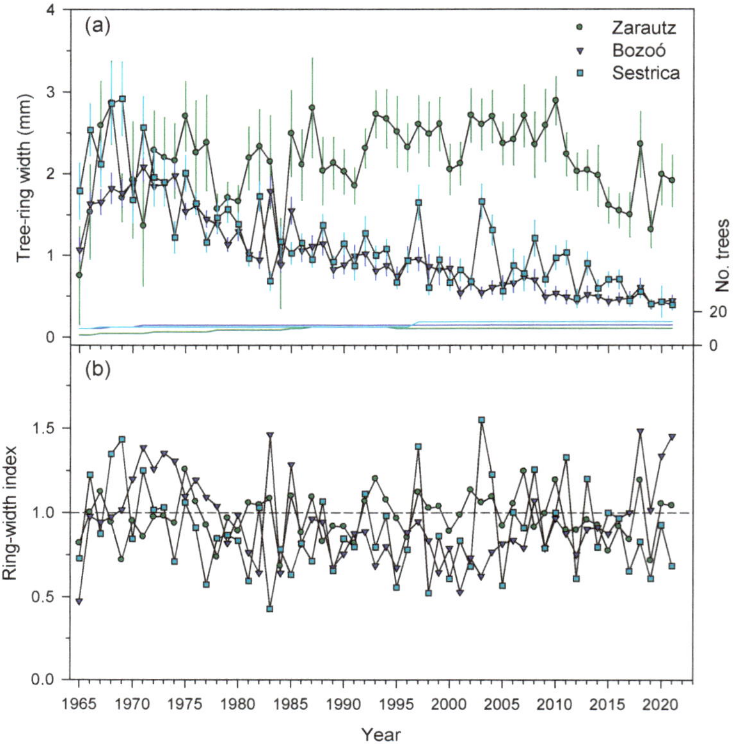

Figure 3. Mean site series of (**a**) ring width data (means ± SE) and (**b**) residual ring width indices of the three study sites (Zarautz, Bozoó and Sestrica). In the upper plot, the sample size (annual number of measured trees, lower lines of different colors) is shown on the right y axis.

3.2. Relationships between Climate Variables, Drought and Growth Indices

The correlations between climate variables and ring width indices showed that cork oak growth was constrained by warm and dry spring and early summer conditions in Bozoó and Sestrica (Figure 4). In Zarautz, cool and wet conditions in the previous October enhanced growth. In Sestrica, high minimum temperatures in April and elevated precipitation in the prior winter, spring and current September improved growth. However, some relationships with precipitation were of the opposite sign in Zarautz, where elevated precipitation in January reduced growth indices. This was related to the positive correlation found between ring width and the NAO indices of the previous year (r = 0.43, p = 0.001) in Zarautz (Figure S2). This association weakened in the 2010s–2020s.

The correlations between the precipitation summed from the previous September to the current May and the series of ring width indices yielded the highest coefficients and were positive and significant at the driest Bozoó and Sestrica sites (Figure 5). Cumulative precipitation explained 12% and 30% of the variability of ring width indices in Bozoó and Sestrica, respectively. The slope was significantly higher (F = 4.43, p = 0.038) in Sestrica than in Bozoó.

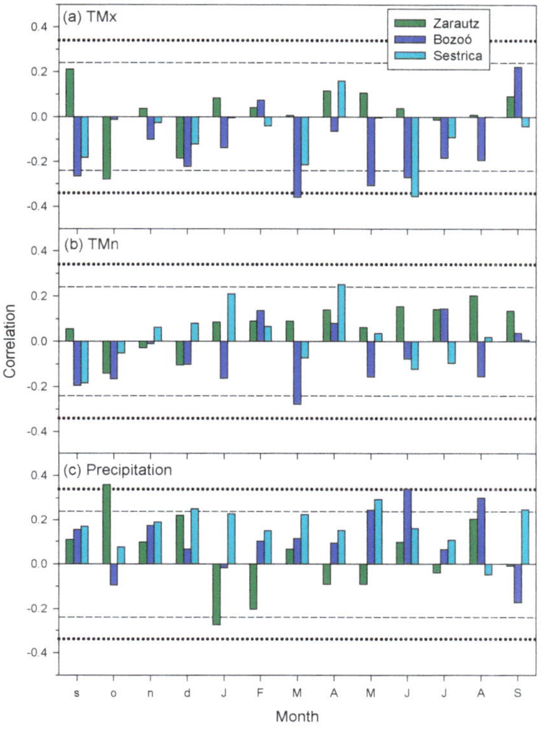

Figure 4. Relationships between monthly climate variables ((**a**), TMx, mean maximum temperature; (**b**), TMn, mean minimum temperature; (**c**), precipitation) and site series of ring width indices of the three study sites. The bars are Pearson correlation coefficients (r), and the dashed and dotted horizontal lines indicate the 0.05 and 0.01 significance levels, respectively. The correlation window spans from previous September to current September, with months of the prior year abbreviated with lowercase letters.

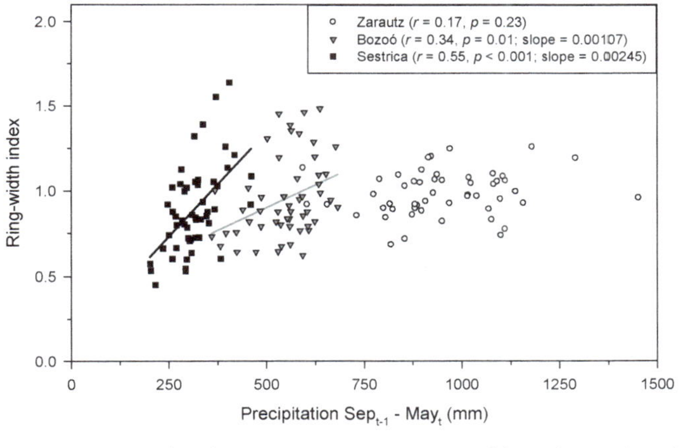

Figure 5. Relationships between precipitation summed from the previous (year t − 1) September to the current (year t) May and ring width indices in the three study cork oak populations. The Pearson correlation coefficients (r), their significance levels (p) and the slopes corresponding to fitted linear regressions are shown in the box.

The highest correlation coefficient found between the values of the SPEI drought index and ring width indices ($r = 0.56$, $p < 0.001$) corresponded to early June considering 9-month-long SPEI scales, and it was found in the driest Sestrica site (Figure 6).

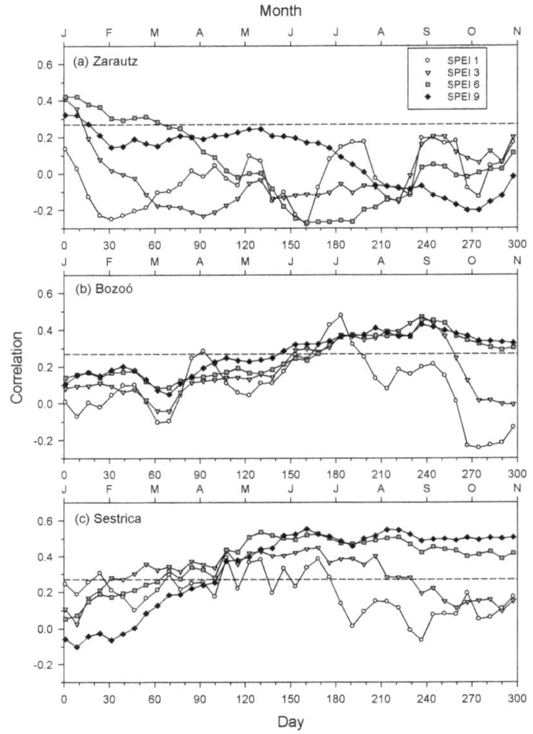

Figure 6. Correlations calculated between SPEI weekly values and the site series of ring width indices considering the common interval 1967–2021 and the temporal window from January to October. Dashed lines show the 0.05 significance levels.

The continental Bozoó site presented the second highest correlation ($r = 0.48$, $p = 0.0003$) in early July for 1-month-long SPEI values. In Sestrica and Bozoó, 6-month SPEI values also showed strong positive correlations with cork oak ring width indices. Finally, in the wettest Zarautz site, the highest correlation ($r = 0.42$, $p = 0.0019$) was found in early January for 6-month-long SPEI values.

3.3. Analyses of Climate–Growth Relationships Based on Climwin

Regarding *climwin* analyses, the highest coefficient of determination (R^2) was found for precipitation in Sestrica, which explained 41% of the variability of ring width indices (Table 3). In this case, precipitation had a positive effect, followed by a linear relationship and its impact on growth indices spanned from previous September until current November (Figure 7). Precipitation showed similar impacts on growth indices in Zarautz and Bozoó, where it explained 34% and 21% of the growth indices variability, respectively, but along different temporal windows, previous September–October and current June–August in that order. The relationships between growth indices and precipitation were mostly linear, but nonlinear models also showed good fits (Figure S3). Maximum temperatures showed negative correlations with the ring width indices, explaining from 13% (Sestrica) to 22% (Zarautz) of growth variability, but with very different temporal windows (Zarautz, September–October; Bozoó, March; Sestrica, June; see Figure 4). Lastly, minimum temperature exerted a negative effect on growth in Bozoó, explaining 20% of the variability and corresponding to January temperature values. In the other sites, the influences of minimum temperature on growth indices were positive, explaining 12% and 14% of growth variability in Zarautz and Sestrica, respectively, and peaking the correlations in August and April, correspondingly. In any case, the relationships between temperature and growth were not significant, according to randomization tests.

Table 3. Variables of *climwin* analyses relating series of ring width indices and monthly climate variables (Tmax, mean maximum temperature; Tmin, mean minimum temperature; Prec, precipitation). The previous year is indicated with "(t − 1)". DOY is the day of the year. The ΔAICc is a subtraction between the model AICc and the AICc value of the null model. The *p* AICc is the probability value obtained in the randomization test.

Site	Climate Variable	Window Open (DOY)	Window Close (DOY)	Coefficient	*p*	R^2	ΔAICc	*p* AICc
Zarautz	Tmax	244 (t − 1)	292 (t − 1)	−0.046	<0.001	0.22	−10.85	0.079
	Tmin	210	228	0.041	0.011	0.12	−4.49	0.661
	Prec	238 (t − 1)	296 (t − 1)	0.043	<0.001	0.34	−19.69	0.002
Bozoó	Tmax	70	85	−0.044	<0.001	0.21	−10.16	0.144
	Tmin	360 (t − 1)	10	−0.048	<0.001	0.20	−9.4	0.167
	Prec	147	238	0.206	<0.001	0.21	−10.28	0.116
Sestrica	Tmax	152	170	−0.036	0.009	0.13	−4.85	0.904
	Tmin	104	126	0.063	0.007	0.14	−5.4	0.431
	Prec	258 (t − 1)	312	0.665	<0.001	0.41	−25.6	<0.001

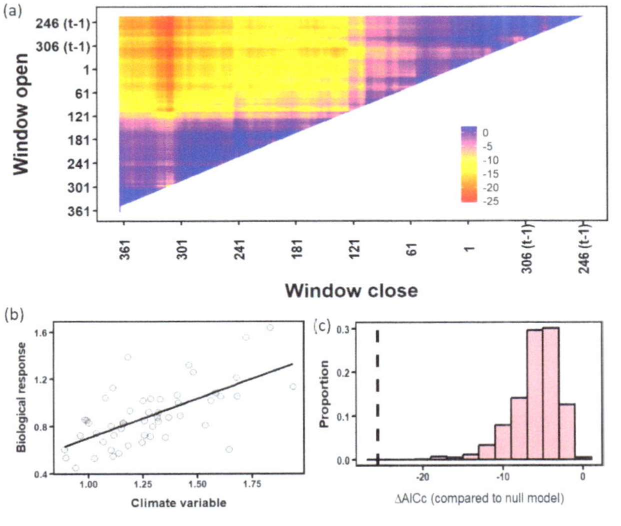

Figure 7. Plots summarizing *climwin* analyses between precipitation (climate variable in the scatter plot) and the site series of ring width indices (biological response in the scatter plot) in the Sestrica site. The upper plot (**a**) shows temporal windows and best fits (see Table 3) as red grids corresponding to lowest ΔAICc values (shown in the histogram). The ΔAICc is a subtraction between the model AICc and the AICc value of the null model. Dates of the previous year are indicated with "(t − 1)". The bottom plots show (**b**) the relationship between the climate of the best supported window and the growth indices, and (**c**) the histogram of ΔAICc values of the models fitted on randomized data with a vertical dashed line showing the ΔAICc of the best model fitted on the observed data.

4. Discussion

As expected, the cork oaks from the driest site (Sestrica) showed the highest responsiveness (correlation coefficient, precipitation–growth slope) to spring–summer drought, but the lowest growth rates were found in the continental site (Bozoó), indicating that cork oak growth was particularly constrained by cold spring conditions at that site. Unexpectedly, we also found that cork oak growth in the wettest site (Zarautz) was limited by low precipitation in the previous autumn, suggesting a marked sensitivity to late-summer drought stress. In addition, elevated precipitation in the previous winter also reduced growth in this humid site, indicating a negative influence of cloudy and rainy conditions on spring growth through lagged effects. The highest sensitivity of radial growth to water shortage in the driest site could have been exacerbated by its steep slopes and E aspect.

From a methodological point of view, we have to emphasize that applying dendrochronology to cork oak is challenging because annual rings are hard to distinguish, cross-date and measure. Nevertheless, we were able to develop robust tree-ring chronologies with the help of anatomical cross-sections developed at some sites. The continentality of some sites (e.g., Bozoó) and the marked and long summer drought of others (e.g., Sestrica) contributed to a clear seasonality and the formation of annual growth rings. This would explain the lowest coherence statistics (rbar, EPS) found in the mesic Zarautz site, located near the Atlantic coast and subjected to temperate conditions.

There have been a few studies on climate–growth relationships in cork oak, particularly in southern Portugal, i.e., in the distribution core area [51–53]. They also found that cork oak growth was enhanced by higher cumulative precipitation from the previous autumn until early spring [52]. These authors also found a positive effect of warmer September conditions on growth [52], a correlation similar to that found with maximum temperatures in the continental Bozoó site, albeit it was not significant in our case. In addition, a positive effect of warmer spring temperatures (as we found in Bozoó) and a negative effect of summer maximum temperatures (as we found in Bozoó and Sestrica) were reported on growth [51]. These thermal influences may be explained by a higher cambial activity in spring triggered by warmer night conditions and a higher evapotranspiration rate induced by warmer day conditions in summer, making cork oak trees more dependent on deeper soil water sources and reducing their photosynthesis rates and meristem activity [14–16,19]. Overall, these studies and our findings support the importance of soil water recharge during autumn and winter prior to the start of the growing season. Such an effect has already been noted in several Mediterranean conifers and could explain the different temporal scales of SPEI-growth correlations [54].

Cork oak is a drought-avoiding species that reduces leaf conductance to water vapor through efficient stomatal control of transpiration [55]. This isohydric strategy is maintained during the dry Mediterranean summer through the uptake of deep underground water despite radial growth rates being very low in summer [14–16,22]. In this study, cork oak growth was notably constrained by 9-month and 1-month-long droughts ending in June (i.e., dry conditions from prior October to current June) or July (i.e., dry conditions from current June to July) in the Sestrica and Bozoó sites, respectively. This agrees with the growth sensitivity of cork oak to spring–summer precipitation observed in these two sites, particularly in Sestrica, where the highest R^2 value (0.41) was found for precipitation in the selected *climwin* models. The contrasting growth responses to different timings and durations of drought also suggest different growth phenologies, with a more delayed growth onset in Bozoó. This idea could be tested through xylogenesis analyses and/or using dendrometers. In Zarautz, growth responded to 6-month-long SPEI values in January (i.e., dry conditions from prior August to current January), indicating that late-summer drought stress in the previous year could constrain cork oak growth in this site through legacy effects (e.g., carbohydrate consumption due to warm conditions and high respiration rates, fine root and bud formation, etc.). In this wet site, the positive correlation of ring width indices with prior November NAO indices is explained because of the negative association between growth indices and the January precipitation. Such a negative relationship could be caused by wet, cloudy winter conditions reducing photosynthetic rates and leading to lower growth rates in spring.

Cork oak radial growth may also be reduced by cork extraction, which can intensify the negative impact of droughts on cork oak [52,53,56,57], but we discarded this effect by selecting stands that have not been subjected to cork harvesting during the past two decades. Furthermore, cork oak growth positively depends on tree basal area and tree diameter [53], which could partially explain the lowest growth rates found in Bozoó, where the trees with the smallest diameter were also sampled. Tree-to-tree competition could also modulate cork oak growth responses to climate, particularly precipitation, but the sampled stands were usually open forests, and the basal area was similar. Masting could also influence cork oak growth responses to climate, but long series of acorn production are not available for the study sites.

Our findings have profound implications for the management and conservation of the studied relict cork oak populations. These forests persist in topoclimatic refugia as isolated populations within adverse regional climate conditions under drought (Sestrica) or cold (Bozoó) stress. In addition to evolutionary factors leading to genetic differences among populations, which we did not analyze, contrasting growth responses to climate may represent key mechanisms implied in population persistence [58]. For instance, the highest responsiveness to spring–summer drought in the driest site (Sestrica) indicates a remarkable capacity to tolerate water shortage, whereas the negative response to prior winter precipitation in the wettest site (Zarautz) may correspond to a reduction in carbohydrate formation due to cloudy–rainy conditions. Nevertheless, the Zarautz site is the most sensitive to drier conditions in the prior autumn, which could lead to growth decline even in such mesic conditions. Lastly, in the most continental Bozoó site, forecasted warmer spring conditions would enhance tree growth more than in the other sites. If this site is considered the "winner" in response to projected and ongoing climate warming, Sestrica could be regarded as the "loser" if further aridity in spring and summer leads to growth decline and dieback once a drought tolerance threshold is surpassed. Nevertheless, no recent dieback episode has been observed at the Sestrica site, to the best of our knowledge, indicating that this population is a valuable genetic source for assisted migration of cork oak under more arid conditions. Finally, the cork oaks at the wettest Zarautz site could also be negatively impacted by hotter droughts in late summer and autumn.

5. Conclusions

The study of the three relict cork oak populations in northern Spain evidenced the plastic and contrasting climate–growth relationships observed under different topoclimatic conditions. Relict tree populations persist in topographic and climatic refugia and may have developed adaptations to withstand climatic stressors. Cork oak growth was severely limited by cold spring conditions in the continental site (Bozoó) and by spring–summer warm–dry conditions during the growing season in that site and also in the xeric site (Sestrica). In contrast, growth was limited by prior autumn droughts and wet winters in the humid site (Zarautz), subjected to more oceanic influence, including cloudier winter–spring conditions. If climate keeps warming and drying, the persistence of the relict cork oak populations located in the continental and xeric sites may be threatened. However, these stands have not shown recent symptoms of crown dieback or increased mortality despite recent hotter droughts in the past four decades. Therefore, trees from isolated relict stands under drought stress, such as those in Sestrica, represent valuable genetic pools for assisted migration under forecasted more arid conditions.

Supplementary Materials: The following supporting information can be downloaded at: https://www.mdpi.com/article/10.3390/f15010072/s1, Figure S1: Associations found between tree age and mean growth rate; Figure S2: Series of ring width indices from Zarautz and the previous November NAO index; Figure S3: Quadratic relationships observed between precipitation (standardized climate variable) and cork oak series of ring width indices (biological response).

Author Contributions: Conceptualization, J.J.C. and A.G.; methodology, J.J.C., A.G., C.V., M.C. and Á.R.-C.; software, J.J.C., A.G., C.V. and Á.R.-C.; validation, J.J.C. and A.G.; formal analysis, J.J.C., Á.R.-C. and A.G.; investigation, J.J.C.; resources, J.J.C.; data curation, J.J.C.; writing—original draft preparation, J.J.C.; writing—review and editing, J.J.C., A.G., C.V., M.C. and Á.R.-C.; visualization, J.J.C.; supervision, J.J.C.; project administration, J.J.C.; funding acquisition, J.J.C. and A.G. All authors have read and agreed to the published version of the manuscript.

Funding: This research was funded by the Spanish Ministry of Science and Innovation, grants numbers PID2021-123675OB-C43 and TED2021-129770B-C21. ARC acknowledges support from the Margarita Salas postdoctoral fellowship (reference RCMS-22-G1T6IW-17-NLHJ80) of the Universidad Politécnica de Madrid.

Data Availability Statement: Data are available on request to the first author.

Acknowledgments: We thank Roberto del Val for his help during field sampling in Sestrica.

Conflicts of Interest: The authors declare no conflicts of interest.

References

1. Giorgi, F.; Lionello, P. Climate change projections for the Mediterranean region. *Glob. Planet. Chang.* **2008**, *63*, 90–104. [CrossRef]
2. Lionello, P.; Scarascia, L. The relation between climate change in the Mediterranean region and global warming. *Reg. Environ Chang.* **2018**, *18*, 1481–1493. [CrossRef]
3. Aronson, J.; Pereira, J.S.; Pausas, J.G. (Eds.) *Cork Oak Woodlands on the Edge: Ecology, Adaptive Management, and Restoration*; Island Press: Washington, DC, USA, 2009.
4. Brasier, C.M. Oak tree mortality in Iberia. *Nature* **1992**, *360*, 539. [CrossRef]
5. David, T.S.; Cabral, M.T.; Sardinha, R. A mortalidade dos sobreiros e a seca. *Finisterra* **1992**, *27*, 17–24.
6. Costa, A.; Pereira, H.; Madeira, M. Analysis of spatial patterns of oak decline in cork oak woodlands in Mediterranean conditions *Ann. For. Sci.* **2010**, *67*, 204. [CrossRef]
7. Gentilesca, T.; Camarero, J.J.; Colangelo, M.; Nolè, A.; Ripullone, F. Drought-induced oak decline in the western mediterranean region: An overview on current evidences, mechanisms and management options to improve forest resilience. *iForest-Biogeosci For.* **2017**, *10*, 796–806. [CrossRef]
8. Sánchez-Cuesta, R.; Ruiz-Gómez, F.J.; Duque-Lazo, J.; González-Moreno, P.; Navarro-Cerrillo, R.M. The environmental drivers influencing spatio-temporal dynamics of oak defoliation and mortality in dehesas of Southern Spain. *For. Ecol. Manag.* **2021**, *485*, 118946 [CrossRef]
9. Sánchez-González, M.; Beltrán, R.S.; Lanzo Palacios, R.; Prades, C. Analysis of cork quality and cork tree health in stands of western Spain. *For. Ecol. Manag.* **2023**, *539*, 121012. [CrossRef]
10. Touhami, I.; Chirino, E.; Aouinti, H.; Khorchani, A.E.; Elaieb, M.T.; Khaldi, A.; Nasr, Z. Decline and dieback of cork oak (*Quercus suber* L.) forests in the Mediterranean basin: A case study of Kroumirie, Northwest Tunisia. *J. For. Res.* **2020**, *31* 1461–1477. [CrossRef]
11. Palma, J.H.N.; Paulo, J.A.; Faias, S.P.; Garcia-Gonzalo, J.; Borges, J.G.; Tomé, M. Adaptive management and debarking schedule optimization of *Quercus suber* L. stands under climate change: Case study in Chamusca, Portugal. *Reg. Environ. Ch.* **2015**, *15* 1569–1580. [CrossRef]
12. Pereira, H. *Cork: Biology, Production and Uses*; Elsevier: Amsterdam, The Netherlands, 2007.
13. Aranda, I.; Castro, L.; Pardos, M.; Gil, L.; Pardos, J.A. Effects of the interaction between drought and shade on water relations, gas exchange and morphological traits in cork oak (*Quercus suber* L.) seedlings. *For. Ecol. Manag.* **2005**, *210*, 117–129. [CrossRef]
14. David, T.S.; Henriques, M.O.; Kurz-Besson, C.; Nunes, J.; Valente, F.; Vaz, M.; Pereira, J.S.; Siegwolf, R.; Chaves, M.M. Gazarini, L.C.; et al. Water-use strategies in two co-occurring Mediterranean evergreen oaks: Surviving the summer drought. *Tree Physiol.* **2007**, *27*, 793–803. [CrossRef] [PubMed]
15. Kurz-Besson, C.; Otieno, D.; Lobo do Vale, R.; Siegwolf, R.; Schmidt, M.; Herd, A.; Nogueira, C.; David, T.S.; David, J.S. Tenhunen, J.; et al. Hydraulic lift in cork oak trees in a savannah-type Mediterranean ecosystem and its contribution to the local water balance. *Plant Soil* **2006**, *282*, 361–378. [CrossRef]
16. Kurz-Besson, C.; Lobo do Vale, R.; Rodrigues, M.L.; Almeida, P.; Herd, A.; Grant, O.M.; David, T.S.; Schmidt, M.; Otieno, D. Keenan, T.F.; et al. Cork oak physiological responses to manipulated water availability in a Mediterranean woodland. *Agric. For Meteorol.* **2014**, *184*, 230–242. [CrossRef]
17. Mendes, M.P.; Ribeiro, L.; David, T.S.; Costa, A. How dependent are cork oak (*Quercus suber* L.) woodlands on groundwater A case study in southwestern Portugal. *For. Ecol. Manag.* **2016**, *378*, 122–130. [CrossRef]
18. Otieno, D.O.; Kurz-Besson, C.; Liu, J.; Schmidt, M.W.T.; Vale do Lobo, R.; David, T.S.; Siegwolf, R.; Pereira, J.S.; Tenhunen, J.D Seasonal variations in soil and plant water status in a *Quercus suber* L. stand: Roots as determinants of tree productivity and survival in the Mediterranean-type ecosystem. *Plant Soil* **2006**, *283*, 119–135. [CrossRef]

19. Vaz, M.; Pereira, J.S.; Gazarini, L.C.; David, T.S.; David, J.S.; Rodrigues, A.; Maroco, J.; Chaves, M.M. Drought induced photosynthetic inhibition and autumn recovery in two Mediterranean oak species (*Quercus ilex* and *Quercus suber*). *Tree Physiol.* **2010**, *30*, 946–956. [CrossRef]
20. Cabral, M.T.; Ferreira, M.C.; Moreira, T.; Carvalho, E.C.; Diniz, A.C. Diagnóstico das causas da anormal mortalidade dos sobreiros a sul do Tejo. *Sci. Gerund* **1992**, *18*, 205–214.
21. Camilo-Alves, C.S.P.; Vaz, M.; Da Clara, M.I.E.; Ribeiro, N.A. Chronic cork oak decline and water status: New insights. *New For.* **2017**, *48*, 753–772. [CrossRef]
22. Costa, A.; Pereira, H.; Oliveira, A. Variability of radial growth in cork oak mature trees under cork production. *For. Ecol. Manag.* **2003**, *175*, 239–246. [CrossRef]
23. Camilo-Alves, C.S.P.; Da Clara, M.I.E.; Ribeiro, N.M.D.A. Decline of Mediterranean oak trees and its association with *Phytophthora cinnamomi*: A review. *Eur. J. For. Res.* **2013**, *132*, 411–432. [CrossRef]
24. Matías, L.; Abdelaziz, M.; Godoy, O.; Gómez-Aparicio, L. Disentangling the climatic and biotic factors driving changes in the dynamics of *Quercus suber* populations across the species' latitudinal range. *Div. Distrib.* **2019**, *25*, 524–535. [CrossRef]
25. Morillas, L.; Leiva, M.J.; Pérez-Ramos, I.M.; Cambrollé, J.; Matías, L. Latitudinal variation in the functional response of *Quercus suber* seedlings to extreme drought. *Sci. Total Environ.* **2023**, *887*, 164122. [CrossRef] [PubMed]
26. Ramírez-Valiente, A.; Valladares, F.; Gil, L.; Aranda, I. Population differences in juvenile survival under increasing drought are mediated by seed size in cork oak (*Quercus suber* L.). *For. Ecol. Manag.* **2009**, *257*, 1676–1683. [CrossRef]
27. Linares, J.C.; Camarero, J.J.; Carreira, J.A. Interacting effects of climate and forestcover changes on mortality and growth of the southernmost European fir forests. *Glob. Ecol. Biogeogr.* **2009**, *18*, 485–497. [CrossRef]
28. Camarero, J.J.; Manzanedo, R.D.; Sánchez-Salguero, R.; Navarro-Cerrillo, R.M. Growth response to climate and drought change along an aridity gradient in the southernmost Pinus nigra relict forests. *Ann. For. Sci.* **2013**, *70*, 769–780. [CrossRef]
29. Cornes, R.; van der Schrier, G.; van den Besselaar, E.J.M.; Jones, P.D. An ensemble version of the E-OBS temperature and precipitation datasets. *J. Geophys. Res. Atmos.* **2018**, *123*, 9391–9409. [CrossRef]
30. Aronson, J.; Pereira, J.S.; Pausas, J.G. (Eds.) Introduction. In *Cork Oak Woodlands on the Edge*; Island Press: Washington DC, USA, 2009.
31. Vicente-Serrano, S.M.; Beguería, S.; López-Moreno, J.I. A Multi-scalar drought index sensitive to global warming: The Standardized Precipitation Evapotranspiration Index—SPEI. *J. Clim.* **2010**, *23*, 1696–1718. [CrossRef]
32. Vicente-Serrano, S.M.; Tomas-Burguera, M.; Beguería, S.; Reig, F.; Latorre, B.; Peña-Gallardo, M.; Luna, M.Y.; Morata, M.; González-Hidalgo, J.C. A High Resolution Dataset of Drought Indices for Spain. *Data* **2017**, *2*, 22. [CrossRef]
33. Caudullo, G.; Welk, E.; San-Miguel-Ayanz, J. Chorological maps for the main European woody species. *Data Brief* **2017**, *12*, 662–666. [CrossRef]
34. Fick, S.E.; Hijmans, R.J. WorldClim 2: New 1km spatial resolution climate surfaces for global land areas. *Int. J. Climatol.* **2017**, *37*, 4302–4315. [CrossRef]
35. Hijmans, R. Raster: Geographic Data Analysis and Modeling; R Package Version 3.6-21. 2023. Available online: https://rspatial.org/raster (accessed on 10 May 2023).
36. Fritts, H.C. *Tree Rings and Climate*; Academic Press: London, UK, 1976.
37. Larsson, L.A.; Larsson, P.O. *CDendro and CooRecorder, v. 9.3.1*; Manual; Cybis Elektronik and Data AB: Saltsjöbaden, Sweden, 2018.
38. Holmes, R.L. Computer-assisted quality control in tree-ring dating and measurement. *Tree-Ring Bull.* **1983**, *43*, 69–78.
39. Gärtner, H.; Lucchinetti, S.; Schweingruber, F.H. A new sledge microtome to combine wood anatomy and tree-ring ecology. *IAWA J.* **2015**, *36*, 452–459. [CrossRef]
40. Briffa, K.R.; Jones, P.D. Basic chronology statistics and assessment. In *Methods of Dendrochronology*; Cook, E.R., Kairiukstis, L.A., Eds.; Kluwer: Dordrecht, The Netherlands, 1990; pp. 137–152.
41. Bunn, A.; Korpela, M.; Biondi, F.; Campelo, F.; Mérian, P.; Qeadan, F.; Zang, C. *dplR: Dendrochronology Program Library in R, R Package Version 1.7.1*; CRAN: Vienna, Austria, 2020.
42. Wigley, T.M.; Briffa, K.R.; Jones, P.D. On the average value of correlated time series, with applications in dendroclimatology and hydrometeorology. *J. Clim. Appl. Meteorol.* **1984**, *23*, 201–213. [CrossRef]
43. Zang, C.; Biondi, F. Treeclim: An R package for the numerical calibration of proxy-climate relationships. *Ecography* **2015**, *38*, 431–436. [CrossRef]
44. Hurrell, J. Decadal trends in North Atlantic Oscillation and relationship to regional temperature and precipitation. *Science* **1995**, *269*, 676–679. [CrossRef]
45. Hurrell, J.; Van Loon, H. Decadal variations in climate associated with the North Atlantic Oscillation. *Clim. Chang.* **1997**, *36*, 301–326. [CrossRef]
46. Rubio-Cuadrado, A.; Camarero, J.J.; Bosela, M. Applying climwin to dendrochronology: A breakthrough in the analyses of tree responses to environmental variability. *Dendrochronologia* **2022**, *71*, 125916. [CrossRef]
47. Bailey, L.; van de Pol, M. Climwin: An R toolbox for climate window analysis. *PLoS ONE* **2016**, *11*, e0167980. [CrossRef]
48. van de Pol, M.; Bailey, L.D.; McLean, N.; Rijsdijk, L.; Lawson, C.R.; Brouwer, L. Identifying the best climatic predictors in ecology and evolution. *Methods Ecol. Evol.* **2016**, *7*, 1246–1257. [CrossRef]
49. Burnham, K.P.; Anderson, D.R. *Model Selection and Multimodel Inference: A Practical Information-Theoretic Approach*; Springer: New York, NY, USA, 2004.
50. R Core Team. *R: A Language and Environment for Statistical Computing*; R Foundation for Statistical Computing: Vienna, Austria, 2023.

51. Costa, A.; Pereira, H.; Oliveira, A. A dendroclimatological approach to diameter growth in cork-oak adult trees under cork production. *Trees Struct. Funct.* **2001**, *15*, 438–443. [CrossRef]
52. Leal, S.; Nunes, E.; Pereira, H. Cork oak (*Quercus suber* L.) wood growth and vessel characteristics variations in relation to climate and cork harvesting. *Eur. J. For. Res.* **2008**, *127*, 3341.
53. Mendes, M.P.; Cherubini, P.; Plieninger, T.; Ribeiro, L.; Costa, A. Climate effects on stem radial growth of *Quercus suber* L.: Does tree size matter? *Forestry* **2019**, *92*, 73–84. [CrossRef]
54. Pasho, E.; Camarero, J.J.; de Luis, M.; Vicente-Serrano, S.M. Impacts of drought at different time scales on forest growth across a wide climatic gradient in North-Eastern Spain. *Agric. For. Meteorol.* **2011**, *151*, 1800–1811. [CrossRef]
55. Nardini, A.; LoGullo, M.A.; Salleo, S. Competitive strategies for water availability in two Mediterranean *Quercus* species. *Plant Cell Environ.* **1999**, *22*, 109–116. [CrossRef]
56. González-Adrados, J.R.; Gourlay, I. Applications of Dendrochronology to *Quercus suber* L. In *Cork Oak and Cork. Proceedings of the European Conference on Cork Oak and Cork*; Pereira, H., Ed.; Centro de Estudos Florestais: Lisboa, Portugal, 1998; pp. 162–166.
57. Costa, A.; Nunes, L.C.; Spiecker, H.; Graça, J. Insights into the responsiveness of cork oak (*Quercus suber* L.) to bark harvesting *Econ. Bot.* **2015**, *69*, 171–184. [CrossRef]
58. Hampe, A.; Jump, A.S. Climate relicts: Past, present, future. *Ann. Rev. Ecol. Evol. Syst.* **2011**, *42*, 313–333. [CrossRef]

forests

MDPI

Article

Impact of Environmental Conditions on Wood Anatomical Traits of Green Alder (*Alnus alnobetula*) at the Alpine Treeline

Andreas Gruber, Gerhard Wieser, Marion Fink and Walter Oberhuber *

Department of Botany, Leopold-Franzens-University of Innsbruck, Sternwartestrasse 15, A-6020 Innsbruck, Austria; andreas.gruber@uibk.ac.at (A.G.); gerhard.wieser@uibk.ac.at (G.W.); marion.fink@student.uibk.ac.at (M.F.)

* Correspondence: walter.oberhuber@uibk.ac.at

Abstract: Due to land use change, green alder (*Alnus alnobetula*), formerly restricted to moist slopes, is now expanding to drier sun-exposed sites at the alpine treeline. The highly productive shrub is forming closed thickets, establishing nitrogen-saturated species poor shrublands. To evaluate wood anatomical adaptations to changing environmental conditions, we analyzed vessel characteristics (mean vessel area, MVA; vessel density, VD; and theoretic conductive area, TCA) and axial parenchyma abundance, as well as their distribution in the annual ring at a moist north-facing and a dry south-facing site at the alpine treeline on Mt. Patscherkofel (Central European Alps, Austria). Results revealed that lower soil water availability and enhanced evaporative demand did not affect MVA while VD and TCA were significantly reduced at the dry south-facing site. This suggests that in green alder, vessel size is a static trait whereas vessel number responds plastic. Limited water availability also triggered a significant increase in axial parenchyma, confirming the important role of xylem parenchyma in water relations. Harsh environmental conditions at the distributional limit of green alder may have affected xylogenesis, leading to a near semi-ring-porous distribution of vessels and an accumulation of parenchyma in the late growing season. We conclude that in a warmer and drier climate, growth limitation and physiological stress may set limits to the distribution of *Alnus alnobetula* at drought-prone sites in the alpine treeline ecotone.

Keywords: alpine treeline ecotone; global change; green alder; phenotypic plasticity; xylem anatomy; xylem parenchyma

Citation: Gruber, A.; Wieser, G.; Fink, M.; Oberhuber, W. Impact of Environmental Conditions on Wood Anatomical Traits of Green Alder (*Alnus alnobetula*) at the Alpine Treeline. *Forests* **2024**, *15*, 24. https://doi.org/10.3390/f15010024

Academic Editors: Yassine Messaoud, Jan Světlík and Giorgio Alberti

Received: 7 November 2023
Revised: 14 December 2023
Accepted: 15 December 2023
Published: 21 December 2023

1. Introduction

Mountain ecosystems are particularly vulnerable and are expected to undergo considerable transformations due to global change [1–3]. However, within the European Alps, land abandonment and decreasing grazing pressure may have a greater impact on the treeline ecotone in the coming decades than rising temperatures [4,5]. There are indications that shrubs may benefit more from land use changes than trees and take over abandoned pastures [6–8]. During the last decades, in particular, green alder (*Alnus alnobetula* (Ehrh.) K. Koch = *Alnus viridis* (Chaix) DC.) has spread rapidly across the Alps [7,9–11]. *A. alnobetula* is a cold-resistant and moderately shade-tolerant species of the subalpine and subarctic zones of the northern hemispheres, growing in mountains, tundra, and river valleys [12]. Due to clonal growth, high seed production, and symbiosis with N_2-fixing actinobacteria and ectomycorrhizal fungi, *A. alnobetula* is spreading rapidly after land abandonment. As green alder forms closed thickets, with canopy heights up to 4 m, it drives former N-poor grassland into nitrogen-saturated-species-poor shrubland and suppresses tree establishment within the treeline ecotone [7].

Although the occurrence of green alder was restricted to north-facing slopes and avalanche gullies, with high water availability [10,13,14], this species is now expanding into moderately steep well-drained subalpine grasslands, as well as sun-exposed sites with shallow soils and impaired water availability [9,15]. Green alder is known to be an anisohydric species, keeping its stomata open even under high vapor pressure deficits [16–18], a strategy that can be risky under drought conditions [19]. Nevertheless, the former restriction of *A. alnobetula* to moist habitats may be due to former land use patterns, while *A. alnobetula* is able to adapt to drier condition [15].

A. alnobetula is a diffuse-porous species, keeping vessel diameters constant throughout the annual ring [20]. However, drought-induced changes in vessel development in diffuse-porous species have been reported before [21–23]. Size, number, and distribution of vessels are closely linked to tree hydraulic conductivity and drought resistance [24,25], but still little is known about intraspecific adaptations in hydraulic properties. The intraspecific studies available to date have shown the effects of water availability on conduit traits [26,27] where decreasing water availability resulted in narrower conduits and higher conduit density in different species [28–31]. However, the adaptation of wood anatomical traits might vary between species and there may be different adaptation strategies for ring-porous and diffuse-porous angiosperms [32,33].

Furthermore, there is increasing awareness that aside from conductive cells, parenchyma tissues in the secondary xylem also play a critical role in the water relations of woody plants [34–37]. It is well known that parenchyma in the secondary xylem stores carbohydrates, which can subsequently be used for growth, establishment of freeze tolerance, and protection against or recovery after infestations [38–41]. Increasing drought stress in the course of global warming emphasizes the role of xylem parenchyma in maintenance of the water transport and embolism repair [33,42,43], as well as water storage and circulation between xylem and phloem [34,36,40,44,45]. Axial parenchyma is reported to be more plastic than ray parenchyma [39,40,46] and its fractions vary with environmental conditions at the intra- and interspecific level [47,48].

Analysis of functional adaptations in the wood anatomy of *A. alnobetula* provides insights into the ecological limits of this pioneer species. We, therefore, evaluated vessel and axial parenchyma properties and distribution in *A. alnobetula* at a south-facing windward and a north-facing leeward site at the alpine treeline, differing in soil water availability and evaporative demand. We tested the hypotheses that reduced soil water availability and enhanced evaporative demand on the south-facing site (i) reduces vessel diameter, (ii) increases vessel density, (iii) changes vessel distribution within the annual ring leading to a semi-ring-porous arrangement of vessels, and (iv) causes an increase in axial parenchyma cells.

2. Materials and Methods

2.1. Study Area and Sample Plots

The study was performed within the treeline ecotone of Mt. Patscherkofel in the Central European Alps, Tyrol, Austria (47°12′ N, 11°27′ E), where *A. alnobetula* stands primarily spread out in avalanche gullies on leeward north-facing slopes between 1950 and 2150 m asl. Even so, during the last decades, *A. alnobetula* stands spread out to windward sites on south- to southeast-facing slopes [9]. Within the study area, the bedrock is dominated by gneisses and schists [49], and the soils are classified as haplic podzols [50,51].

The mean annual temperature at the meteorological station on top of Mt. Patscherkofel was 0.8 ± 0.7 °C during the period from 1991 to 2020, with February being the coldest month (−6.6 °C) and July the warmest (8.9 °C). Despite the high altitude, air temperature can frequently reach maxima around 20 °C during summer. Mean annual precipitation was 889 ± 128 mm with precipitation maxima occurring in summer (long-term mean 371 ± 74 mm, June to August) and winters being the driest season (132 ± 60 mm, December to February). The study area is also characterized by the frequent occurrence of strong southerly Föhn-type winds [52], which strongly influences snow depth and snow distribution—[53] and hence, the duration period of the permanent snow cover. At south facing slopes, snow depth is generally < than 1 m and only small patches of snow are present till early April; while on leeward north-facing slopes, a closed snow cover with snow depths up to 3 m frequently persists from the end of October till the end of May.

Within the study area, we selected two study plots at the upper edge of the treeline ecotone (Figure 1, Table 1): a north-exposed leeward (hereafter N-site) and a south-exposed windward site (hereafter S-site). Although the two plots were only 150 m apart in linear distance, they differed considerably with respect to slope exposure and soil depth (Table 1), soil water availability, and evaporative demand (Table 2, Figure 2). The canopy height of *A. alnobetula* was higher on the N-site, although shrubs were older on the S-site (Table 1).

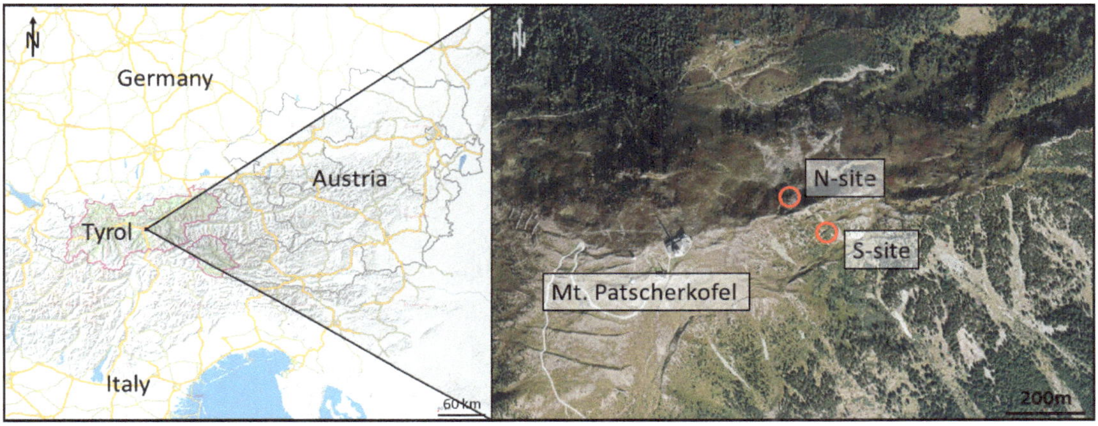

Figure 1. Geographical location of the N-site and S-site in the Central Austrian Alps. Source: tirisMaps Land Tirol, (https://maps.tirol.gv.at/synserver?user=guest&project=tmap_master, accessed on 6 November 2023).

Table 1. Site description (N-site = north-facing leeward site; S-site = south-exposed windward site) and characteristics of the selected *A. alnobetula* stands. Mean values ± standard deviation (SD) are shown.

	Elevation (m asl)	Aspect	Slope (°)	Soil Depth (cm) Mean ± SD	Canopy Height (m) Mean ± SD	Stand Age (yrs) Mean ± SD
N-site	2150	N	35	12 ± 3	270 ± 80	15 ± 7
S-site	2140	SE	30	7 ± 3	140 ± 40	20 ± 8

Table 2. Climatic parameters (mean values ± standard deviation for July and August) obtained during the growing season 2022 and mean ring width on the S- and N-site. Statistical significance (*p*-values) of the differences between the study plots are indicated.

	T_{air} (°C) [1]	Solar Radiation (Wm^{-2})	T_{soil} (°C) [2]	Soil Moisture (% vol.)	Ring Width (μm)
S-site	10.8 ± 4.0	229.5 ± 82.7	10.3 ± 1.0	20.1 ± 2.7	378 ± 171
N-site	10.4 ± 3.6	180.7 ± 67.1	9.8 ± 1.1	23.0 ± 3.2	714 ± 228
p	0.421	<0.001	0.021	<0.001	<0.001

[1] T_{air} = air temperature, [2] T_{soil} = soil temperature.

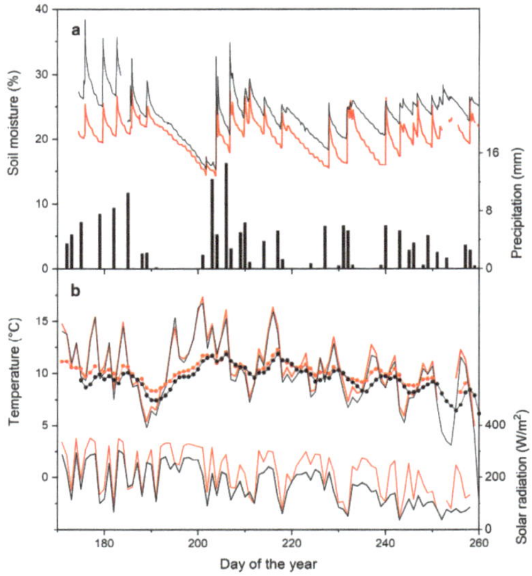

Figure 2. (**a**) Seasonal variation of soil moisture (solid lines) and precipitation (bars). (**b**) Air and soil temperature (solid and dotted lines, respectively) and solar radiation (solid lines) at both study plots. Black and red lines indicate the N- and S-site, respectively.

2.2. Environmental Data

Environmental conditions were recorded at both study sites from June through September 2022. At each study plot, air temperature and relative air humidity (CS215 Temperature and Relative Humidity Sensor) and solar radiation (SP1110 Pyranometer Sensor) (all sensors, Campbell Scientific, Shepshed, UK) were monitored 2 m above ground, while soil temperature (T 107 temperature probe, Campbell Scientific) and moisture in 10 cm soil depth were monitored using four ThetaProbes ML2 (Delta-T Devices Ltd., Burwell, UK). Precipitation (ARG100 Rain Gauge) was recorded at the S-site. All the environmental data were recorded with a Campbell CR1000 data logger (Campbell Scientific, Shepshed, UK) programmed to record 30-minute averages of measurements taken every minute.

Daily mean air temperature averaged throughout the growing season did not differ significantly between the two study plots (Figure 2, Table 2). During the growing season daily mean solar radiation and soil temperature in 10 cm soil depth, respectively, were significantly higher on the windward S-site than on the leeward N-site (Table 2). Conversely soil water content in 10 cm soil depth was significantly lower on the S-site than on the N-site (Table 2), indicating that drier conditions prevail on the S-site compared with the N-site. This assumption is supported by significantly lower annual radial increments and a lower canopy height at the S-site compared with the N-site (Table 1).

2.3. Sample Collection, Preparation, and Wood Anatomical Analysis

Stem discs were sampled from 3 different stocks on the N-site (mean diameter of stem discs: 2.13 ± 0.21 cm) and 4 stocks and the S-site (mean diameter of stem discs 2.08 ± 0.17 cm). Discs were taken 150 cm from the shoot tip as there is evidence that xylem anatomy in trees and shrubs is changing from tip to base [54–57], with the narrowest conduits found at the tip of the shoots. Stem discs were air-dried, and subsequently rectangular pieces were cut from the discs to prepare the probes for microtome sectioning. After the samples were soaked in glycerin for one day to soften the wood, microsections of 10 μm thickness were produced using a sledge microtome (WSL-core microtome, WSL Birmensdorf, Switzerland). The microsections were stained with safranin and astrablue to differentiate between lignified (red) and unlignified (blue) cells. The stained cross sections were observed under the microscope (Olympus Typ BX50, Olympus Corporation, Tokyo, Japan), and images were taken at 10× magnification with an HD-microscope camera (ProgRes GryphaxR, Jenoptik, Jena, Germany).

Wood anatomic variables were evaluated from the microscopic images using the open-source image analysis program Fiji (ImageJ2). Vessel number, vessel area, and diameter were evaluated for 10 annual rings (2011–2020). In total wood anatomy of 70 annual rings from 7 samples was determined. In total, 5677 and 4402 vessels were measured at the N- and the S-site, respectively. From the collected data, mean vessel area (MVA), vessel density (VD, number of vessels mm^{-2}), and theoretic conductive area (TCA, i.e., percentage of vessel area to total area) were calculated for each sample Kernel density estimations were used to compare vessel area distributions between sites. All individual vessel measurements were included in this analysis. To compare vessel area distributions related to wood surface area, all measured data were assigned to size classes and the number of vessels per size classes and xylem area was calculated.

To quantify phenological differences between vessels formed in different phases during the growing season, we determined the position within the annual ring for each vessel. Using this positioning data, we analyzed differences in the number and sizes of vessels in the first and second 50% of each annual ring (i.e., early and late growing season). MVA, VD, and TCA were calculated to gain insight into the plasticity of vessel formation throughout the growing period. In addition, vessel area distribution was analyzed for the first and the second 50% of the growth rings. Because some of the inner growth rings were curved, which made reliable positioning difficult, only the outermost 6-year rings (2015–2020) were used for the processing of the position data.

In a second series of measurements, we evaluated the number, area, and position of the diffuse parenchyma within the annual ring. Using these data, we calculated the percentage of the total area occupied by the axial parenchyma (PA). We also used the positioning data to calculate the percentage of PA for the first and the second 50% of the annual ring to get insight into the timing of the formation of diffuse parenchyma in the course of the growing period.

2.4. Data Analysis

Shapiro–Wilk tests and Q-Q plots were used to check for normal distribution of data. Most anatomical and climate datasets were not normally distributed, so we used nonparametric statistical tests for analyses. Because temperature data were consistently normally distributed, differences in air and soil temperature at the N-site and S-site were tested for significance using paired *t*-tests. For analyzing differences in MVA, VD, and TCA at the two sites, Mann–Whitney U tests were applied. For comparison of vessel and parenchyma properties in the first and second half of the annual rings and for analysis of soil moisture and solar radiation data, we used paired samples Wilcoxon tests. All statistical analyses were performed using SPSS Statistics, Version 26 (IBM, New York, NY, USA).

3. Results

3.1. Wood Anatomy

On an entire growing season basis, MVA was not significantly different between the N-site and S-site (Figure 3, Table 3). VD ($p = 0.037$) and TCA ($p = 0.017$), by contrast, were significantly higher on the N-site than on the S-site (Figure 3).

Vessel area distribution calculated by Kernel density estimation (Figure 4) showed similar patterns at both sites. Vessels around 250 μm^2 had the highest frequency at both measuring sites, then tailing out at with increasing vessel area (Figure 4). Nevertheless, wood samples at the S-site showed a higher proportion of smaller vessels between 200 μm^2 and 800 μm^2 and vessels between 1200 and 2000 μm^2 when compared with the N-site (Figure 3). Numbers of vessels per surface area were higher at the N-site in size classes up to 1200 μm^2. Values for vessels with areas greater than 1200 μm^2 were quite equal at both sites. Vessels up to 2500 μm^2 were frequently found in samples from both sampling sites.

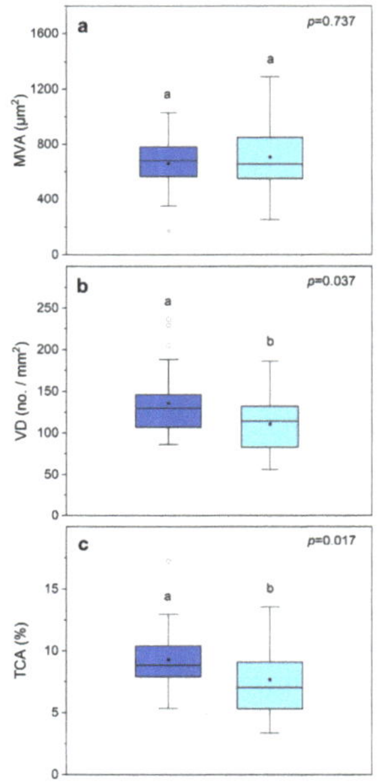

Figure 3. Boxplots of (**a**) mean vessel area (MVA), (**b**) vessel density (VD), and (**c**) theoretic conductive area (TCA). N-site: dark blue; S-site: light blue. Different letters indicate $p < 0.05$ among sites. p-values are shown in the upper right corner.

Table 3. Mean vessel area (MVA), vessel density (VD), theoretical conductive area (TCA), and percentage of axial parenchyma (PA) at the S- and N-site for the total annual ring (AR_{tot}) and the first (AR_{first}) and the second half (AR_{second}) of the annual ring. Mean values ± standard deviations are shown.

	MVA (μm²)	VD (no./mm⁻²)	TCA (%)	PA (%)
N-site (AR_{tot})	659.5 ± 177.4	135.6 ± 39.5	9.3 ± 2.4	10.1 ± 2.0
S-site (AR_{tot})	708.4 ± 223.5	110.9 ± 32.9	7.7 ± 2.9	16.5 ± 5.2
N-site (AR_{first})	741.6 ± 187.5	166.1 ± 65.6	11.5 ± 3.1	4.0 ± 2.1
N-site (AR_{second})	658,8 ± 153.7	133.3 ± 55.4	8.3 ± 2.3	14.9 ± 3.5
S-site (AR_{first})	823.6 ± 241.8	109.7 ± 37.8	8.8 ± 3.4	9.5 ± 4.9
S-site (AR_{second})	701.1 ± 161.9	84.9 ± 34.0	6.5 ± 4.3	19.6 ± 5.7

MVA, VD, and TCA were significantly higher in the first 50% of the annual ring compared with the second half at both sites (p-values, Figure 5). In the first half of the annual rings, the N-site and the S-site showed no significant differences in MVA while VD and TCA were significantly higher in N-site ($p \leq 0.006$). The same accounted for the second half of the annual ring ($p \leq 0.002$). Kendal density curves and the calculated numbers of vessels per area for the first and second half of the annual ring (Figure 6) provided a more detailed insight into the differences in vessel production throughout the growing season. In the first half of the annual ring, mean Kendal density curves were bimodal in shape at both sampling sites. There were years where the bimodality of distribution was quite pronounced, even in the second half of the annual ring, while it was weak or missing in other annual rings. The first half of the annual ring showed a higher density of midsize and large vessels. Vessels in the second half of the annual ring showed higher density in small vessels up to a size of 500 and 750 μm² at the N-site and S-site, respectively. The number of vessels per size class and area was higher in the first 50% of the annual ring throughout all size classes, with the greatest differences found for vessel areas between 20 and 1000 μm². At the S-site, the numbers of vessels per size class were overall significantly smaller compared with the N-site.

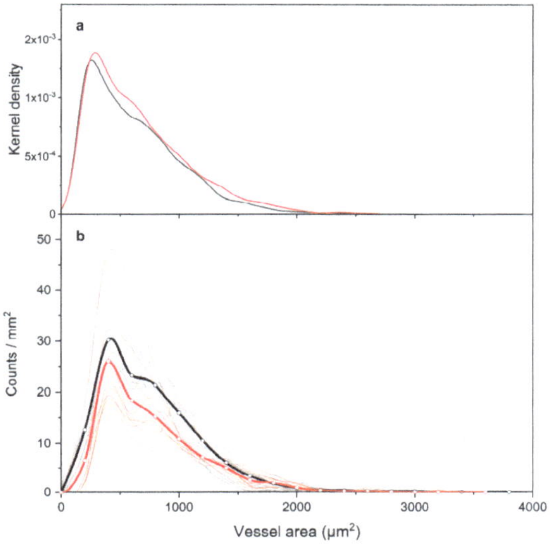

Figure 4. (**a**) Kernel Gauss distribution of vessel area for the N-site (black line) and S-site (red line). (**b**) Counts of vessels of different size classes per cross-section area. N-site: black line, S-site: red line; grey and thin red lines represent single measuring years (2011–2020) at the N-site and S-site, respectively.

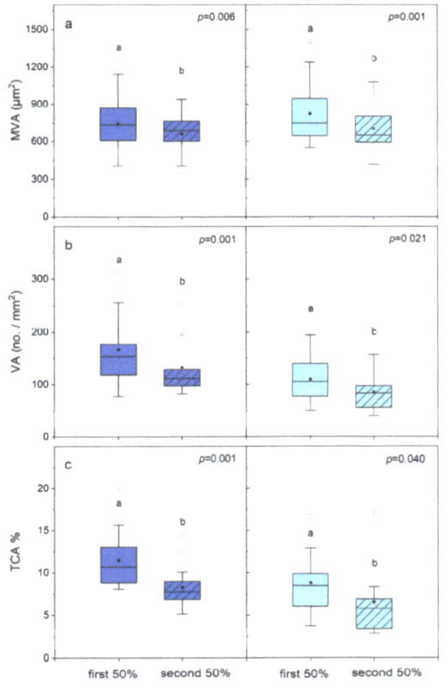

Figure 5. Boxplots of (**a**) mean vessel area (MVA), (**b**) vessel density (VD), and (**c**) theoretic conductive area (TCA), in the first and second half of the annual ring. Left panels: N-site in dark blue; right panels: S-site in light blue. Different letters indicate $p < 0.05$ between the first and second 50% of the annual ring. p-values are shown in the upper right corner.

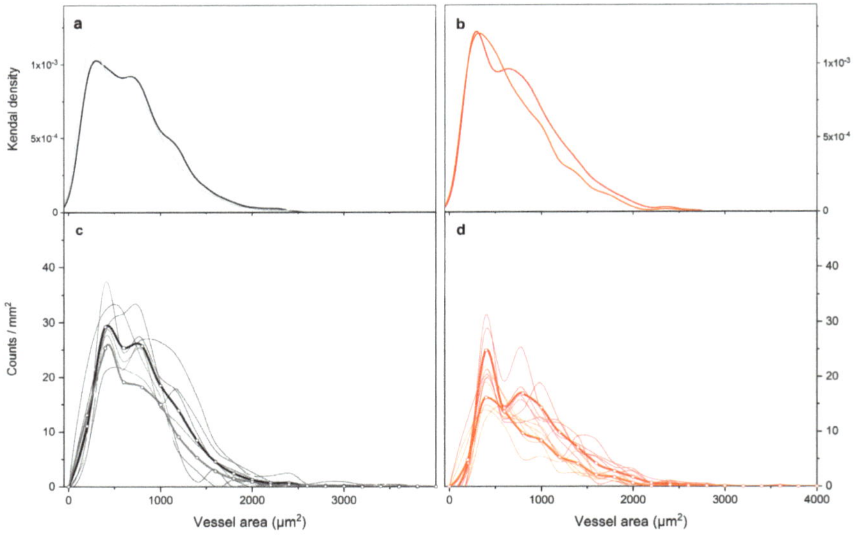

Figure 6. Kernel Gauss distribution of vessel area at the (**a**) N-site and (**b**) S-site and counts of vessels of different size classes per cross-section area at the (**c**) N-site and (**d**) S-site. Black lines and dark red lines represent values for the first half of the annual ring; grey lines and orange lines represent values for the second half of the annual ring. In (**c**,**d**), thin lines represent single measuring years (2015–2020).

3.2. Percentage of Axial Parenchyma

The percentage of axial parenchyma in the total annual ring was significantly higher ($p < 0.001$) at the S-site (16.5 ± 5.2%) when compared with the N-site (10.1 ± 2.0%) (Figure 7). At both sites, the percentage of axial parenchyma was significantly higher in the second half of the annual ring, ($p < 0.001$ for N- and S-site), and the highest percentages of axial parenchyma (19.6 ± 5.7%) were found in the second half of the annual ring at the S-site. Furthermore, for both sections of the annual ring, the percentage of axial parenchyma was significantly higher at the S-site ($p < 0.001$ and $p = 0.002$ for the first and second half of the annual ring) compared with the N-site.

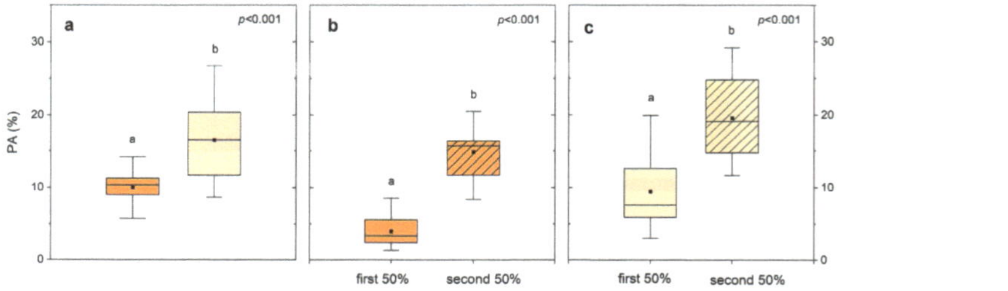

Figure 7. (**a**) Boxplots of the percentage of axial parenchyma (PA) at the N-site (orange) and the S-site (light orange). (**b**) Boxplots of the percentage of axial parenchyma at the N-site in the first and the second half of the annual ring. (**c**) Boxplots of the percentage of axial parenchyma at the S-site in the first and the second half of the annual ring. Different letters indicate $p < 0.05$ among groups. p-values are shown in the upper right corner.

4. Discussion

Analysis of site-specific functional adaptations of A. alnobetula enables us to determine the effects of global warming on the limits of the expansion of this pioneer species. Intraspecific studies provide insights into the impact of environmental factors on species-specific anatomical traits, disentangling genotypic and environmental effects contributing to xylem development is a difficult task. There is evidence that provenances can have more impact on xylem anatomy than environmental conditions [58–60]. Therefore, anatomical variations in plants growing at faraway sites or along wide-ranging environmental gradients may result from genetic variations. However, in the presented study, the examined plants were growing in close proximity, and *A. alnobetula* has expanded from the north-facing to the south-facing site only during the last decades [9], ruling out differences in provenience. Even so, the two sites differed significantly with respect to environmental conditions. The S-site experienced significantly higher solar radiation and was more exposed to the prevailing southerly winds while soil water availability was significantly lower than at the N-site (Figure 2, Table 2; c.f. [53]). Moreover, in anisohydric *A. alnobetula*, stomata remain open even under high evaporative demand [16–18], which, in combination with frequently occurring strong winds reducing boundary layer resistance (e.g., [61,62]), may lead to drought stress during periods of reduced soil water availability at the S-site. This is confirmed by a lower canopy height (Table 1), a significantly lower radial growth (Table 2; [53]), and premature leaf wilting, occurring at the S-site in mid-August after a dry period in summer 2022. At the N-site, by contrast, leaf wilting did not occur until early October.

Xylem cell division and differentiation are driven by environmental and internal factors [63]. As a key factor, temperature controls cambial cell division and plays a major role in cell wall lignification [64,65]. Conversely, as a turgor-driven procees cell enlargement is mainly controlled by cell water status [66,67] and sugar availability acting as an osmotic agent [68]. Cell enlargement and cambial cell division are seriously hampered under drought, often leading to a decline in radial growth [69] and vessel diameter [70–72]. Air temperature did not significantly differ between the two sampling sites (Table 2), and due to a high surface roughness, *A. alnobetula* experienced strong aerodynamic coupling to the free atmosphere [16]. On the other hand, significantly higher solar radiation at the S-site, especially during cloudless periods, enhanced solar heating of the ground. Nevertheless, as samples were taken in the upper branch section, temperature effects are less likely to explain observed differences in anatomical traits between the N- and the S-site.

Fast-growing species like *Populus* sp. show high phenotypic plasticity and are used as model species to examine environmental impacts on xylogenesis [73]. In poplar, drought often initiates a reduction in the vessel area [23,29,74,75]. In our study, however, MVA and HD did not differ significantly between the S- and N-site. However, tracheid enlargement might be based on turgor control in gymnosperms [66], while in more complex angiosperms, cell differentiation is probably actively regulated by endogen mechanisms rather than being a result of passive processes [76]. Anatomical traits are often under strong genetic control [58,73] suggesting species-specific adaptations to environmental impacts. In beech, for example, like in our study, no change in MVA under drought was detected [22,77,78]. Despite significant differences in environmental conditions, plant height, and annual ring width, Kernel density estimations of vessel areas (Figure 4a) showed no differences between the two sampling sites, confirming our assumption that vessel size is a static trait in green alder. Nevertheless, the control of cell formation and the patterns of cell distribution and size are poorly understood and the role of auxin and other morphogens are still under debate [24,79,80].

The number of vessels per size class and wood surface area (Figure 4b) showed a lower absolute number of small and midsize vessels (up to 1200 μm^2) at the dry S-site, which is in accordance with the significantly lower VD and TCA at the S-site. On the contrary, an increase in VD and TCA under drought conditions has been reported in several studies [21,58,77]. However, Arnic et al. [21] and Giagli et al. [22] pointed out that drier summer conditions resulted in narrower annual rings and, consequently, higher VD and TCA. Environmental and provenance effects on anatomical traits often result from stem height and ring width [58]. In beech, VD was positively related to the above-ground biomass and increased with tree height [77]. Plant height and crown size might often explain anatomical properties like VD and TCA better than climatic conditions [81], and leaf area has been found to be correlated with stem hydraulic conductivity [74]. In our study, the green alder stand at the N-site reached greater height, had a denser crown and, thus, higher leaf area compared with the S-site. Considering this, higher stem hydraulic conductivity was needed at the N-site to adjust the balance between the water supply and transpiring surface, which explains higher VD and TCA.

A. alnobetula is classified as a diffuse-porous species, with constant vessel diameters throughout the annual ring [20]. Nevertheless, anatomical parameters and density curves in our study estimate a near semi-ring-porous vessel distribution at both sampling sites. Environmentally induced changes in vessel distribution in diffuse-porous species have been reported—e.g., a near semi-ring-porous distribution of vessels in dry years and continuous decrease in vessel size under dry conditions have been reported for diffuse-porous beech [21–23]. As in diffuse-porous species, wide and narrow vessels can be formed during the entire growing period, short-term adaptations to changing environmental conditions are possible in these species [82,83]. The bimodal shape of density curves over several years might be another reference for frequent semi-ring-porous vessel distribution at our treeline site [84]. Schreiber et al. [29] linked the bimodal distribution of vessel diameters in trembling aspen, which was found at a boreal site, to rough environmental conditions and a short growing period. Therefore, the harsh conditions at the altitudinal limit of *A. alnobetula* and not water deficiency might be the reason for hampered vessel formation in the second half of the growing season.

In addition to vascular tissue, axial parenchyma seems to play a key role in hydraulic optimization [33,44]. Xylem parenchyma is known to modulate xylem flow and hydraulic resistance through osmotic exudation into the xylem [34,39] and is involved in in embolism repair [34,85]. Moreover, water stored in xylem parenchyma cells [86,87] buffers a decline in water potential to sustain water transport under drought stress [34,88]. Aritsara et al. [44] found that species with more axial parenchyma were close to their hydraulic limits and [38] hypothesized that axial parenchyma fractions potentially keep vessels hydrated during drought periods. We, therefore, assume that the significantly higher amount of axial parenchyma at the S-site might be an adaptation to drought conditions.

However, there are still gaps in knowledge, especially when it comes to diffuse axial parenchyma in temperate species as most of the available studies concentrate on paratracheal parenchyma and species from subtropical and tropical regions (e.g., [38,44,89–91]). An accumulation of diffuse apotracheal parenchyma in latewood has been described for several conifer species [92] and European beech and pedunculate oak [20,93]. Nevertheless, in our study, the accumulation of diffuse parenchyma is very pronounced and not restricted to latewood. Therefore, we suggest that the higher fraction of parenchyma cells in the second half of the growing season, when vessel production is declining, indicates an enhanced investment in carbohydrate storage triggered by harsh environmental conditions at the upper distributional limit of the species.

5. Conclusions

In contrast to our first and second hypotheses, lower soil water availability and enhanced evaporative demand did not affect the vessel diameter. However, vessel density and the theoretic conductive area were reduced at the drier south-facing site. This is confirmed by the higher number of vessels per size class at equal distribution, suggesting that vessel size is a static trait in green alder. Vessel numbers, by contrast, demonstrate plastic response. In accordance with our third hypothesis, limited water availability triggered a significant increase in axial parenchyma. This confirms the important role of xylem parenchyma when it comes to drought stress. Harsh environmental conditions at the distributional limit o green alder have affected xylogenesis, leading to a semi-ring-porous distribution of vessels and accumulation of parenchyma in the late growing season. We conclude from the present study that reduced annual increment, early leaf fall, and an accumulation of parenchyma indicate growth limitation and physiological stress, setting limits to the distribution of *A. alnobetula* in the treeline ecotone in a future drier environment.

Author Contributions: W.O., G.W. and A.G. conceived the study. W.O. coordinated the overall project. A.G., W.O. and M.F. collected, prepared, and analyzed the data. W.O., G.W., A.G. and M.F. interpreted and discussed the data. A.G. wrote the manuscript and G.W. and W.O. provided editorial advice. All authors have read and agreed to the published version of the manuscript.

Funding: This research was funded by the Austrian Science Fund (FWF), P34706-B. For the purpose of open access, the author has applied a CC BY public copyright license to any author accepted manuscript version arising from this submission.

Data Availability Statement: The wood anatomical data presented in this study are openly available on Zenodo at: 10.5281/zenodo.10405899.

Acknowledgments: We thank Bernhard Nairz and Michaela Schweinschwaller for their help with anatomical measurements.

Conflicts of Interest: The authors declare no conflict of interest.

References

1. Verrall, B.; Pickering, C.M. Alpine vegetation in the context of climate change: A global review of past research and future directions. *Sci. Total Environ.* **2020**, *748*, 141344. [CrossRef] [PubMed]
2. Gottfried, M.; Pauli, H.; Futschik, A.; Akhalkatsi, M.; Barančok, P.; Benito Alonso, J.L.; Coldea, G.; Dick, J.; Erschbamer, B.; Fernández Calzado, M.R.; et al. Continent-wide response of mountain vegetation to climate change. *Nat. Clim. Chang.* **2012**, *2*, 111–115. [CrossRef]
3. Grabherr, G.; Gottfried, M.; Pauli, H. Climate change impacts in alpine environments. *Geogr. Compass* **2010**, *4*, 1133–1153. [CrossRef]
4. Gruber, A.; Oberhuber, W.; Wieser, G. Treeline-Quo Vadis? An Ecophysiological Approach. *Forests* **2022**, *13*, 857. [CrossRef]
5. Gehring-Fasel, J.; Gusian, A.; Zimmermann, N.E. Tree line shifts in the Swiss Alps: Climate change or land abandonment. *J. Veg. Sci.* **2007**, *18*, 571–582. [CrossRef]
6. Francon, L.; Roussel, E.; Lopez-Saez, J.; Saulnier, M.; Stoffel, M.; Corona, C. Alpine shrubs have benefited more than trees from 20th century warming at a treeline ecotone site in the French Pyrenees. *Agric. For. Meteorol.* **2023**, *329*, 109284. [CrossRef]
7. Bühlmann, T.; Körner, C.; Hiltbrunner, E. Shrub expansion of *Alnus viridis* drives former montane grassland into nitrogen saturation. *Ecosystems* **2016**, *19*, 968–985. [CrossRef]
8. Dullinger, S.; Dirnböck, T.; Grabherr, G. Patterns of shrub invasion into high mountain grasslands of the Northern Calcareous Alps, Austria. *Arct. Antarct. Alp. Res.* **2003**, *35*, 434–441. [CrossRef]
9. Oberhuber, W.; Wieser, G.; Bernich, F.; Gruber, A. Radial stem growth of the clonal shrub Alnus alnobetula at treeline is constrained by summer temperature and winter desiccation and differs in carbon allocation strategy compared to co-occurring Pinus cembra. *Forests* **2022**, *13*, 440. [CrossRef]
10. Bühlmann, T.; Hiltbrunner, E.; Körner, C. Alnus viridis expansion contributes to excess reactive nitrogen release, reduces biodiversity and constrains forest succession in the Alps. *Alp. Bot.* **2014**, *124*, 187–191. [CrossRef]
11. Anthelme, F.; Villaret, J.C.; Brun, J.J. Shrub encroachment in the Alps gives rise to the convergence of sub-alpine communities on a regional scale. *J. Veg. Sci.* **2007**, *18*, 355–362. [CrossRef]
12. Hantemirova, E.; Marchuk, E. Phylogeography and genetic structure of a subarctic-alpine shrub species, *Alnus alnobetula* (Ehrh.) K. Koch s. l., inferred from chloroplast DNA markers. *Tree Genet. Genomes* **2021**, *17*, 18. [CrossRef]
13. Richard, L. Ecologie de l'Aulne Vert (*Alnus viridis* Chaix): Facteurs climatiques et édaphiques. *Doc. Carte. Veg. Alpes.* **1968**, *6*, 107–158.
14. Schröter, C. *Das Pflanzenleben der Alpen: Eine Schilderung der Hochgebirgsflora*; Albert Raustein: Zürich, Switzerland, 1908.
15. Caviezel, C.; Hunziker, M.; Kuhn, N.J. Green alder encroachment in the European Alps: The need for analyzing the spread of a native-invasive species across spatial data. *Catena* **2017**, *159*, 149–158. [CrossRef]
16. Van den Bergh, T.; Körner, C.; Hiltbrunner, E. *Alnus* shrub expansion increases evapotranspiration in the Swiss Alps. *Reg. Environ. Chang.* **2018**, *18*, 1375–1385. [CrossRef]
17. McDowell, N.; Pockman, W.T.; Allen, C.D.; Breshears, D.D.; Cobb, N.; Kolb, T.; Yepez, E.A. Mechanisms of plant survival and mortality during drought: Why do some plants survive while others succumb to drought? *New Phytolog.* **2008**, *178*, 719–739. [CrossRef]

18. Körner, C.; Jussel, U.; Schiffer, K. Transpiration, diffusionswiderstand und wasserpotential in verschiedenen schichten eines grünerlenbestandes. In *Ökologische Analysen von Almflächen im Gasteiner Tal. Veröffentlichungen des Österreichischen MaB-Hochgebirgsprogramms Hohe Tauern, Band 2*; Cernusca, A., Ed.; Universitätsverlag Wagner: Innsbruck, Austria, 1978; pp. 81–98.
19. Sade, N.; Gebremedhin, A.; Moshelion, M. Risk-taking plants: Anisohydric behavior as a stress-resistance trait. *Plant Signal. Behav.* **2012**, *7*, 767–770. [CrossRef]
20. Schweingruber, F.H.; Börner, A.; Schulze, E.D. *Atlas of Stem Anatomy in Herbs, Shrubs and Trees*; Springer Science & Business Media Berlin/Heidelberg, Germany, 2011; Volume 1.
21. Arnič, D.; Gričar, J.; Jevšenak, J.; Božič, G.; von Arx, G.; Prislan, P. Different wood anatomical and growth responses in European beech (*Fagus sylvatica* L.) at three forest sites in Slovenia. *Front. Plant Sci.* **2021**, *12*, 669229. [CrossRef]
22. Giagli, K.; Gričar, J.; Vavrčík, H.; Menšík, L.; Gryc, V. The effects of drought on wood formation in *Fagus sylvatica* during two contrasting years. *IAWA J.* **2016**, *37*, 332–348. [CrossRef]
23. Schume, H.; Grabner, M.; Eckmüllner, O. The influence of an altered groundwater regime on vessel properties of hybrid poplar *Trees* **2004**, *18*, 184–194. [CrossRef]
24. Hacke, U.G.; Lachenbruch, B.; Pittermann, J.; Mayr, S.; Domec, J.-C.; Schulte, P.J. *The Hydraulic Architecture of Conifers BT—Functional and Ecological Xylem Anatomy*; Springer International Publishing: Cham, Switzerland, 2015.
25. Tyree, M.T.; Zimmermann, M.H. *Xylem Structure and the Ascent of Sap*; Springer Series in Wood Science; Springer Berlin/Heidelberg, Germany, 2013.
26. Chambi-Legoas, R.; Tomazello-Filho, M.; Laclau, J.P.; Chaix, G. Potassium fertilization enhances xylem plasticity and growth recovery of Eucalyptus grandis trees in response to drastic changes in water availability. *For. Ecol. Manag.* **2023**, *528*, 120656 [CrossRef]
27. Chenlemuge, T.; Schuldt, B.; Dulamsuren, C.; Hertel, D.; Leuschner, C.; Hauck, M. Stem increment and hydraulic architecture of a boreal conifer (*Larix sibirica*) under contrasting macroclimates. *Trees* **2015**, *29*, 623–636. [CrossRef]
28. Schuldt, B.; Knutzen, F.; Delzon, S.; Jansen, S.; Müller-Haubold, H.; Burlett, R.; Leuschner, C. How adaptable is the hydraulic system of European beech in the face of climate change-related precipitation reduction? *New Phytol.* **2016**, *210*, 443–458. [CrossRef] [PubMed]
29. Schreiber, S.G.; Hacke, U.G.; Hamann, A. Variation of xylem vessel diameters across a climate gradient: Insight from a reciprocal transplant experiment with a widespread boreal tree. *Funct. Ecol.* **2015**, *29*, 1392–1401. [CrossRef]
30. Montwé, D.; Spiecker, H.; Hamann, A. An experimentally controlled extreme drought in a Norway spruce forest reveals fast hydraulic response and subsequent recovery of growth rates. *Trees* **2014**, *28*, 891–900. [CrossRef]
31. Sterck, F.J.; Zweifel, R.; Sass-Klaassen, U.; Chowdhury, Q. Persisting soil drought reduces leaf specific conductivity in Scots pine (*Pinus sylvestris*) and pubescent oak (*Quercus pubescens*). *Tree Physiol.* **2008**, *28*, 529–536. [CrossRef]
32. Bader, M.K.F.; Scherrer, D.; Zweifel, R.; Körner, C. Less pronounced drought responses in ring-porous than in diffuse-porous temperate tree species. *Agric. For. Meteorol.* **2022**, *327*, 109184. [CrossRef]
33. Kiorapostolou, N.; Da Sois, L.; Petruzzellis, F.; Savi, T.; Trifilò, P.; Nardini, A.; Petit, G. Vulnerability to xylem embolism correlates to wood parenchyma fraction in angiosperms but not in gymnosperms. *Tree Physiol.* **2019**, *39*, 1675–1684. [CrossRef]
34. Slupianek, A.; Dolzblasz, A.; Sokołowska, K. Xylem parenchyma—Role and relevance in wood functioning in trees. *Plants* **2021** *10*, 1247. [CrossRef]
35. Hearn, D.J.; Poulsen, S.; Spicer, R. The evolution of growth forms with expanded root and shoot parenchymatous storage is correlated across the eudicots. *Int. J. Plant Sci.* **2013**, *174*, 1049–1061. [CrossRef]
36. Sevanto, S.; Hölttä, T.; Holbrook, N.M. Effects of the hydraulic coupling between xylem and phloem on diurnal phloem diameter variation. *Plant Cell Environ.* **2011**, *34*, 690–703. [CrossRef] [PubMed]
37. Höltta, T.; Vesala, T.; Sevanto, S.; Peramaki, M.; Nikinmaa, E. Modeling xylem and phloem water flows in trees according to cohesion theory and Munch hypothesis. *Trees* **2006**, *20*, 67–78. [CrossRef]
38. Morris, H.; Gillingham, M.A.; Plavcová, L.; Gleason, S.M.; Olson, M.E.; Coomes, D.A.; Fichtler, E.; Klepsch, M.M.; Martínez Cabrera, H.I.; McGlinn, D.J.; et al. Vessel diameter is related to amount and spatial arrangement of axial parenchyma in woody angiosperms. *Plant Cell Environ.* **2018**, *41*, 245–260. [CrossRef] [PubMed]
39. Morris, H.; Plavcová, L.; Cvecko, P.; Fichtler, E.; Gillingham, M.A.; Martínez-Cabrera, H.I.; Jansen, S. A global analysis of parenchyma tissue fractions in secondary xylem of seed plants. *New Phytol.* **2016**, *209*, 1553–1565. [CrossRef] [PubMed]
40. Spicer, R. Symplasmic networks in secondary vascular tissues: Parenchyma distribution and activity supporting long-distance transport. *J. Exp. Bot.* **2014**, *65*, 1829–1848. [CrossRef] [PubMed]
41. Höll, W. Distribution, fluctuation and metabolism of food reserves in the wood of trees. In *Cell and Molecular Biology of Wood Formation*; Savidge, R., Barnett, J., Napier, R., Eds.; BIOS Scientific Publishers: Oxford, UK, 2000; pp. 347–362.
42. Secchi, F.; Zwieniecki, M.A. Sensing embolism in xylem vessels: The role of sucrose as a trigger for refilling. *Plant Cell Environ* **2011**, *34*, 514–524. [CrossRef] [PubMed]
43. Salleo, S.; Trifilò, P.; Esposito, S.; Nardini, A.; Gullo, M.A.L. Starch-to-sugar conversion in wood parenchyma of field-growing Laurus nobilis plants: A component of the signal pathway for embolism repair? *Funct. Plant Biol.* **2009**, *36*, 815–825. [CrossRef]
44. Aritsara, A.N.A.; Razakandraibe, V.M.; Ramananantoandro, T.; Gleason, S.M.; Cao, K.F. Increasing axial parenchyma fraction in the Malagasy Magnoliids facilitated the co-optimisation of hydraulic efficiency and safety. *New Phytol.* **2021**, *229*, 1467–1480 [CrossRef]

45. Mencuccini, M.; Hölttä, T.; Sevanto, S.; Nikinmaa, E. Concurrent measurements of change in the bark and xylem diameters of trees reveal a phloem-generated turgor signal. *New Phytol.* **2013**, *198*, 1143–1154. [CrossRef]
46. Pratt, R.B.; Jacobsen, A.L. Conflicting demands on angiosperm xylem: Tradeoffs among storage, transport and biomechanics. *Plant Cell Environ.* **2017**, *40*, 897–913. [CrossRef]
47. Alves, E.S.; Angyalossy-Alfonso, V. Ecological trends in the wood anatomy of some Brazilian species. 2. Axial parenchyma, rays and fibres. *IAWA J.* **2002**, *23*, 391–418. [CrossRef]
48. Wheeler, E.A.; Baas, P. A survey of the fossil record for dicotyledonous wood and its significance for evolutionary and ecological wood anatomy. *IAWA J.* **1991**, *12*, 275–318. [CrossRef]
49. Tollmann, A. *Geologie von Österreich*; Band 1. Die Zentralalpen; Deuticke: Wien, Austria, 1977.
50. FAO. *World Reference Base for Soil Resources*; FAO: Rome, Italy, 1998.
51. Neuwinger, I. Böden der subalpinen und alpinen Stufe in den Tiroler Alpen. *Mitteilungen Ostalpin. Dinarischen Ges. Veg.* **1970**, *11*, 135–150.
52. Richner, H.; Hächler, P. Understanding and forecasting Alpine foehn. In *Mountain Weather Research and Forecasting: Recent Progress and Current Challenges*; Chow, F.K., De Wekker, S.F.J., Snyder, B.J., Eds.; Springer: Berlin/Heidelberg, Germany, 2013; pp. 219–260.
53. Oberhuber, W.; Dobler, A.-L.; Heinzle, T.; Scandurra, F.; Gruber, A.; Wieser, G. Climate overrides the influence of microsite conditons on radial growth of the tall multi-stemmed shrub *Alnus alnobetula* at treeline. *Plants* **2023**, *12*, 1708. [CrossRef] [PubMed]
54. Olson, E.M.; Anfodillo, T.; Gleason, M.S.; McCulloh, K.A. Tip-to-base xylem conduit widening as an adaptation: Causes, consequences, and empirical priorities. *New Phytol.* **2020**, *229*, 1877–1893. [CrossRef] [PubMed]
55. McCulloh, K.A.; Sperry, J.S. Patterns in hydraulic architecture and their implications for transport efficiency. *Tree Physiol.* **2005**, *25*, 257–267. [CrossRef] [PubMed]
56. Anfodillo, T.; Carraro, V.; Carrer, M.; Fior, C.; Rossi, S. Convergent tapering of xylem conduits in different woody species. *New Phytol.* **2006**, *169*, 279–290. [CrossRef] [PubMed]
57. Meinzer, F.C.; Clearwater, M.J.; Goldstein, G. Water transport in trees: Current perspectives, new insights and some controversies. *Environ. Exp. Bot.* **2001**, *45*, 239–262. [CrossRef]
58. Hietz, P.; Rungwattana, K.; Scheffknecht, S.; George, J.P. Effects of Provenance, Growing Site, and Growth on Quercus robur Wood Anatomy and Density in a 12-Year-Old Provenance Trial. *Front. Plant Sci.* **2022**, *13*, 795941. [CrossRef]
59. George, J.P.; Theroux-Rancourt, G.; Rungwattana, K.; Scheffknecht, S.; Momirovic, N.; Neuhauser, L.; Hietz, P. Assessing adaptive and plastic responses in growth and functional traits in a 10-year-old common garden experiment with pedunculate oak (*Quercus robur* L.) suggests that directional selection can drive climatic adaptation. *Evol. Appl.* **2020**, *13*, 2422–2438. [CrossRef]
60. Bayramzadeh, V.; Funada, R.; Kubo, T. Relationships between vessel element anatomy and physiological as well as morphological traits of leaves in *Fagus crenata* seedlings originating from different provenances. *Trees* **2008**, *22*, 217–224. [CrossRef]
61. Schymanski, S.J.; Or, D. Wind increases leaf water use efficiency. *Plant Cell Environ.* **2016**, *39*, 1448–1459. [CrossRef] [PubMed]
62. Körner, C. *Alpine Plant Life: Functional Plant Ecology of High Mountain Ecosystems*; Springer: New York, NY, USA, 2003.
63. Rathgeber, C.B.K.; Perez-de-Lis, G.; Fernandez-de-Una, L.; Fonti, P.; Rossi, S.; Treydte, K.; Ponton, S. Anatomical, developmental and physiological bases of tree-ring formation in relation to environmental factors. In *Stable Isotopes in Tree Rings: Inferring Physiological, Climatic and Environmental Responses*; Siegwolf, R.T.W., Brooks, J.R., Roden, J., Saurer, M., Eds.; Springer International: Cham, Switzerland, 2022; pp. 61–99.
64. Chaffey, N.J. (Ed.) *Wood Formation in Trees—Cell and Molecular Biology Techniques*; Taylor and Francis: London, UK; New York, NY, USA, 2002; p. 364.
65. Donaldson, L.A. Lignification and lignin topochemistry—An ultrastructural view. *Phytochemistry* **2001**, *57*, 859–873. [CrossRef] [PubMed]
66. Cabon, A.; Fernandez-de-Una, L.; Gea-Izquierdo, G.; Meinzer, F.C.; Woodruff, D.R.; Martınez-Vilalta, J.; de Caceres, M. Water potential control of turgor driven tracheid enlargement in Scots pine at its xeric distribution edge. *New Phytol.* **2020**, *225*, 209–221. [CrossRef] [PubMed]
67. Fromm, J. Xylem development in trees: From cambial divisions to mature wood cells. In *Cellular Aspects of Wood Formation*; Springer: Berlin/Heidelberg, Germany, 2013; pp. 3–39.
68. Koch, K. Sucrose metabolism: Regulatory mechanisms and pivotal roles in sugar sensing and plant development. *Curr. Opin. Plant Biol.* **2004**, *7*, 235–246. [CrossRef] [PubMed]
69. Gruber, A.; Strobl, S.; Veit, B.; Oberhuber, W. Impact of drought on the temporal dynamics of wood formation in *Pinus sylvestris*. *Tree Physiol.* **2010**, *30*, 490–501. [CrossRef] [PubMed]
70. Balducci, L.; Cuny, H.E.; Rathgeber, C.B.K.; Deslauriers, A.; Giovannelli, A.; Rossi, S. Compensatory mechanisms mitigate the effect of warming and drought on wood formation. *Plant Cell Environ.* **2016**, *39*, 1338–1352. [CrossRef] [PubMed]
71. Rossi, S.; Simard, S.; Rathgeber, C.B.K.; Deslauriers, A.; De Zan, C. Effects of a 20-day-long dry period on cambial and apical meristem growth in *Abies balsamea* seedlings. *Trees* **2009**, *23*, 85–93. [CrossRef]
72. Nonami, H.; Boyer, J.S. Turgor and growth at low water potentials. *Plant Physiol.* **2008**, *89*, 798–804. [CrossRef]
73. Hacke, U.G.; Spicer, R.; Schreiber, S.G.; Plavcová, L. An ecophysiological and developmental perspective on variation in vessel diameter. *Plant Cell Environ.* **2017**, *40*, 831–845. [CrossRef]
74. Plavcova, L.; Hacke, U.G. Phenotypic and developmental plasticity of xylem in hybrid poplar saplings subjected to experimental drought, nitrogen fertilization, and shading. *J. Exp. Bot.* **2012**, *63*, 6481–6491. [CrossRef] [PubMed]

75. Awad, H.; Barigah, T.; Badel, E.; Cochard, H.; Herbette, S. Poplar vulnerability to xylem cavitation acclimates to drier soil conditions. *Physiol. Plant.* **2010**, *139*, 280–288. [CrossRef] [PubMed]
76. Noyer, E.; Stojanović, M.; Horáček, P.; Pérez-de-Lis, G. Toward a better understanding of angiosperm xylogenesis: A new method for a cellular approach. *New Phytol.* **2023**, *239*, 792–805. [CrossRef] [PubMed]
77. Hajek, P.; Kurjak, D.; von Wühlisch, G.; Delzon, S.; Schuldt, B. Intraspecific variation in wood anatomical, hydraulic, and foliar traits in ten European beech provenances differing in growth yield. *Front. Plant Sci.* **2016**, *7*, 791. [CrossRef] [PubMed]
78. Oladi, R.; Bräuning, A.; Pourtahmasi, K. "Plastic" and "static" behavior of vessel-anatomical features in Oriental beech (*Fagus orientalis* Lipsky) in view of xylem hydraulic conductivity. *Trees* **2014**, *28*, 493–502. [CrossRef]
79. Buttò, V.; Rozenberg, P.; Deslauriers, A.; Rossi, S.; Morin, H. Environmental and developmental factors driving xylem anatomy and micro-density in black spruce. *New Phytol.* **2021**, *230*, 957–971. [CrossRef] [PubMed]
80. Hartmann, F.; Rathgeber, C.; Fournier, M.; Moulia, B. Modelling wood formation and structure: Power and limits of a morphogenetic gradient in controlling xylem cell proliferation and growth. *Ann. For. Sci.* **2017**, *74*, 1–15. [CrossRef]
81. Rosell, J.A.; Olson, M.E.; Anfodillo, T. Scaling of xylem vessel diameter with plant size: Causes, predictions, and outstanding questions. *Curr. For. Rep.* **2017**, *3*, 46–59. [CrossRef]
82. Schweingruber, F.H. *Wood Structure and Environment*; Springer: Berlin/Heidelberg, Germany, 2007.
83. Bosshard, H. *Holzkunde. Band 1: Mikroskopie und Makroskopie des Holzes*; Birkhauser: Basel, Switzerland, 1982.
84. Carlquist, S. *Comparative Wood Anatomy*; Springer: Berlin/Heidelberg, Germany, 2001.
85. Schenk, H.J.; Jansen, S.; Hölttä, T. Positive pressure in xylem and its role in hydraulic function. *New Phytol.* **2021**, *230*, 27–45. [CrossRef]
86. Jupa, R.; Plavcová, L.; Gloser, V.; Jansen, S. Linking xylem water storage with anatomical parameters in five temperate tree species. *Tree Physiol.* **2016**, *36*, 756–769. [CrossRef]
87. Holbrook, N.M. Stem water storage. In *Plant Stems: Physiology and Functional Morphology*; Academic Press: San Diego, CA, USA, 1995; pp. 151–174.
88. Meinzer, F.C.; Johnson, D.M.; Lachenbruch, B.; McCulloh, K.A.; Woodruff, D.R. Xylem hydraulic safety margins in woody plants: Coordination of stomatal control of xylem tension with hydraulic capacitance. *Funct. Ecol.* **2009**, *23*, 922–930. [CrossRef]
89. Kawai, K.; Minagi, K.; Nakamura, T.; Saiki, S.T.; Yazaki, K.; Ishida, A. Parenchyma underlies the interspecific variation of xylem hydraulics and carbon storage across 15 woody species on a subtropical island in Japan. *Tree Physiol.* **2022**, *42*, 337–350. [CrossRef] [PubMed]
90. Janssen, T.A.; Hölttä, T.; Fleischer, K.; Naudts, K.; Dolman, H. Wood allocation trade-offs between fiber wall, fiber lumen, and axial parenchyma drive drought resistance in neotropical trees. *Plant Cell Environ.* **2019**, *43*, 965–980. [CrossRef] [PubMed]
91. Zheng, J.; Martínez-Cabrera, H.I. Wood anatomical correlates with theoretical conductivity and wood density across China: Evolutionary evidence of the functional differentiation of axial and radial parenchyma. *Ann. Bot.* **2013**, *112*, 927–935. [CrossRef]
92. Carlquist, S. Living cells in wood. 1. Absence, scarcity and histology of axial parenchyma as keys to function. *Bot. J. Linn. Soc.* **2015**, *177*, 291–321. [CrossRef]
93. Gasson, P. Some implications of anatomical variations in the wood of pedunculate oak (*Quercus robur* L.), including comparisons with common beech (*Fagus sylvatica* L.). *IAWA J.* **1987**, *8*, 149–166. [CrossRef]

forests

Article

Climate-Sensitive Diameter Growth Models for White Spruce and White Pine Plantations

Mahadev Sharma

Ontario Forest Research Institute, Ontario Ministry of Natural Resources and Forestry, 1235 Queen St. East, Sault Ste. Marie, ON P6A 2E5, Canada; mahadev.sharma@ontario.ca; Tel.: +1-(519)-572-7285

Abstract: Global change in the climate is affecting tree/forest growth. There have been many studies that analyzed climate effects on tree growth. Results presented in these studies showed that the climate had both positive and negative effects on tree growth. The nature (positive/negative) and magnitude of the effects and the climate variables affecting growth depended on tree species. Climate-sensitive diameter growth models are not available for white pine (*Pinus strobus* L.) and white spruce (*Picea glauca* (Moench) Voss) plantations. These models are needed to project forest growth and yield and develop forest management plans. Therefore, diameter growth models were developed for white pine and white spruce plantations by incorporating climate variables. Four hundred white pine and white spruce trees (200 per species) were sampled from 80 (40 per species) even-aged monospecific plantations (five trees per plantation) across Ontario, Canada. Diameter–age pairs were obtained from these trees using stem analysis. A nonlinear mixed-effects modeling approach was used to develop diameter growth models. To make the models climate sensitive, model parameters were expressed in term of climate variables. Inclusion of climate variables significantly improved model fit statistics and predictive accuracy. For evaluation, diameters (inside bark) at breast height were estimated for three geographic locations (east, west, and south) across Ontario for an 80-year growth period (2021–2100) under three climate change (emissions) scenarios (representative concentration pathway or RCP 2.6, 4.5, and 8.5 watts m^{-2}). For both species, the overall climate effects were negative. For white spruce, the maximum pronounced difference in projected diameters after the 80-year growth period was in the west. At this location, compared to the no climate change scenario, projected spruce diameters under RCPs 2.6 and 8.5 were thinner by 4.64 (15.99%) and 3.72 (12.80%) cm, respectively. For white pine, the maximum difference was in the south. Compared to the no climate change scenario, projected pine diameters at age 80 under RCPs 2.6 and 8.5 at this location were narrower by 4.54 (13.99%) and 7.60 (23.43%) cm, respectively. For both species, climate effects on diameter growth were less evident at other locations. If the values of climate variables are unavailable, models fitted without climate variables can be used to estimate these diameters for both species.

Keywords: climate change; tree growth; DBH growth models; boreal tree species; nonlinear regression; mixed-effects model

Citation: Sharma, M. Climate-Sensitive Diameter Growth Models for White Spruce and White Pine Plantations. *Forests* **2023**, *14*, 2457. https://doi.org/10.3390/f14122457

Academic Editors: Yassine Messaoud, Jan Světlík and Giorgio Alberti

Received: 24 November 2023
Revised: 11 December 2023
Accepted: 15 December 2023
Published: 17 December 2023

1. Introduction

Accurate information about forest stand development over time is essential for forest management planning [1–3]. This information is obtained using forest growth and yield models that are driven by certain measurable variables (e.g., diameter at breast height (dbh) and total height of a tree, site index, tree/stand age, basal area). These models are mainly classified into two categories: stand- and tree-scale [2]. Stand-scale models rely on stand-level attributes such as stand age, stand density, and site index for growth and yield calculations. On the other hand, tree-scale models rely on individual tree attributes (e.g., dbh, height) [3]. Stand-scale models are less comprehensive than tree-scale models for providing stand structure and its development over time [1]. Tree-scale models provide detailed information about stand dynamics and structure, including stand volume distribution by size classes [2].

Information about individual tree diameters and their growth is needed to determine current and projected product recoveries from trees growing in a stand under a range of management alternatives. Therefore, diameter growth models are key components of tree-scale growth and yield models. The models can be used to estimate individual tree growth rates, which can be summed to obtain stand-scale estimates [3].

Diameter growth is affected by climate change [4–11]. The nature (positive/negative) and magnitude of climate effects depend on tree species and geographic locations [3,12]. Pokharel and Frose [13] analyzed climate effects on basal area growth of black spruce (*Picea mariana* (Mill.) B.S.P.), jack pine (*Pinus banksiana* Lamb.), balsam fir (*Abies balsamea* (L.) Mill.), and trembling aspen (*Populus tremuloides* Michx.) grown in natural stands in eastern Canada. They reported that basal area of all these tree species was significantly affected by climate.

Maxime and Hendrik [5] examined climate effects on diameter growth of common beech (*Fagus sylvatica* L.) and silver fir (*Abies alba* Mill.) grown in France. They reported that climate significantly affected diameter growth of these tree species. In another study, Subedi and Sharma [12] found positive effects of climate change on jack pine diameter growth but negative effects on black spruce grown in plantations across Ontario, Canada. For both species, the effects (positive or negative) were more pronounced in the west than the east. Similarly, Matisons et al. [8] reported that the radial growth of European (common) beech was sensitive to climate, with sensitivity depending on the method of establishment and social status of trees.

Sharma [3] recently analyzed climate effects on diameter growth of red pine (*Pinus resinosa Ait.*) plantations grown in eastern Canada. He reported that red pine diameter growth was significantly affected by climate, with positive effects in the southeast and southwest and negative effects in the central west of Ontario. No effect was evident in the far west. The magnitude (positive or negative) of the effect varied by geographic location. Oboite and Comeau [14] also found varying effects of climate on diameter growth of lodgepole pine (*Pinus contorta* Douglas ex Loudon), jack pine, trembling aspen, white spruce (*Picea glauca* (Moench) Voss), and balsam poplar (*Populus balsamifera* L.) trees grown in western Canada. Similarly, Bayat et al. [15] analyzed climate effects on diameter increment of various tree species of the Hyrcanian Forest of Northern Iran.

Most of the studies described above examined whether diameter/basal area growth of different tree species in various locations were affected by climate change. Since these studies reported significant effects (positive/negative of varying magnitude) of climate on diameter/basal area growth of various tree species, climate factors need to be accounted for to accurately project future tree growth [16]. Accurate growth projections are crucial for developing credible forest management plans. Therefore, in addition to analyzing climate effects on diameter growth of different tree species, Subedi and Sharma [12], Sharma [3], and Oboite and Comeau [14] also developed diameter growth models by incorporating climate variables. Similarly, Bayat et al. [15] modeled diameter growth in terms of climate variables.

White pine (*Pinus strobus* L.) and white spruce grow throughout much of Ontario. These species are the most commonly planted commercial tree species after jack pine and black spruce. The objectives of this study were to (1) examine climate effects on diameter growth of plantation-grown white pine and white spruce trees, (2) develop climate-sensitive diameter growth models for these species, and (3) analyze climate effects on their future diameter growth under three climate change scenarios.

2. Materials and Methods

2.1. Tree Data

For each species, forty even-aged monoculture plantations were selected from across Ontario, Canada (Figure 1) to sample trees for this study. In total, 400 white spruce and white pine trees (200 per species) were sampled from these plantations. White spruce and white pine trees were sampled in fall 2021 and 2022, respectively. For both species the latitude of sample sites ranged from 42.38° N to 50.83° N and longitude varied from 75.06° W to 94.07° W. Similarly, the variation in elevation was from 83.0 m to 534.0 m. A circular temporary sample plot of 400 m^2 (11.28 m radius) was established, representing the tree population in the stand at each plantation site to sample the trees for each species. If required, the plot size was increased (up to 600 m^2) to ensure at least 40 planted live trees were in the sample plots.

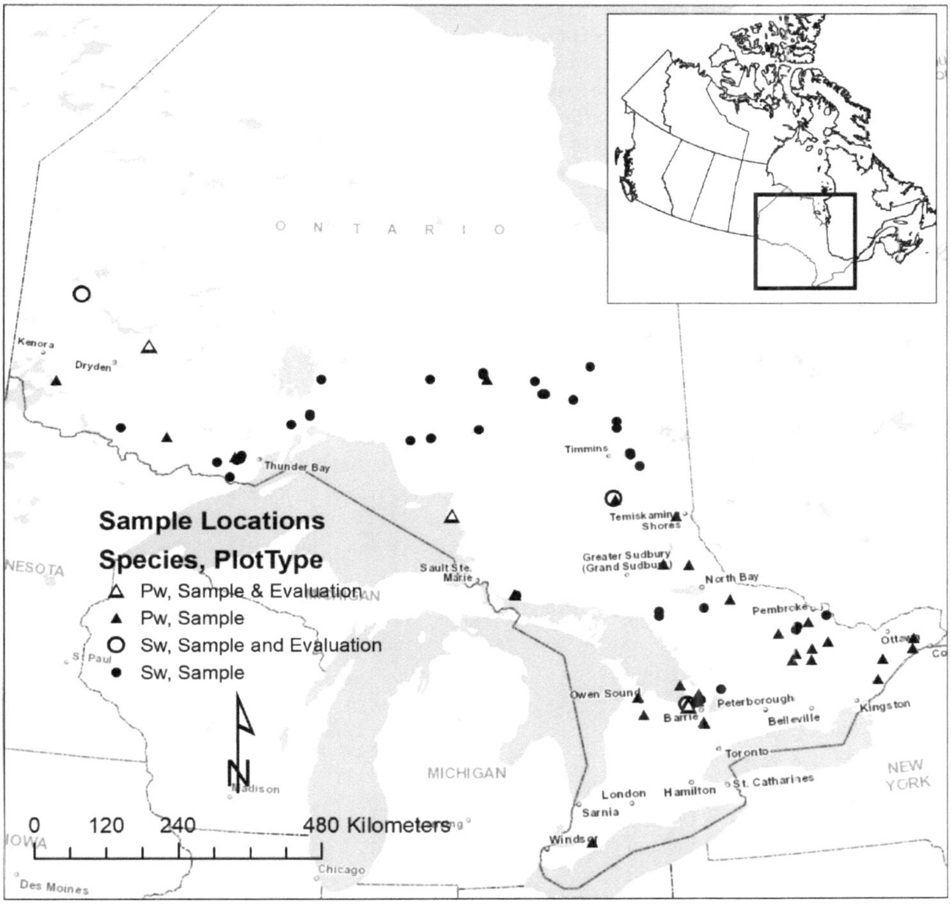

Figure 1. Distribution of white spruce (Sw) and white pine (Pw) plantation sites sampled across Ontario, Canada.

All live trees were measured in each sample plot, and stem density (trees ha^{-1}) and total basal area (BA ha^{-1}) were calculated by consecutively numbering all trees growing in the plot for each species. Total basal area of each tree species from each plot was divided into 5 BA classes. One tree was randomly selected from each BA class for each species at each plot and destructively sampled. Five trees without visible deformities (e.g., broken tops, forked, dead, injuries) were sampled from each plot (one tree from each BA class) for each species. This sampling resulted in 200 trees for each species.

Each sample tree was cut at 0.15, 0.5, and 0.9 m heights below breast height and one disk was sampled at 1.3 m (breast height). The rest of the height (tree height from breast height to tip) was divided by 10, with sections cut at the resulting interval. This approach resulted in 13 sections per tree for each species. The largest outside and inside bark diameters and those perpendicular to them passing through the pith were measured at each stem height where sections were cut. Mean inside and outside bark diameters were obtained by averaging these diameters at that stem height. Since this study involved diameter at breast height (DBH) growth, only the disks cut and measurements taken at breast height from each tree of each species were used for analysis. The sampling and other data collection protocols used in [3] were applied in this study. For details about the procedures used to measure mean annual radial growth by stem analysis and calculate mean annual diameter growth for each tree, refer to [3].

Although all trees of a particular tree species at a specific site were planted during the same year, erratic early height growth means not all of them reached breast height the same year. However, climate variable values are tied to a particular calendar year. Therefore, diameter growth of 5 trees from a site could not be combined to obtain a site-scale growth series to analyze climate effects on diameter growth. Hence, climate effects on diameter growth of a tree in a particular calendar year were analyzed using annual/seasonal values of climate variables from the same calendar year that the tree reached breast height. This analysis resulted in 200 diameter–age growth series for white pine plantations. For white spruce, one tree was missing from the final data set, resulting in 199 trees in the growth series.

The growth period used to analyze climate effects included the time between when the sample tree reached breast height and when it was sampled, ending in 2021 for white spruce and 2022 for white pine trees. Summary statistics of trees and stand characteristics used in this study are listed in Table 1.

Table 1. Summary statistics for measured tree and stand characteristics of plantation-grown white spruce and white pine trees and climate variables from across Ontario used in this study. DBH = diameter at breast height; BA = basal area ha^{-1}; trees ha^{-1} = density; TPGS = growing season total precipitation; MTGS = growing season mean temperature; CMI = climatic moisture index; Std Dev = standard deviation.

Attribute	N	Mean	Std Dev	Minimum	Maximum
	White spruce				
DBH (outside bark) (cm)	199	24.83	7.03	10.10	48.80
DBH (inside bark) (cm)	199	23.20	6.82	9.48	45.53
Total height (m)	199	19.59	3.08	12.30	26.75
Breast height age (yr)	199	48.22	7.71	28.00	69.00
BA (m^2 ha^{-1})	40	41.65	11.16	22.76	81.50
Trees ha^{-1}	40	1134.14	451.22	533.33	2625.00
	White pine				
DBH (outside bark) (cm)	200	27.78	8.84	11.50	55.10
DBH (inside bark) (cm)	200	25.42	8.01	10.82	49.23
Total height (m)	200	21.09	4.59	8.60	34.90
Breast height age (yr)	200	51.31	15.67	21.00	88.00
BA (m^2 ha^{-1})	40	44.00	12.15	23.09	78.84
Trees ha^{-1}	40	975.08	451.72	366.67	2425.00
	Climate variables				
TPGS (mm)	9680	459.43	95.86	108.20	960.40
MTGS (°C)	9680	13.41	1.04	9.65	17.20
CMI for June (cm)	9680	−1.29	3.58	−10.57	13.87
Sum of growing months (April to August) CMIs (cm)	9680	−1.78	8.26	−27.35	29.94
Annual CMI (sum of 12 months CMIs) (cm)	9680	30.61	13.69	−15.47	94.72

2.2. Climate Data

Climate variable values for each sample site were estimated using Canadian climate models [17]. Estimates of annual/seasonal values of these variables were calculated for each year, starting when the sample tree reached breast height and ending in 2022. In this study, 68 climate-related variables were used including mean, minimum, and maximum values of total precipitation and air temperatures estimated for each month and quarter of the year and annually (see [3] for a detailed description of climate variable calculations).

In addition, climatic moisture index (CMI) was computed by subtracting potential evapotranspiration (PET) from mean monthly precipitation (MMP) for each month of each year (see [18]). Climate variable values were taken from [16]. Summary statistics of climate variables used in this study (climate variables that significantly explained diameter growth of trees in white spruce and white pine plantations) are shown in Table 1.

2.3. Diameter Growth Models

In general, basal area/diameter growth models are developed using composite models [19–22]. These models include individual tree size, competition effects (tree vigor), and a measure of site productivity [14,23]. Recently, growth functions commonly used to model stand height growth have also been used to model diameter growth.

Sharma et al. [24] used a variant (algebraic-difference type) of the Chapman–Richards function to model diameter growth of Norway spruce (*Picea abies* (Linnaeus) H. Karsten). Similarly, Sharma [3] used the McDill and Amateis growth model (see [25]) to describe diameter growth of red pine in plantations across Ontario, Canada. Comparing this model to the variant of the Chapman–Richards function used by Sharma et al. [24] for red pine plantation diameter growth data, he found the McDill and Amateis model had better fit statistics and predictive ability. The mathematical form of the McDill and Amateis model was:

$$D = \frac{\beta_0}{1 - \left(1 - \frac{\beta_0}{D_0}\right)\left(\frac{A_0}{A}\right)^{\beta_1}} + \varepsilon \quad (1)$$

where D and D_0 are diameters at ages A and A_0, respectively, β_0 and β_1 are the parameters to be estimated, and ε is the model error term. In this model, β_0 and β_1 determine the asymptote and shape, respectively, of the curve.

Equation (1) can be easily modified to analyze climate effects on diameter growth by expressing its parameters in terms of climate variables [3]. Therefore, Equation (1) was used as the base function to describe diameter growth of trees in white spruce and white pine plantations. Climate effects on diameter growth of these tree species were analyzed by expressing the asymptote and the shape parameters as a function of climate variables. Three site-related variables (longitude, latitude, and elevation) were added to climate variables to account for site effects on tree diameter growth in white spruce and white pine plantations.

2.4. Methods

Diameter growth data of white spruce and white pine trees were obtained by measuring annual ring widths along the radius of the disks sampled at breast height from each tree at each plot (site). These data have hierarchical structures (i.e., rings within trees, trees within plots/sites). Among trees, diameter measurements can be considered independent, but within a tree they are correlated. Within-tree correlation (autocorrelation) can be addressed by using a mixed-effects modeling technique [26]. Therefore, in this study a diameter growth model was fit using a mixed-effects modeling technique. Random effects at stand and tree scale were added to both asymptote and rate parameters.

Climate- and site-related variables were incorporated into diameter growth models by dividing them into 3 categories: temperature, precipitation, and site-specific. First, all temperature-related variables were introduced one at a time and fitted using the NLINMIXED procedure in SAS. The variable that significantly explained the variation in diameter growth (alpha = 0.05) and produced the lowest Akaike information criterion (AIC) value was selected as the first climate (temperature) variable to be included in the model.

Precipitation-related variables were then introduced one by one in the presence of the first temperature-related variable. The one that was significant in the regression and resulted in the lowest AIC value was selected to be included in the model. All other climate- and site-related variables were introduced individually in the presence of the first 2 variables. Similarly, derived variables obtained by making quadratic and/or exponential transformations of climate- (temperature and precipitation) and site-related variables were introduced in the model. Variables that significantly explained the variation in diameter growth and reduced the AIC value were added to the model.

Random effects were added to fixed-effects parameters at site and tree scales as required. Estimated values of residuals (observed − predicted) from the diameter growth model were calculated for all 1-year growth periods for each diameter growth series. Heteroscedasticity in the data was checked by plotting these residuals against predicted diameter growth.

Climate effects on future diameter growth were evaluated by randomly selecting 3 sites from eastern, central, and western (1 site from each area) Ontario (Figure 1). Inside bark diameters of white spruce and white pine trees were estimated using the model with projected values of climate variables for each area for each species under 3 climate change scenarios. These scenarios include emissions trajectories with 2.6, 4.5, and 8.5 watts m^{-2} of warming projected for the end of the century [27]. These scenarios were chosen as these represent the mildest, an intermediate, and an extreme case of projected climate change scenarios. These trajectories are known as representative concentration pathways (RCPs). Since projected values of climate variables were available for 80 years beginning in 2021, diameters were estimated for an 80-year growth period. These diameters were plotted against breast height age (BHA).

3. Results

To check whether Equation (1) is appropriate to model diameter growth of trees in white spruce and white pine plantations, this equation was fit to diameter growth data collected from these species (Table 2). Nonlinear regression in SAS was used to fit the equation for both species. Annual diameter growth was determined with the parameters estimated using regression. Initial values used to estimate the diameters were average diameter values at age 1 (0.52 and 0.72 cm, for white spruce and white pine, respectively).

Table 2. Parameter estimates and fit statistics (σ^2 = mean squared error and AIC = Akaike's information criterion) for the base model (Equation (1)) fitted to diameter growth data collected from trees in white spruce and white pine plantations across Ontario. NA = not applicable.

Parameters	White Spruce		White Pine	
	Estimates	SE	Estimates	SE
β_0	43.9032	0.36930	51.8574	0.64230
β_1	1.2177	0.00581	1.0898	0.00613
σ^2	0.18866	0.00054	0.05236	0.00075
AIC	−4309	NA	−1073	NA

Estimated diameters were plotted against breast height age (BHA) for both species (Figure 2). Observed diameters across BHA were also overlaid in the plot. Diameter growth profiles generated using Equation (1) closely followed the trend of observed values for both species. These results confirmed that Equation (1) was appropriate to model diameter growth of both white spruce and white pine grown in plantations.

Equation (1) was modified to include climate- and site-related variables. Parameters of Equation (1) were expressed in terms of climate- and site-related variables. The equation was fit to diameter growth data by expressing each parameter in terms of temperature-related variables individually for both species. Although many temperature-related variables were significant in the regression, the rate parameter (β_1) expressed in terms of the mean growing season temperature (MTGS) resulted in the best fit (lowest AIC value) for both tree species.

(a)

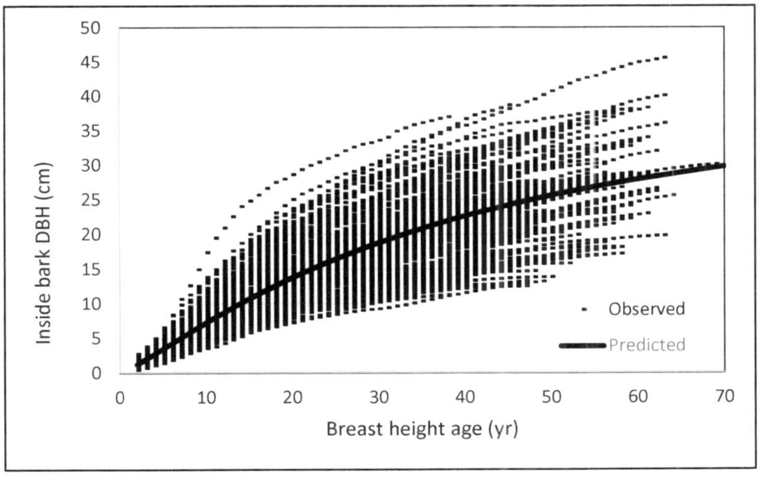

(b)

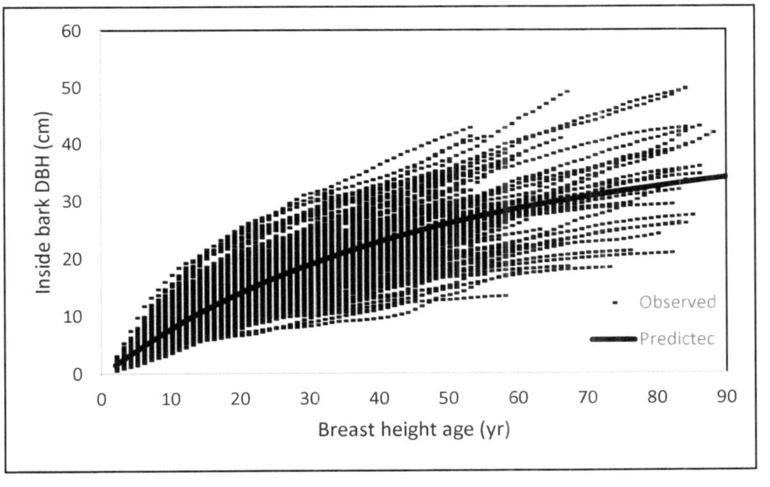

Figure 2. Inside bark diameters at breast height (DBH) of all trees across breast height age (observed) and diameter profiles (predicted) generated using Equation (1) for plantation-grown (**a**) white spruce and (**b**) white pine in Ontario fitted without climate variables.

Precipitation-related variables were then introduced in the model in the presence of MTGS. The asymptote (β_0) expressed in terms of annual climatic moisture index (ACMI), defined as the sum of the monthly moisture indices, and GMCMI, defined as the sum of April to August (growing months) moisture indices, explained the variations in diameter growth more than other variables from this category for white spruce and white pine, respectively. Other temperature and precipitation variables were then introduced in the presence of MTGS and ACMI and GMCMI.

In the presence of MTGS and the sum of climatic moisture indices, total growing season precipitation (TPGS) and June climatic moisture index (JCMI) were also significant in expressing the rate parameters for both species. No site-related variables (elevation, latitude, longitude) were significant for both species. For transformations and interactions, only the quadratic transformation of JCMI was significant for white pine. The AIC value decreased significantly for both species when climate variables were introduced.

The diameter growth models that include climate variables can be written as:

White spruce

$$D_{ij} = \frac{\beta_0 + \beta_2 ACMI}{1 - \left(1 - \frac{\beta_0+\beta_2 ACMI}{D_{ik(k\neq j)}}\right)\left(\frac{A_0}{A}\right)^{\beta_1+\beta_3 MTGS+\beta_4 TPGS+\beta_5 JCMI}} + \varepsilon_{ij} \quad (2)$$

White pine

$$D_{ij} = \frac{\beta_0 + \beta_2 GMCMI}{1 - \left(1 - \frac{\beta_0+\beta_2 GMCMI}{D_{ik(k\neq j)}}\right)\left(\frac{A_0}{A}\right)^{\beta_1+\beta_3 MTGS+\beta_4 TPGS+\beta_5 JCMI+\beta_6 JCMI^2}} + \varepsilon_{ij} \quad (3)$$

where D_{ij} is the diameter at breast height of ring j and tree i, β_2 to β_6 are fixed-effects parameters associated with climate-related variables, ε_{ij} is the error term associated with the jth ring of tree i, and other variables/parameters are as defined earlier. Random effects were then introduced to fixed-effects parameters. Only random effects associated with intercepts in the expressions for asymptote and the rate parameter were significant at tree level. However, introduction of these random effects resulted in very unstable and inconsistent parameter estimates for both species.

For example, estimates for β_3 and β_4 were both negative for the model without random effects but turned positive when the random effects were introduced for white spruce. Additionally, the estimated standard error for β_3 was zero for the model with random effects for this tree species. Moreover, diameters estimated using the parameters with random effects were negative for some trees. Results for white spruce were also inconsistent. Therefore, random effects were not included in the final model for either species. Table 1 includes the summary statistics for climate variables (TPGS, MTGS, JCMI, GMCMI, and ACMI) that were significant in the regression.

Table 3 displays the estimated values of parameters for Equations (2) and (3). The asymptote was expressed as a linear function of ACMI and GMCMI for white spruce and white pine, respectively. Similarly, the rate parameter was a linear function of MTGS, TPGS and JCMI for white spruce, and of MTGS and TPGS and a quadratic function of JCMI for white pine. The asymptote for the white spruce is negatively correlated with ACMI. On the other hand, it is positively correlated with GMCMI for white pine.

Table 3. Parameter estimates and fit statistics (σ^2 = mean squared error and AIC = Akaike's information criterion) for diameter growth models incorporating climate variables (Equations (2) and (3)) fitted to diameter growth data collected from white spruce and white pine trees in plantations from across Ontario. NA = not applicable.

Parameters	White Spruce		White Pine	
	Estimates	SE	Estimates	SE
β_0	49.7726	0.93160	52.9072	0.7130
β_1	1.5247	0.06365	1.3601	0.06869
β_2	−0.1585	0.02286	0.3975	0.06065
β_3	−0.03690	0.004823	−0.01308	0.00482
β_4	0.000381	0.000054	−0.00015	0.00004
β_5	0.005878	0.001331	0.00687	0.00133
β_6	NA	NA	−0.00071	0.00025
σ^2	0.03596	0.000531	0.05161	0.00074
AIC	−4464	NA	−1203	NA

The rate parameter was negatively associated with MTGS for both species. However the association of the rate parameter with TPGS was positive for white spruce and negativ for white pine. JCMI was positively correlated with the rate parameter for both species. Its quadratic term ($JCMI^2$) was only significant for white pine and its association with the rate parameter was negative. The magnitude of the coefficient for the linear term of JCMI for white pine was almost 10 times larger than the one for its quadratic term. This result implies that the positive effect of JCMI on the rate parameter diminishes as the value of JCMI increases.

Residual plots were made by predicting inside bark diameters of all trees using Equations (2) and (3) for white spruce and white pine, respectively, at all breast height ages (Figure 3). These plots indicated that heteroscedasticity was not a concern in fitting the models. However, diameters larger than 35 cm seemed slightly underestimated for both species. Residual plots were also made against tree breast height age for both species. All residuals were clustered around the zero line and there was no bias at any point across the breast height age for both species.

(a)

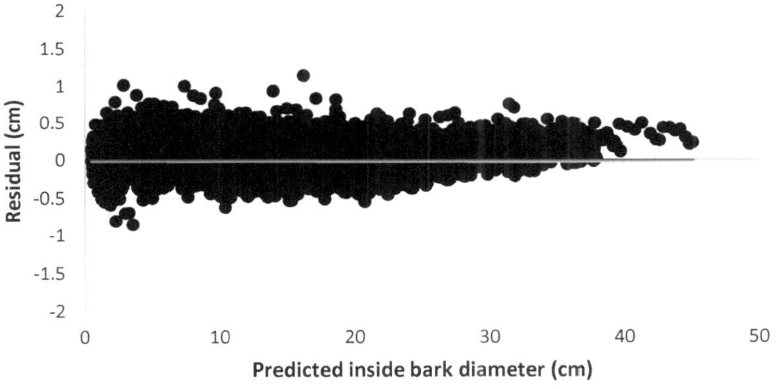

(b)

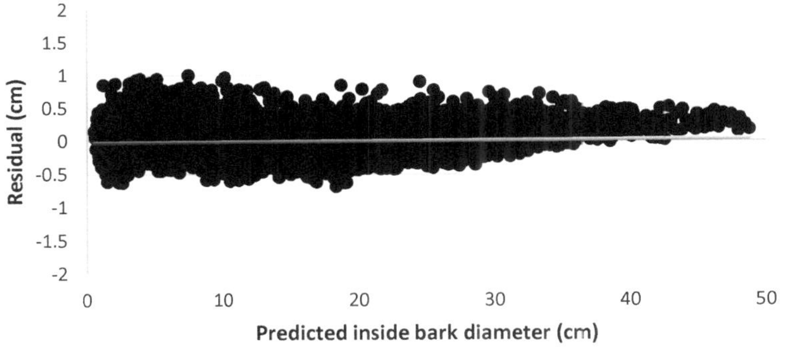

Figure 3. Residuals (observed—predicted) of inside bark diameter at breast height (DBH) estimated using Equation (2) for plantation-grown white spruce (**a**) and Equation (3) for white pine (**b**) in Ontario plotted against predicted inside bark diameters.

Equations (2) and (3) were further analyzed by computing bias across breast height age and diameter classes for both species. Predicted diameters were subtracted from their observed counterparts to calculate residuals. Bias was obtained for each age and diameter class by averaging residuals in those classes. Standard deviation of the bias was also obtained for each age and diameter class. These biases and standard deviations are provided in Tables 4 and 5 for white spruce and white pine, respectively. The maximum bias was 0.1316 cm for diameters larger than 45 cm for white pine, with a standard deviation of 0.2302 cm. Most of the biases were less than 0.09 cm for both species.

Table 4. Bias (observed—predicted) and its standard deviation (Std Dev), minimum, and maximum of the residuals for diameter at breast height (DBH) class and age class that resulted from fitting Equation (2) for white spruce trees in Ontario, Canada. (N = number of sample trees).

DBH (cm)	N	Bias	Std Dev	Min	Max
<15	626	0.1070	0.1282	−0.4102	0.3816
15–20	1661	−0.0880	0.1521	−0.5379	0.6449
20–25	2390	−0.0473	0.1626	−0.7024	0.7316
25–30	2374	0.0056	0.1835	−0.8580	1.0140
30–35	1137	−0.0473	0.1941	−0.5168	0.9998
40–45	721	0.1141	0.2197	−0.5915	1.1369
>45	280	0.1235	0.1875	−0.5004	0.8356
Age (Years)					
<10	1791	0.0713	0.2197	−0.8580	1.1369
10–20	1990	0.0131	0.2016	−0.5495	0.8356
20–30	1987	−0.0849	0.1601	−0.4884	0.6388
30–40	1883	−0.0549	0.1570	−0.3814	0.6272
40–50	1123	−0.0083	0.1537	−0.2738	0.7589
50–60	388	0.0332	0.1546	−0.2242	0.7089
>60	27	0.0257	0.0883	−0.1506	0.2247

Table 5. Bias (observed—predicted) and its standard deviation (Std Dev), minimum, and maximum of the residuals for diameter at breast height (DBH) class and age class that resulted from fitting Equation (3) for white pine trees in Ontario, Canada. (N = number of sample trees).

DBH (cm)	N	Bias	Std Dev	Min	Max
<15	474	−0.1274	0.1826	−0.6106	0.7236
15–20	1049	−0.0660	0.1982	−0.5861	1.0820
20–25	1918	−0.0558	0.2097	−0.8363	1.2187
25–30	1878	0.0368	0.2285	−0.7897	1.0537
30–35	1822	−0.0141	0.2151	−0.7197	1.0113
40–45	1266	0.0559	0.2213	−0.8960	1.4841
>45	1254	0.1316	0.2302	−0.5750	1.1761
Age (years)					
<10	1800	0.1128	0.2892	−0.8960	1.4841
10–20	1999	−0.0711	0.2131	−0.7197	0.8433
20–30	1900	−0.0953	0.1755	−0.4930	0.8103
30–40	1630	0.0548	0.1708	−0.4117	0.9912
40–50	1240	0.0394	0.1823	−0.2799	0.6525
50–60	532	−0.1264	0.2180	−0.2043	1.1761
60–70	291	0.0894	0.1684	−0.1952	0.6819
70–80	229	0.0814	0.1578	−0.1612	0.5978
>80	40	0.1126	0.1686	−0.1161	0.4775

Climate effects on future diameter growth were evaluated by estimating inside bark diameters at breast height using Equations (2) and (3) for white spruce and white pine respectively. Estimates were made for three sites in Ontario under three emissions scenarios (RCPs 2.6, 4.5, and 8.6 watts m^{-2} trajectories). Average diameters of sample trees from those sites measured at BHA 1 year were used as initial values for these estimations. These diameters were plotted against BHA for each climate change scenario for each site for each species (Figures 4 and 5).

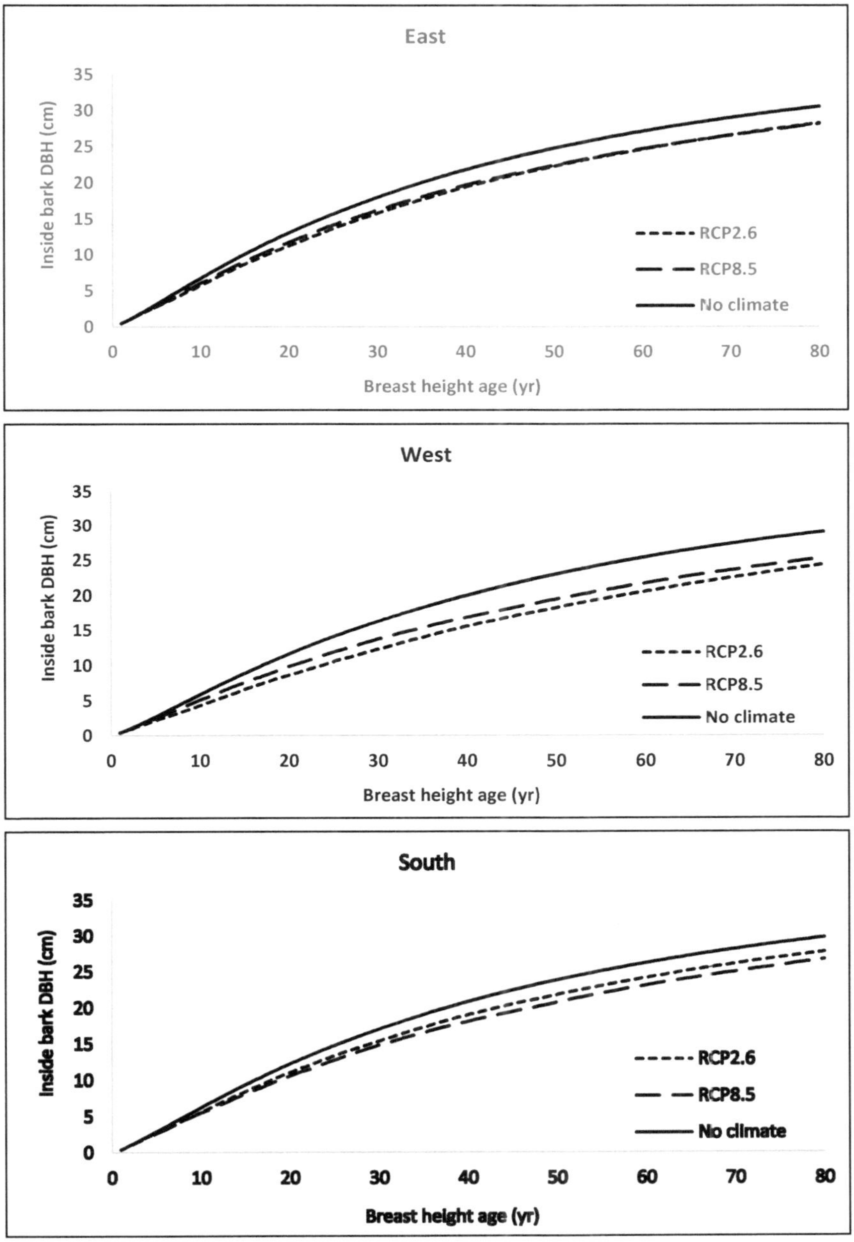

Figure 4. Diameter growth profiles for plantation-grown white spruce trees generated using the models without climate variables (Equation (1)) (no climate) and with climate variables (Equation (2)) under representative concentration pathway (RCP) 2.6 and 8.5 for eastern, western, and southern parts of Ontario, Canada.

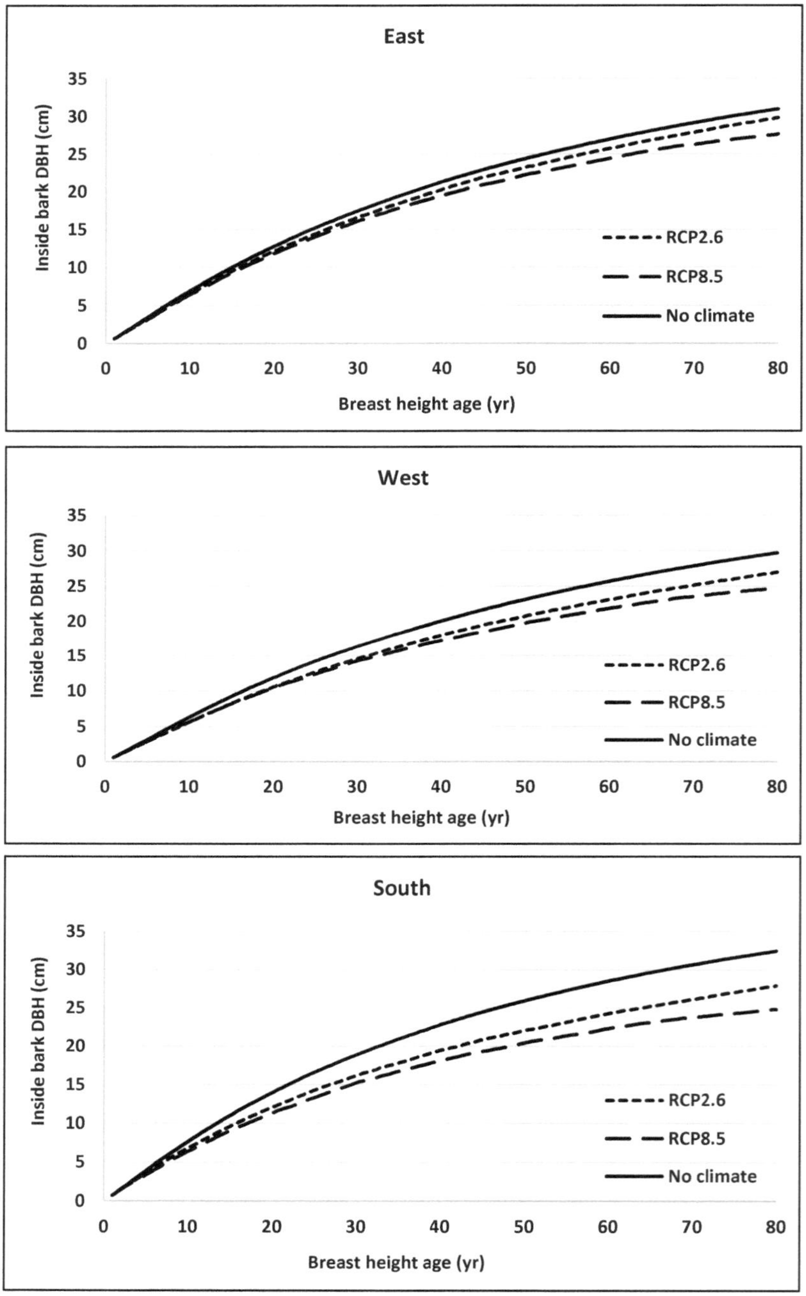

Figure 5. Diameter growth profiles for plantation-grown white pine trees generated using the models without climate variables (Equation (1)) (no climate) and with climate variables (Equation (3)) under representative concentration pathway (RCP) 2.6 and 8.5 for eastern, western, and southern parts of Ontario, Canada.

Projected diameter growths under RCP 2.6 were very similar to those under RCP 4.5 for both tree species. Therefore, RCP 4.5 growth profiles were not included in Figures 4 and 5. Annual and seasonal values of all climatic variables were used in Equations (2) and (3) to estimate future diameter growth for white spruce and white pine, respectively. Diameter growth was projected from 2021 through 2100 (80-year growth period) for all climate change scenarios for both species. Diameters were also estimated using Equation (1) without climate variables.

The climate effects were negative and varied with species and location. The maximum difference in projected diameters at age 80 for white spruce was in the west. Projected diameters under RCPs 2.6 and 8.5 were thinner by 4.64 (15.99%) and 3.72 (12.80%) cm, respectively, compared to those under no climate change scenario. In the east, diameter differences under RCP 2.6 and 8.5 scenarios were minimal. However, the difference in diameters with and without climate change was significant. Under climate change scenarios, diameters at age 80 were thinner than those under no climate change by 2.30 cm (8.00%). In the south, diameters at age 80 under RCP 2.6 and 8.5 were thinner than those under no climate scenario by 2.02 (6.79%) and 3.07 (10.33%) cm, respectively.

Climate effects on diameter growth were more consistent for white pine than for white spruce. These effects were more pronounced under RCP 8.5 than under RCP 2.6 for all three locations for this tree species. The maximum difference in projected diameters was in the south and the minimum difference was in the east. At age 80, projected diameters under RCPs 2.6 and 8.5 in the south were narrower than those under no climate change scenario by 4.54 (13.99%) and 7.60 (23.43%) cm, respectively. In the east, these diameters under RCPs 2.6 and 8.5 were thinner than those under the no climate change scenario by 1.17 (3.79%) and 3.36 (10.85%) cm, respectively. In the west, however, projected diameters under RCPs 2.6 and 8.5 were narrower than those under the no climate change scenario by 2.76 (9.28%) and 4.96 (16.71%) cm, respectively. Summary statistics of projected climate variables under three climate change scenarios for the 80-year (2021–2100) growth period used in this study are presented in Table 6.

Table 6. Summary statistics of projected climate variables significant in explaining the variation of diameter growth of white pine and white spruce plantations of sample sites used in this study for the 80-year (2021–2100) growth period. TPGS = growing season total precipitation; MTGS = growing season mean temperature; CMI = climatic moisture index; Std Dev = standard deviation.

Attribute	N	Mean	Std Dev	Minimum	Maximum
	RCP 2.6				
TPGS (mm)	6400	570.85	147.86	225.90	1257.30
MTGS (°C)	6400	14.97	1.10	11.89	18.84
CMI for June (cm)	6400	−3.27	4.25	−12.69	15.59
Sum of growing months (April to August) CMIs (cm)	6400	0.60	7.68	−23.49	28.94
Annual CMI (sum of 12 months CMIs) (cm)	6400	33.01	18.33	−24.68	111.98
	RCP 4.5				
TPGS (mm)	6400	569.71	128.33	234.00	1038.10
MTGS (°C)	6400	15.36	1.11	11.90	19.26
CMI for June (cm)	6400	−4.72	4.45	−14.97	13.36
Sum of growing months (April to August) CMIs (cm)	6400	−1.93	7.09	−22.32	31.17
Annual CMI (sum of 12 months CMIs) (cm)	6400	28.84	17.48	−31.15	103.06
	RCP 8.5				
TPGS (mm)	6400	587.72	135.52	242.70	1225.80
MTGS (°C)	6400	16.64	1.74	12.21	22.66
CMI for June (cm)	6400	−6.13	5.00	−20.35	13.91
Sum of growing months (April to August) CMIs (cm)	6400	−3.23	9.84	−38.15	30.26
Annual CMI (sum of 12 months CMIs) (cm)	6400	22.46	19.88	−52.11	92.22

4. Discussion

Climate has both positive and negative effects on tree growth. The nature (positive/negative) and magnitude of the effects and the climate variables affecting growth depend on tree species and location. Even for a given species, the nature and magnitude of the effects can vary from one location to another [28]. Goldblum and Rigg [29] found that temperature and precipitation had positive effects on the growth of white spruce and sugar maple (*Acer saccharum* Marsh.), but temperature had no significant effects on diameter growth of balsam fir grown in the boreal forest near the coast of Lake Superior in Ontario, Canada. Similarly, Pokharel and Froese [13] reported that including average annual temperature in modeling basal area growth of natural stand-grown trembling aspen, balsam fir, jack pine, and black spruce trees improved the fit statistics and predictive accuracy.

Subedi and Sharma [12] examined climate effects on diameter growth of jack pine and black spruce trees grown in plantations in Ontario. Overall, they found effects of climate were positive for jack pine but negative for black spruce. The climate variables that explained the variation in diameter growth were total precipitation of growing season, precipitation of wettest quarter, and the mean temperature of the growing season for both species. Negative climate effects on black spruce were more evident than the positive climate effects on jack pine trees. They used average values of climate variables over a 30-year growth period (1971–2000) to examine climate effects.

Sharma [3] used seasonal/annual values of climate variables to examine climate effects on diameter growth of red pine in plantations. He reported that diameter growth was affected by climate and varied by location. The overall effect was positive and pronounced more in southern than in eastern Ontario. It was neutral in the far west. However, the effects were negative and less pronounced in the central west. The climate variables that significantly explained variation in the diameter growth of red pine trees in plantations were the total growing season precipitation (TPGS) and the range of mean diurnal temperature (MDTR). Diurnal temperature range is the difference between the maximum and minimum temperatures on the same day.

In this study, the overall climate effects on diameter growth of white spruce and white pine trees in plantations were negative. The magnitude of negative effects depended on tree species and location. The effects were more pronounced for white pine than for white spruce trees at all locations. For white spruce, the asymptote decreased as the sum of monthly CMIs for the whole year (January to December) (ACMI) increased. For white pine, the sum of monthly CMIs of growing months (April-August) (GMCMI) explained more variation in the asymptote than ACMI and the asymptote increased as the value of GMCMI increased.

The rate parameter increased as growing season total precipitation and JCMI increased but decreased as mean temperature of growing season increased. On the other hand, the rate parameter increased as JCMI increased to a certain limit (the coefficient of the quadratic term of JCMI was negative) and decreased if the values of both total precipitation and mean temperature of the growing season were elevated.

In other studies, Matisons et al. [8] and Adhikari et al. [11] found June weather corelated with radial growth of European beach and post oak (*Quercus stellata* Wangenh.) grown in Europe and Oklahoma, United States, respectively. Similarly, Oboite and Comeau [14] reported positive effects of CMI on diameter growth of lodgepole pine, jack pine, trembling aspen, white spruce, and balsam poplar grown in western Canada. They also reported that a longer frost-free period (FFP) had positive effects on the diameter growth of balsam poplar and trembling aspen trees. However, the FFP had negative effects on the diameter growth of lodgepole and jack pine trees. On the other hand, Bayat et al. [15] found that the impact of climate change on diameter growth of several tree species in the Hyrcanian Forest was not very pronounced. They reported that the diameter growth under climate change decreased by 7% at the end of 2070 as compared to the beginning of the growth period.

Climate effects varied among species. Even for the same species, effects varied by location even within a province (Ontario). The projected values of diameter growth under different climate change scenarios showed that climate had positive effects on diameter growth of jack pine [12] but negative effects on that of black spruce [12], white spruce, and white pine (this study). Climate had both positive and negative effects on red pine. It was positive in the east and west but negative in the south. Effects on black spruce were most evident, followed by those on red pine, white pine, and white spruce. Among red pine, white spruce, and white pine, white spruce growth was least affected. Effects on red pine and white pine growth were similar, with red pine affected slightly less.

Competitive interactions can modify the growth responses to climate change [28]. Similarly, the response may vary from its northern to southern boundary [3]. It was not possible to cover the native range and stand density in sampling white pine and white spruce trees. Additionally, Sharma [30] reported that the climate effects on height growth of black spruce in mixed stands depended on the tree species that it grew with. Therefore, caution should be applied in utilizing the diameter growth models presented here to determine the climate effects on growth of all white pine and white pine populations.

5. Conclusions

Climate-sensitive diameter growth models were developed for white pine and white spruce plantations. The McDill–Amateis growth function was used as the base function to model diameter growth of these tree species. To make the model climate sensitive, the asymptote and rate parameter of the function were expressed in terms of climate variables. The climate variable that explained the variation in the asymptote of white spruce trees was the sum of monthly climatic moisture index (CMI) for the whole year. For white pine, variation in the asymptote was explained by the sum of monthly CMI of the growing months.

For the rate parameter, mean temperature and total precipitation of the growing season and June CMI significantly explained the variation for both species. Climate effects were evaluated at three geographic locations across Ontario (east, west, and south) for each species under three climate change (emissions) scenarios. The overall effects of climate were negative, with magnitude depending on tree species and growing location. The negative effect of climate on diameter growth was more pronounced for white pine than for white spruce.

Inside bark diameters at breast height of white spruce and white pine plantations can be estimated using the models presented here. Projected seasonal/annual values of climate variables under the most accurate emissions (climate change) scenario are needed for accurate estimations. Forest management plans developed using climate-sensitive models should provide forest managers with more accurate information about forest growth than traditional models. If the values of climate variables are unavailable, the models fitted without climate variables can be used to estimate the diameters of the tree species used in the study.

Funding: Forest Future Trust Ontario Knowledge Transfer and Tool Development (FFT-KTTD) Program (Project # 14B-2021).

Data Availability Statement: Summary statistics of the data used in this study are presented in Table 1.

Acknowledgments: The author is grateful to Lisa Buse (Ontario Forest Research Institute) for editing an earlier version of manuscript and Chris Stratton and Todd Little, Biodiversity and Monitoring Section, MNRF, for coordinating data collection work and producing sample location map, respectively.

Conflicts of Interest: The author declares no conflict of interest.

References

1. Vanclay, K. *Modelling Forest Growth and Yield: Application to Mixed Tropical Forests*; CAB International: Wallingford, UK, 1994.
2. Avery, T.E.; Burkhart, H.E. *Forest Measurements*; McGraw-Hill: New York, NY, USA, 2002; 456p.

3. Sharma, M. Modelling climate effects on diameter growth of red pine trees in boreal Ontario, Canada. *Trees For. People* **2021**, *4*, 100064. [CrossRef]
4. Chhin, S.; Hogg, E.H.; Lieffers, V.J.; Huang, S. Potential effects of climate change on the growth of lodgepole pine across diameter size classes and ecological regions. *For. Ecol. Manag.* **2008**, *256*, 1692–1703. [CrossRef]
5. Maxime, C.; Hendrik, D. Effects of climate on diameter growth of co-occurring Fagus sylvatica and Abies alba along an altitudinal gradient. *Trees* **2011**, *25*, 265–276. [CrossRef]
6. Toledo, M.; Poorter, L.; Pena-Clara, M.; Balkazar, A.A.J.; Leano, C.; Licona, J.C.; Llanque, O.; Zuidema, V.V.P.; Bongers, F. Climate is a stronger driver of tree and forest growth rates than soil and disturbance. *J. Ecol.* **2011**, *99*, 254–264. [CrossRef]
7. Hökkä, H.; Salminen, H.; Ahti, E. Effect of temperature and precipitation on the annual diameter growth of Scots pine on drained peatlands and adjacent mineral soil sites in Finland. *Dendrochronologia* **2012**, *30*, 157–165. [CrossRef]
8. Matisons, R.; Puriņa, L.; Adamovičs, A.; Robalte, L.; Jansons, Ā. European beech in its northeasternmost stands in Europe: Varying climategrowth relationships among generations and diameter classes. *Dendrochronologia* **2017**, *45*, 123–131. [CrossRef]
9. Rubio-Cuadrado, Á.; Bravo-Oviedo, A.; Mutke, S.; Del Río, M. Climate effects on growth differ according to height and diameter along the stem in Pinus pinaster Ait. *iForest* **2018**, *11*, 237–242. [CrossRef]
10. Jiao, L.; Xue, R.; Qi, C.; Chen, K.; Liu, X. Comparison of the responses of radial growth to climate change for two dominant coniferous tree species in the eastern Qilian Mountains, northwestern China. *Int. J. Biometeorol.* **2021**, *65*, 1823–1836. [CrossRef]
11. Adhikari, A.; Masters, R.E.; Adams, H.D.; Will, R.E. Radial growth responses of post oak (Quercus stellata) to climate variability and management in southeastern Oklahoma, USA. *Can. J. For. Res.* **2022**, *52*, 209–219. [CrossRef]
12. Subedi, N.; Sharma, M. Climate-diameter growth relationships of black spruce and jack pine trees in boreal Ontario, Canada. *Glob. Chang. Biol.* **2013**, *19*, 505–516. [CrossRef]
13. Pokharel, B.; Froese, R.B. Representing site productivity in the basal area increment model for FVS-Ontario. *For. Ecol. Manag.* **2009**, *258*, 657–666. [CrossRef]
14. Oboite, O.F.; Comeau, P.G. Climate sensitive growth models for predicting diameter growth of western Canadian boreal tree species. For. *Int. J. For. Res.* **2021**, *94*, 363–373. [CrossRef]
15. Bayat, M.; Knoke, T.; Heidari, S.; Hamidi, S.K.; Burkhart, H.; Jaafari, A. Modeling Tree Growth Responses to Climate Change: A Case Study in Natural Deciduous Mountain Forests. *Forests* **2022**, *13*, 1816. [CrossRef]
16. Yeh, H.Y.; Wensel, L.C. The relationship between tree diameter growth and climate for coniferous species in northern California. *Can. J. For. Res.* **2000**, *30*, 1463–1471. [CrossRef]
17. McKenney, D.W.; Hutchinson, M.F.; Papadopol, P. Customized spatial climate models for North America. *Bull. Am. Meteorol. Soc.* **2011**, *92*, 1161–1622. [CrossRef]
18. Hogg, E.H. Climate and the southern limit of the western Canadian boreal forest. *Can. J. For. Res.* **1994**, *24*, 1835–1845. [CrossRef]
19. Wykoff, W.R. A basal area increment model for individual conifers in the northern Rocky Mountains. *For. Sci.* **1990**, *36*, 1077–1104.
20. Murphy, P.A.; Shelton, M.G. An individual-tree basal area growth model for loblolly pine stands. *Can. J. For. Res.* **1996**, *26*, 327–331. [CrossRef]
21. Weiskittel, A.R.; Garber, S.M.; Johnson, G.P.; Maguire, D.A.; Monserud, R.A. Annualized diameter and height growth equations for Pacific Northwest plantation-grown Douglas-fir, western hemlock, and red alder. *For. Ecol. Manag.* **2007**, *250*, 266–278. [CrossRef]
22. Leites, L.P.; Robinson, A.P.; Crookston, N.L. Accuracy and equivalence testing of crown ratio models and assessment of their impact on diameter growth and basal area increment predictions of two variants of the Forest Vegetation Simulator. *Can. J. For. Res.* **2009**, *39*, 655–665. [CrossRef]
23. Subedi, N.; Sharma, M. Individual-tree diameter growth model for black spruce and jack pine plantations in northern Ontario. *For. Ecol. Manag.* **2011**, *261*, 2140–2148. [CrossRef]
24. Sharma, R.P.; Vacek, Z.; Vacek, S.; Jansa, V.; Kucera, M. Modelling individual tree diameter growth for Norway spruce in the Czech Republic using a generalized algebraic difference approach. *J. For. Sci.* **2017**, *63*, 227–238. [CrossRef]
25. Burkhart, H.E.; Tome, M. *Modeling Forest Trees and Stands*; Springer: Dordrecht, The Netherlands, 2012; 457p.
26. Pinheiro, J.C.; Bates, D.M. *Mixed-Effects Models in S and S-PLUS*; Springer: New York, NY, USA, 2000; 528p.
27. Vuuren, D.P.; Edmonds, J.A.; Kainuma, M.; Riahi, K.; Weyan, J. A special issue on the RCPs. *Clim. Chang.* **2011**, *109*, 1. [CrossRef]
28. Buechling, A.; Martin, P.H.; Martin, P.H.; Canham, C.D. Climate and competition effects on tree growth in Rockey Mountain forests. *J. Ecol.* **2017**, *105*, 1636–1647. [CrossRef]
29. Goldblum, D.; Rigg, L.S. Tree growth response to climate change at the deciduous-boreal forest ecotone, Ontario, Canada. *Can. J. For. Res.* **2005**, *35*, 2709–2718. [CrossRef]
30. Sharma, M. Climate effects on black spruce and trembling aspen productivity in natural origin mixed stands. *Forests* **2022**, *13*, 43[illegible] [CrossRef]

forests

MDPI

Article

Turkey Oak (*Quercus cerris* L.) Resilience to Climate Change: Insights from Coppice Forests in Southern and Central Europe

Michaela Šimková [1], Stanislav Vacek [1], Václav Šimůnek [1,*], Zdeněk Vacek [1], Jan Cukor [1,2], Vojtěch Hájek [1], Lukáš Bílek [1], Anna Prokůpková [1], Igor Štefančík [3], Zuzana Sitková [3] and Ivan Lukáčik [4]

1 Department of Silviculture, Faculty of Forestry and Wood Sciences, Czech University of Life Sciences Prague, Kamýcká 129, CZ-165 00 Prague-Suchdol, Czech Republic

2 Forestry and Game Management Research Institute, Strnady 136, CZ-252 02 Jíloviště, Czech Republic

3 National Forest Centre, Forest Research Institute, T. G. Masaryka 2175/22, 960 01 Zvolen, Slovakia

4 Department of Silviculture, Faculty of Forestry, Technical University in Zvolen, T.G. Masaryka 24, 960 53 Zvolen, Slovakia

* Correspondence: simunekv@fld.czu.cz

Abstract: Turkey oak (*Quercus cerris* L.) is a thermophilic oak species that is gaining importance in the context of ongoing climate change because of its better resistance to climatic extremes and drier conditions. Therefore, this article focuses on Turkey oak's role and growth properties in the coppice forests of Southern Europe (Italy, Bulgaria) compared to similar site conditions in Central Europe (Slovakia, Czechia). The aims are to evaluate the basic dendrometry indicators, stand biodiversity, growth dynamics, and the effect of climatic factors on tree-ring increment on specific site chronologies. We found that the tree density in coppices of 50–60 years varied between 475 and 775 trees ha^{-1}, and the stand volume ranged from 141 to 407 m^3 ha^{-1}. The complex stand diversity of all plots ranged from a monotonous to uniform structure. The size of tree-ring growth was closely related to indicators of stand density. The lowest influence of climatic factors on tree-ring growth was found in sites in Italy and the highest in Slovakia. The primary limiting factor for growth was the lack of precipitation during the growing season, especially in June and July. In contrast, temperature had a marginal effect on radial growth compared to precipitation. The radial growth in research plots in Southern Europe goes through longer 6 to 8-year growth cycles, and in Central European sites, it goes through shorter cycles of 2.4 to 4.8 years, which confirms better growth conditions in this region. The studied coppice stands exhibit a stable reaction to climate change. Yet, regarding cyclical growth, the Central European stands benefit from an advantageous climate and grow better than in Southern Europe. As part of the changing environmental conditions, Turkey oak is becoming an important tree species that can achieve high production potential even in drier habitats due to its regeneration characteristic as coppice and may play a critical role in its northerly introduction in Europe.

Keywords: dendrochronology; cyclical dynamics; stand structure; biodiversity; timber production

Citation: Šimková, M.; Vacek, S.; Šimůnek, V.; Vacek, Z.; Cukor, J.; Hájek, V.; Bílek, L.; Prokůpková, A.; Štefančík, I.; Sitková, Z.; et al. Turkey Oak (*Quercus cerris* L.) Resilience to Climate Change: Insights from Coppice Forests in Southern and Central Europe. *Forests* **2023**, *14*, 2403. https://doi.org/10.3390/f14122403

Academic Editors: Yassine Messaoud, Jan Světlík and Giorgio Alberti

Received: 24 October 2023
Revised: 1 December 2023
Accepted: 5 December 2023
Published: 9 December 2023

1. Introduction

Forest ecosystems are experiencing ongoing climate change worldwide, and the changes are evident, especially in European forests [1–4]. Standard forest management practices appear unsuitable for providing sustainable and stable timber production and ecosystem services such as carbon sequestration [5–7], improved water dynamics in forest stands, drought mitigation [8,9], biodiversity conservation [10], and many other ecosystem functions. Therefore, forest managers are looking for alternatives to standard management practices, which have existed for decades or even centuries, aiming to provide essential benefits, such as timber production, thereby generating income for forest owners in the first place [11–13].

Coppice forests are an alternative to standard forest management practices. However, it is an almost forgotten management method in some European regions [14]. Coppicing represents the oldest form of systematic and sustainable use of these forests [15,16]. It is a very flexible system that requires low energy consumption and skilled labor and has adapted to the needs of rural societies that provide logs for fuel, charcoal, agriculture, and small businesses [17,18]. The owners or users of these forests build on local ecological knowledge to help maintain and increase the resilience of this social-ecological forest management system [19–21].

The management is based on the stump sprouting of some tree species after felling [22]. Coppice is established by clonal stems forming interconnected groups or predominantly multi-stemmed clusters of individuals that have arisen by vegetative propagation [23,24]. The rotation period of the stand is usually short (about 15–30 years) and, therefore, significant structural changes occur during coppicing [25]. Ancient coppices are a specific type of habitat that reflects long-term human influence and contains high species biodiversity [26]. They preserve local tree ecotypes and, in some locations, are the only remnants of original trees with a natural species composition, even though the stand structure has changed [27]. Coppices show enormous variability and adaptability in the tree and herb layers and in their growth processes [28–30]. This management method preserves the biodiversity of plants, including rare species of ground flora [31], and is, therefore, of considerable interest to nature conservation [32–34].

Growth of coppice begins when a single-trunk tree is felled. Then, multiple shoots start to sprout, forming a multi-stemmed tree. Repeated felling results in multi-stemmed trees Several of these multi-stemmed trees in one area create a coppice forest [35]. Compared to seed regeneration, coppice's initial growth is much faster [36] thanks to the well-developed root system of harvested trees [37]. A sprout can grow up to 1 m per year, depending on many factors, such as tree species, habitat conditions, the stump's age, or the timing of the logging operation [38].

The coppice stands are characterized by rapid changes in thermal, light, and hydrological regimes [24]. The dynamics of regenerative forests thus offer a highly heterogeneous environment within a relatively small area [39]. As a result of regular harvesting interventions, all phases of forest succession occur periodically [35,40], enabling the coexistence of species with different strategies—light-demanding and shade-tolerant [41]. In the initial stages of the coppice cycle, open areas benefit light-demanding species. As the stand density increases and the canopy closes, it limits the ground vegetation growth [35]. These stands are dense, but as competition increases, gaps are formed in later stages due to the death of some sprouts or logging operations [42,43]. In addition to ground vegetation coppice management influences lichens, fungi, beetles, saproxylic insects, [24,44], and birds [45].

Currently, coppice forests are most widespread throughout the Mediterranean, covering 23 million ha [35,46,47]. For example, in Italy, coppice forests cover 3.663 million hectares [35], and of these, both evergreen and deciduous forests of *Quercus* spp. encompass an area of approximately 1.6 million ha [17]. In Slovakia, coppices now cover 110,000 ha compared to 1950 when they covered 196,000 ha, so there is evidence of a decreasing trend of this management practice [48]. The same trend is evident in Czechia, where most of the coppice forests were transformed into high forests [49] (a forest of generative origin with a usual production period of at least 100 years) [50]. In Czechia, coppice forests covered only 109,900 ha in 2013, whereas in 1845, they covered 1,457,400 ha [49]. These transformations from coppice forests are also evident in other European countries, especially since the second half of the 20th century when this forest management declined significantly [41].

Coppice forests in the past were mainly preserved in the form of stumps, which were either transferred to high forests or left untouched [35,39,51]. The primary goal of converting coppices to a high forest is to restore the original physiognomic, structural and spatial diversity of close-to-nature stands [52,53]. Coppices can be converted to high forest in basically two ways: passively by aging (without intervention) or actively by thinning [54]. On the other hand, the current trend is the opposite and is more inclined to coppice forest conservation and restoration in many areas [24]. Among the countries that still actively use coppice forest management are, for example, Romania, Bulgaria, Austria, France, Italy, and Spain [55]. In the last two decades, there has been renewed interest in coppice regarding their ecological functions and the provision of numerous ecosystem services [17,56]. In addition, regeneration in coppice is less damaged by game than artificial planting due to the increasing numbers of ungulates in Europe [57,58]. Interest in coppice forest management has also increased within the last decade due to the rising importance of their resilience to climate change [56].

Traditionally, coppice forest stands consist of broadleaved tree species with a high potential for coppice management [24]. Oaks are among the most suitable species utilized for this purpose, such as native sessile oak (*Quercus petraea*) [33]. Ongoing climate change creates suitable conditions for thermophilic oak species, including Turkey oak (*Quercus cerris* L.), which originally expanded mainly to Southeastern Europe but currently spread to Central Europe. Existing stands of Turkey oak, principally coppices in Europe, are most likely the result of 5000 years of human activity on these stands [59,60]. Coppices of Turkey oak stands in the hills of Italy cover 675,532 ha (18.4%) of the forest area [16]. Coppices in Italy have proven to be the most important cultivation system, widespread primarily in private deciduous forests [61]. In Bulgaria, Turkey oak stands cover an area of 258,400 ha (7.0%) [62], and Turkey oak is also common in Slovakia, where it covers 50,773 ha (2.6%) [63]. Moreover, thermophilic tree species, such as Turkey oak, can play a significant role in the context of ongoing climate change because they can better withstand climatically demanding and drier conditions [4,64]. The synergism of these two factors can be helpful for its successful introduction into new areas outside its natural range of distribution at present, when many native tree species, such as Norway spruce (*Picea abies* [L.] Karst.), are experiencing a large-scale decline in Europe [65].

Turkey oak growth and its reaction to ongoing climate change in Central and Southern Europe have not yet been described in detail. Closer studies of this type are necessary for understanding the adaptability of this tree species to climate change and the possibility of introduction outside its native areas. However, for the first step in determining growth processes, it is crucial to thoroughly analyze the stand structure (diversity, horizontal, and vertical structure) and production parameters (tree density, stocking, stand volume, etc.), which significantly affect the response of trees to climate change and cyclical events [4,66]. Therefore, the main objectives of this paper are (i) to evaluate the production potential, structure, and biodiversity of Turkey oak coppices in research sites in Italy, Bulgaria, Slovakia, and Czechia, (ii) determine growth conditions, cyclical dynamics, and the effect of climate factors (temperature and precipitation) on tree-ring growth in site chronologies in Southern and Central Europe, (iii) analyze the relationship between stand structure, production parameters, and tree-ring growth, and finally (iv) evaluate the growth dynamics during ongoing climate change.

2. Methodology

2.1. Study Area

Areas of interest in Southern and Central Europe (Italy, Bulgaria, Slovakia, and Czechia) include monospecific coppices of Turkey oak. In Italy and Bulgaria, these are foothills to mountainous locations, and in Slovakia and Czechia, they are lowlands. In individual countries, coppices aged 50–60 years were selected in areas where this silvicultural method has a thousand-year tradition. Coppicing is a silvicultural system based on systematically repeating the vegetative regeneration of sprouts, whereas high forest is regenerated generatively and the production period is at least 100 years [50]. Forest management on PRPs is based on individual thinning with respect to health status and target diameter at breast height (DBH). All studied stands belong to the association *Quercion pubescenti-petraeae*. The location of the permanent research plots (PRPs) is shown in Figure 1, and a basic overview of PRPs in Table 1. Long-term annual air temperature ranges from 8.8 to 13.7 °C, and the average sum of precipitation is 497–1022 mm in the study areas (Table 2). According to the worldwide Köppen classification, PRPs in Czechia and Slovakia belong to the climate categorization Cfb—temperate oceanic climate with cool summers, mild winters, and a relatively narrow annual temperature range; PRPs in Bulgaria belong to Dfb—warm summer, humid continental climate, with substantial seasonal temperature differences; PRPs in Italy belong to Csa—hot summer, Mediterranean climate, with dry summers and mild, wet winters.

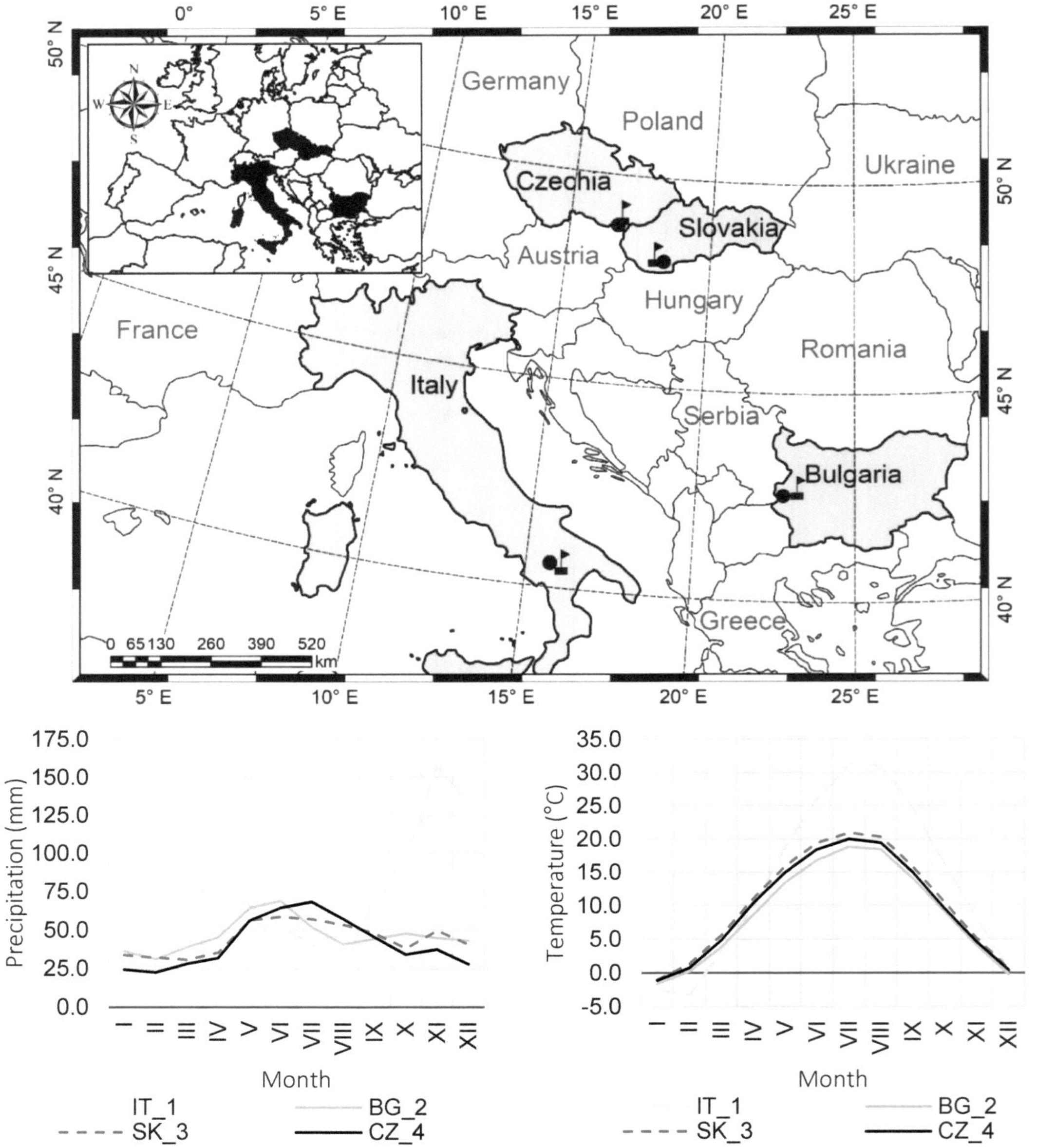

Figure 1. Localization of permanent research plots in coppices of Turkey oak in areas of interest in Europe (**up**); the precipitation (**down left**) and air temperature (**down right**) during the calendar year from I (January) to XII (December) in Italy (IT_1), Bulgaria (BG_2), Slovakia (SK_3), and Czechia (CZ_4).

Table 1. Overview of basic site and stand characteristics of Turkey oak coppices in areas of interest in Europe.

Plot Name	Country	Coordinates	Altitude	Exposition	Slope	Geological Bedrock	Soil	Age	DBH	Height	Stand Volume
			(m)		(°)				(cm)	(m)	(m^3 ha^{-1})
IT_1	Italy	40°32′53.025″ N 15°43′38.775″ E	1000	W	4.3	limestone, marl	rendzina	60	31	22	407
BG_2	Bulgaria	42°31′13.619″ N 22°33′4.277″ E	1060	SW	5.7	sandstone	cambisol	50	19	16	141
SK_3	Slovakia	48°4′14.677″ N 18°21′58.449″ E	220	SE	2.9	loess clay	cambisol	60	29	21	275
CZ_4	Czechia	48°44′31.877″ N 16°47′35.951″ E	200	S	0.0	marl	cambisol	60	28	22	342

Notes: W—west, SW—southwest, SE—southeast, S—south, DBH—diameter breast height.

Table 2. Overview of basic meteorological characteristics of research areas.

Plot	Meteo. Station Name	GPS of Meteo. Station	Station Altitude (m a.s.l.)	Distance to Plot (km)	Annual Temperature (°C)	Seasonal Temperature (°C)	Annual Precipitation (mm)	Seasonal Precipitation (mm)
IT_1	Abriola	40°31′8″ N 15°47′38″ E	1225	6.5	13.7	26.3	1022	258
BG_2	Divlya	42°28′43″ N 22°41′34″ E	720	12.5	8.8	16.3	552	268
SK_3	Hurbanovo	47°52′00″ N 18°12′00″ E	115	25.8	10.4	18.5	532	273
CZ_4	Lednice	48°47′35″ N 16°47′58″ E	177	5.7	9.8	17.6	497	291

2.2. Data Collection

From 2021 to 2022, four PRPs with a size of 25 × 25 m (0.0625 ha) were inventoried. The structure of the tree layer was measured using FieldMap technology (IFER-Monitoring and Mapping Solutions Ltd., Jílové u Prahy, Czech Republic). Each stem was regarded as an individual tree, for both single stems and polycormons (stem with more shoots). The diameter at breast height, position of all individuals with a DBH ≥ 4 cm, total height, height of the green crown base, and crown projection area (at least in four mutually perpendicular directions) were measured. The height of the live crown base was measured at the point where branches formed a continuous whorl of a crown. The crown radii in four directions were measured at a right angle to each other through the centroid of the crown by FieldMap hardware [67]. Boundary trees with more than half of their DBH lying inside a PRP were included. The diameter at breast height was measured with a Mantax Blue caliper (Haglöf, Långsele, Sweden) with an accuracy of 1 mm, while DBH was averaged from two measurements. Individual tree height and the height of the live crown base were recorded with a Vertex laser hypsometer (Haglöf, Långsele, Sweden) with an accuracy of 0.1 m.

For the analysis of the radial growth of Turkey oak, core samples were obtained from the trees with a Pressler auger (Haglöf, Långsele, Sweden) at a height of 1.3 m in the direction up/down the slope. Dendrochronological samples were taken from the visibly healthy trees with no signs of damage in the trunk or crown. The sampled trees were dominant and co-dominant according to the Kraft classification [68]. The selection was examined randomly (RNG function, Excel) according to the distribution in the stand, which describes growth response (compared to sub-dominant and suppressed trees on each research plot [69]. The number of samples had to be sufficient for the EPS indicator described in Data Processing. A total of 104 samples were taken for dendrochronological analysis. The individual numbers of samples per area are described in the table of the site dendrochronology description of the research plots in the Results. Ring widths were measured to the nearest 0.01 mm with an Olympus binocular on a LINTAB measuring table and recorded with the TsapWin program [70].

2.3. Data Processing

2.3.1. Stand Structure and Analysis

The basic structure, diversity, and production characteristics of the tree layer were evaluated by the SIBYLA 5.1 software [71]. The input data were measured by individual dendrometric characteristics of trees (tree species, height, DBH, crown width, live crown base, and age), including coordinates. The PointPro 2.1 program (CZU, Prague, Czechia) was used to calculate the characteristics describing the horizontal structure [72] of tree individuals on the PRPs. The aggregation index was derived from all distances between the two nearest neighbors, the number of trees in the plot, plot area, and the perimeter of the plot [72]. The significance test of deviations from the values expected for a random arrangement of points was performed using Monte Carlo simulations. In the results, statistically significant values (exceeding the confidence interval) are marked with an asterisk. Next, structural diversity was evaluated based on the vertical Arten-profile index [73], diameter and height differentiation [74], crown differentiation, and total stand diversity (Table 3) [75]. The Arten-profile index was calculated using the basal area of tree species in individual stand layers [76]. Diameter and height differentiations are related to the ratio between the larger and the smaller diameter/height of all nearest neighboring trees [74]. The stand diversity index was calculated with regard to complex biodiversity [75]. Total diversity is composed of the following components of diversity: tree species diversity, diversity of vertical structure, diversity of tree spatial distribution, and diversity of crown differentiation. The input variables are the number of tree species, maximum and minimum tree species proportion, maximum and minimum tree height, maximum and minimum tree spacing, minimum height to crown base, and minimum and maximum crown diameter [77].

Table 3. The indices describing stand structure and their common interpretation.

Criterion	Quantifiers	Label	Reference	Evaluation
Horizontal structure	Aggregation pattern	R (C&Ei)	[72]	mean value R = 1; aggregation R < 1; regularity R > 1
Vertical structure	Arten-profile index	*A* (Pri)	[73]	range 0–1; balanced vertical structure *A* < 0.3; selection forest *A* > 0.9
Structure differentiation	Vertical diversity	*S* (J&Di)	[75]	low *S* < 0.3, medium *S* = 0.3–0.5, high *S* = 0.5–0.7, very high *S* > 0.7
	Diameter dif. Height dif.	TM_d (Fi) TM_h (Fi)	[74]	range 0–1; low *TM* < 0.3; very high differentiation *TM* > 0.7
	Crown dif.	*K* (J&Di)	[75]	low *K* < 1.0, medium *K* = 1.0–1.5, high *K* = 1.5–2.0, very high *K* > 2
Complex diversity	Stand diversity	*B* (J&Di)	[75]	monotonous structure *B* < 4; uniform structure B = 4–5.9; non-uniform structure *B* = 6–7.9; diverse structure B = 8–8.9; very diverse structure *B* > 9

Notes: Monotonous structure = stands composed solely of a single tree species; vertically undifferentiated tree canopy; low variation in tree crown diameters; systematic spatial arrangement of trees. Uniform structure = stands composed of one to two tree species; vertical structure of the tree canopy formed by a single layer, occasional identification of a second layer; random horizontal structure of trees. Non-uniform structure = stands composed of up to four tree species with varied mixed proportions; vertical structure consisting of two to three tree layers; average crown size reaching 50%; random to weak clustering tree spatial pattern. Diverse structure = stands composed of an average of five canopy-forming tree species, with two to three having similar mixture proportions; irregularly moderately multilayered vertical structure, rarely differentiated; spatial arrangement of trees classified as heterogeneous with a tendency to cluster. Very diverse structure = forests characterized primarily by high biological diversity; vertically structured profiles forming multiple tree layers, containing up to seven canopy forming tree species, of which at least three to four have relatively equal representation; highly varied crown widths; spatial arrangement of trees perceived as clustered [75].

The stand volume was calculated according to [78]. The relative stand density index (SDI) [79], the canopy closure (CC) [80], and the crown projection area (CPA) were observed for each PRP. The relative SDI was calculated as the ratio of the actual value of the stand density index to its maximum value. The stand density index represents the theoretical number of trees per hectare if the mean quadratic diameter of the stand component is equal to 25 cm [79].

2.3.2. Dendrochronological Processing and Analysis

Dendrochronological data were analyzed in software R (version 4.3.1) [81] using the packages "dplr" [82,83] and "pointRes" [84]. Detrending of each measured sample was carried out by negative exponential detrending with a spline of 2/3 of the age of each tree using "dplr" instructions [85]. The detrended tree-ring growth data are averaged as ring-width index (RWI) that further describes site chronology for the research plot. An analysis of the pointer years through relative growth change was performed [86]. The pointer years reflect the number of standard deviations from the local mean of the average ring-width series in the previous four years. The pointer years identify event years where the pointer year > 0.75 standard deviation of the previous four years. The threshold of the percentage of trees in a negative or positive event year was used [87]. The pointer years and percentage mean annual growth deviation are distinguished by the most common event in the year class [84].

An expressed population signal (EPS) was carried out for the detrended data series. The EPS represents the reliability of a chronology as a fraction of the joint variance of the theoretical infinite tree population. The EPS was employed to represent the limit for using the dendrochronological data series concerning the climatic data. The significant EPS threshold for using the dendrochronological data is EPS > 0.85 [85]. The signal-to-noise ratio (SNR) indicates chronological signal strength. The SNR is a statistical metric that evaluates the strength of the targeted signal within a dataset of the series compared to the background noise level. A higher SNR value indicates a more robust climatic signal relative to noise. Inter-series correlations (R-bar) were calculated for the dendrochronological data series. The R-bar quantifies the similarity of tree-ring patterns among various samples. It represents the average pairwise correlation coefficient between individual trees within a chronology. A higher R-bar value signifies increased coherence among the tree-ring patterns [88]. First-order autocorrelation (Ar1) was also carried out. The Ar1 describes the degree of cross-correlation between a data point and the preceding one in a time series of tree-ring series. The EPS, SNR, R-bar, and Ar1 were calculated by the instructions to "dplr" [85] based on common dendrochronological theories [88,89].

2.3.3. Tree Rings and Climatic Analysis

The average tree-ring series of Turkey oak from research plots IT_1, BG_2, SK_3, and CZ_4 was correlated with climate data, namely precipitation and temperatures; 1968–2022 from weather stations in Italy (Potenza—720 m a.s.l.), Bulgaria (Divlya—600 m a.s.l.), Slovakia (Hurbanovo—115 m a.s.l.), and Czechia (Lednice—177 m a.s.l.) according to individual months and years. The DendroClim 2002 software [90] was used to model the radial growth depending on the climatic characteristics.

Spectral analyses of the data were performed using Statistica 13 software. The calculation was accomplished using the "Single Fourier (Spectral) Analysis" function, utilizing the "Periodogram" plot by "Period" output. The sine and cosine functions are mutually independent, also known as orthogonal. Therefore, we can aggregate the squared coefficients for each frequency to create the periodogram. The values in the periodogram can be understood in relation to the variance, representing the sums of the squares and of the data at the corresponding frequency or period [91]. While our datasets are in yearly intervals, the "period" in spectral analysis describes the length of the yearly interval cycles. The intensity of the cycles of our datasets indicates "periodogram values", which are expressed as the density of cycles per observation. This allows the identification of dominant frequencies or periods (cycles) in the data.

Seasonal temperature was determined by calculating the arithmetic mean of the monthly values within these seasonal months. For the calculation of seasonal precipitation, the sum of monthly precipitation totals during the respective seasonal periods was used.

The intentional selection of this seasonal window was intended to reduce variability at the start and end of the growing season. Thus, the seasonal data assessed within this timeframe accurately represent the shared vegetation period across all research plots.

A principal component analysis (PCA) was performed in the CANOCO 5 program [92] to evaluate the relationships between the stand structure, production parameters, radial growth, and research plots. This tool was used to reduce the dimensionality of a dataset while preserving the most important patterns, information, or relationships between the variables [93]. Prior to analysis, the data were standardized, centralized, and logarithmized. The results of PCA were presented in the forms of species and environmental variables ordination diagram. The input data to the PCA included the following parameters: stand volume, stem volume, basal area, diameter, slenderness coefficient, height, tree density, total diversity, diameter structure, vertical structure, and horizontal structure indices. The total number of variations was 48 (samples × species).

3. Results

3.1. Stand Characteristics, Production, and Diversity

Dendrochronological characteristics of the research plots in Table 4 reveal that all data exhibit significant EPS values for climate comparison with the ring-width index (RWI) of Turkey oak, indicating an EPS higher than 0.85 across the entire examined sample period. The sample count (No. trees) per plot varied from 24 to 29, with IT_1 = 24, BG_2 = 25, SK_3 = 26, and CZ_4 = 29 sample units. The mean ring width (RW) ranges from 2.09 to 3.13 across individual plots, with BG_2 having the lowest RW at 2.09 mm and SK_3 the highest at 3.13 mm. The Ar1 indicator suggests that research plots in BG_2 and IT_1 (Ar1 = 0.79; 0.60) exhibit higher to moderate levels of autocorrelation, indicating a substantial correlation between values in one year and those in the previous year. In contrast, plots in CZ_4 and SK_3 have (Ar1 = 0.41; 0.57), indicating a moderate to slightly lower degree of autocorrelation.

Table 4. Dendrochronological characteristics of Turkey oak stands on permanent research plots.

PRP	No. Trees	Mean RW (mm)	SD RW (mm)	Mean Min–Max (mm)	Age Min–Max	Ar1	R-Bar	EPS	SNR
IT_1	24	2.69	1.03	1.66–4.65	36–66	0.60	0.34	0.90	8.71
BG_2	25	2.09	1.11	1.46–2.97	35–55	0.79	0.43	0.93	12.87
SK_3	26	3.13	1.45	0.95–5.31	33–69	0.57	0.46	0.90	8.89
CZ_4	29	2.23	1.06	1.60–3.09	40–63	0.41	0.55	0.97	29.60

Notes: No. trees—number of used core samples, mean RW—mean ring width, SD RW—standard deviation of ring width, mean min–max—mean ring-width range from the smallest to biggest tree, Age min–max—age range of the youngest and oldest sample tree, Ar1—first order autocorrelation, R-bar—inter-series correlation, EPS—expressed population signal, SNR—signal-to-noise ratio.

The number of live Turkey oak trees from 2021 to 2022 ranged between 475 and 775 trees per ha with an SDI of 0.52–0.82 (Table 5). The highest mean DBH (30.5 cm) and tree volume (0.708 m^3) was from inventoried stands in Italy, while the lowest values were observed in Bulgaria (19.0 cm, 0.182 m^3). In general, as tree density increases, tree dendrometric parameters decrease. The basal area ranged from 22.0 (BG_2) to 41.6 m^2 ha^{-1} (IT_1), and the stand volume was from 141 m^3 ha^{-1} (BG_2) to 407 m^3 ha^{-1} (IT_1). The mean annual increment was from 2.82 m^3 ha^{-1} in Bulgaria to 6.78 m^3 ha^{-1} in Italy.

In terms of the horizontal structure, the spatial pattern of trees was prevailingly random or clustered in Bulgaria (Table 6). The vertical structure was quite variable (A 0.266–0.530), i.e., balanced on IT_1 to moderately differentiated on other PRPs. The diameter and differentiation index varied and reached low values, with the highest variability in Italy. Concerning the overall stand diversity, IT_1, BG_2, and SK_3 showed a uniform structure (B 4.507–5.075), whereas CZ_4 showed a monotonous structure (B 3.917). Similarly, the lowest crown differentiation was observed in the Czech PRP compared to high crown diversity in Bulgarian stands.

Table 5. Overview of stand parameters of Turkey oak coppices.

PRP	DBH	h	f	v	N	G	V	hd	MAI	CC	SDI
	(cm)	(m)		(m^3)	(tree ha^{-1})	(m^2 ha^{-1})	(m^3 ha^{-1})		(m^3 ha^{-1} $year^{-1}$)	(%)	
IT_1	30.5	22.33	0.434	0.708	575	41.6	407	0.732	6.78	88.6	0.82
BG_2	19.0	16.22	0.395	0.182	775	22.0	141	0.854	2.82	77.1	0.52
SK_3	28.8	20.92	0.425	0.579	475	30.9	275	0.726	4.58	80.8	0.62
CZ_4	28.2	21.80	0.402	0.547	625	39.1	342	0.773	5.70	87.4	0.79

Notes: DBH—mean quadratic diameter at breast height, h—mean height, f—form factor, v—average tree volume, N—number of trees, G—basal area, V—stand volume, hd—slenderness ratio, MAI—mean annual increment, CC—canopy closure, SDI—stand density index.

Table 6. Biodiversity of stands with Turkey oak on all PRPs.

PRP	R (C&Ei)	A (Pri)		S (J&Di)		TM_d (Fi)		TM_h (Fi)		K (J&Di)		B (J&Di)	
IT_1	0.896	0.266	↘↘	0.783	↗	0.314	↘	0.156	↘↘	1.219	→	4.507	↘
BG_2	0.676 *	0.385	↘	0.594	→	0.252	↘↘	0.147	↘↘	1.761	↗	5.075	↘
SK_3	0.931	0.530	→	0.758	↗	0.220	↘↘	0.188	↘↘	1.490	→	4.693	↘
CZ_4	1.206	0.426	↘	0.304	↘	0.163	↘↘	0.082	↘↘	0.714	↘	2.105	↘↘

Notes: R—aggregation index, A—Arten-profile index TM_d—diameter differentiation index TM_h—height differentiation index, S—vertical diversity index, K—crown differentiation index, B—stand variability index statistically significant ($p < 0.05$) for horizontal structure (A—aggregation, R—regularity); arrows: ↘↘—low, ↘—low-medium, →—medium, ↗—high. * statistically significant aggregation spatial pattern ($\alpha = 0.05$) for horizontal structure (R index).

The diameter structure shown in Figure 2 demonstrates greater diameter variability with a broader distribution on the diameter spectrum in the case of Italy. The oak stands resembled the Gaussian curve in shape (typical of the same-age stand), which was the least flattened in the case of Bulgaria. Overall, oak was the most common in diameter classes, ranging from 18–23 (Bulgaria) to 28–33 cm (Czechia).

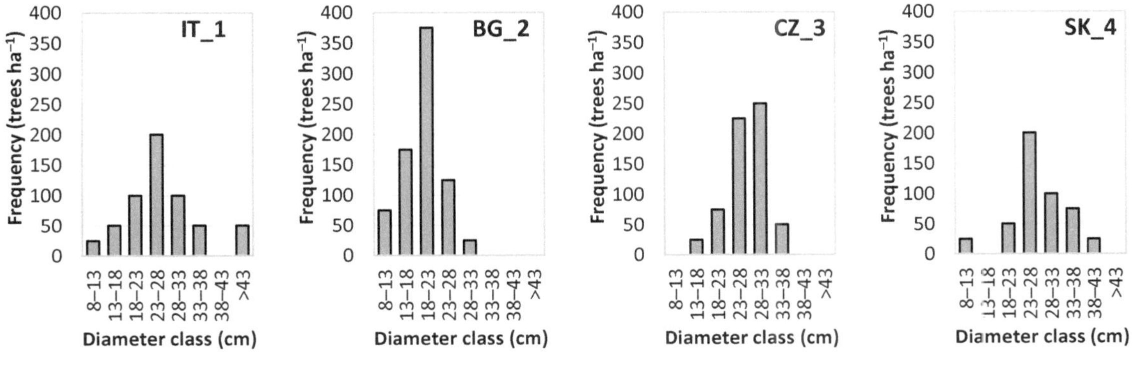

Figure 2. Diameter structure of oak forest stands according to countries.

3.2. Tree-Ring Growth

The ring-width increment of Turkey oak in the PRPs is different for each research plot (Figure 3). In terms of size, none of the PRPs have an increment lower than 0.5 or higher than 1.7 in RWI values. All research plots show irregular growth from one another Overall, the tree-ring chronologies of Turkey oak were not subject to significant long-term fluctuations in the ring-width index over the study period. However, in the short term, the Central European PRPs fluctuate more in growth, as seen in the more irregular growth in CZ_4 and SK_3 from year to year. In contrast, the RWI of the PRPs in Southern Europe—IT and BG—show higher growth stability over the study period.

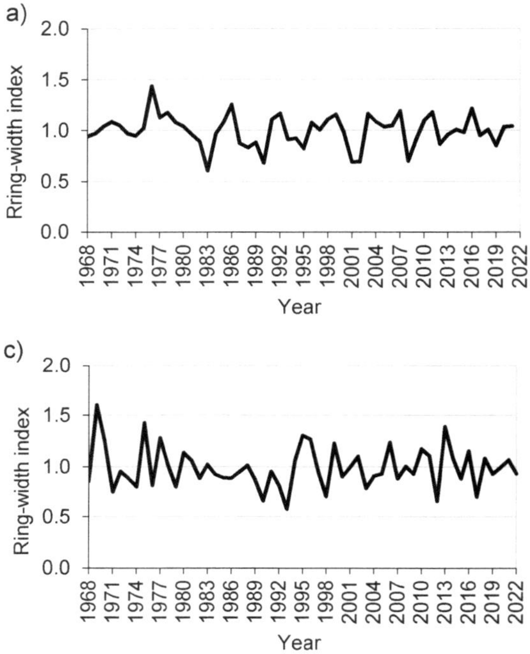

Figure 3. Standardized ring-width index chronology of Turkey oak in the period 1968–2022: (**a**) IT_1 (Italy); (**b**) BG_2 (Bulgaria); (**c**) SK_3 (Slovakia); (**d**) CZ_4 (Czechia).

The research plots in SK_3 and CZ_4 demonstrate a more fluctuating growth pattern in terms of annual oscillations of RWI from year to year. The annual RWI reveals that, in the case of SK_3, the RWI ranged from 0.65 in 2012 to 1.38 in 2013. Similar variability was observed in CZ_4, with notable fluctuations in the RWI, such as in 1978 (RWI = 0.80) to 1979 (RWI = 1.47). Significantly smaller RWI fluctuations were observed in plots in IT_1 and BG_2. This variation in plots from southern regions of Europe exhibits considerably smaller annual differences. For instance, from 1983 to 1984, IT_1 experienced a higher RWI increase, but only from 0.60 to 0.96. Similar small fluctuations are seen in BG_2, where notable RWI increases occurred, for example, in 1987 to 1988, with the RWI rising from 0.86 to 1.13.

Mean growth deviations in RWI are described in Figure 4 for research plots in IT_1 BG_2, SK_3, and CZ_4. Two climatically significant years were recorded for BG_2 in 1979 and 1980. In CZ_4, the negative climatic years were 1973 and 2017. From the viewpoint of negative climatic years, there is no noticeable difference between the PRPs in Southern and Central Europe.

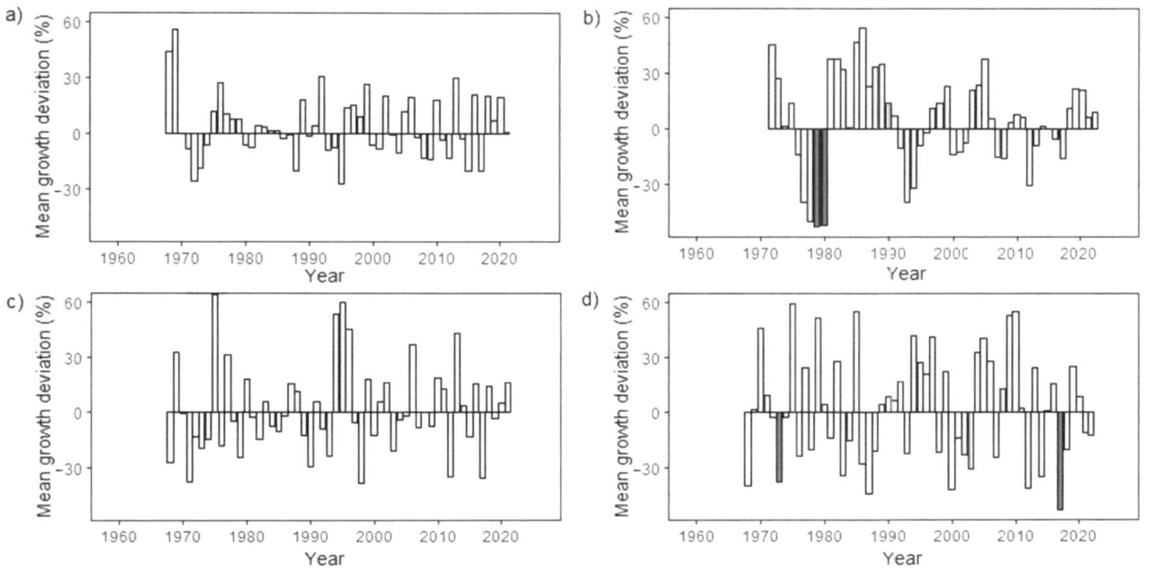

Figure 4. Percentage of mean growth deviation of Turkey oak for the period 1960–2022 with the pointer years (highlighted in dark grey); (**a**) IT_1 (Italy); (**b**) BG_2 (Bulgaria); (**c**) SK_3 (Slovakia); (**d**) CZ_4 (Czechia).

In terms of mean deviations and their oscillations, it is evident that plots in the southern regions of Europe—IT_1 and BG_2—exhibit a smoother transition between negative decline and growth deviations. In contrast, plots in Central Europe—SK_3 and CZ_4—show more regular oscillations in the transition from negative to positive, with these values oscillating mainly from year to year.

3.3. Turkey Oak's RWI with Monthly Precipitation and Temperature

The correlation coefficients in Figure 5 show the course of RWI against the monthly averages of temperatures and precipitation in the research plots. The radial growth of oak in sites in Italy and Bulgaria was least affected by monthly temperature and precipitation development (two significant months). Contrarily, in plots of Central Europe (SK_3 and CZ_4), the climatic factors studied had a significant influence on increment development (four significant months). Overall, the radial growth of Turkey oak correlates significantly ($p < 0.05$) positively with the course of monthly precipitation and negatively with the course of monthly temperatures. Unlike BG and IT, Central European PRPs are significantly more correlated with precipitation when compared to temperature. In comparison, precipitation and temperature for the Italian and Bulgarian PRPs are equally weighted. Generally, temperatures from June to August indicate a negative effect on growth, and precipitation correlates from March to June. These are the primary limiting factors for the radial growth of Turkey oak. Overall, temperatures exert a predominately negative influence on the RWI across all areas during the season, while precipitation has a significantly more positive impact on the RWI.

The results also indicate that precipitation is more frequently correlated in Central Europe on plots in SK_3 and CZ_4. In contrast, fewer correlations with lower values are observed in plots from the southern parts of Europe in IT_1 and BG_2.

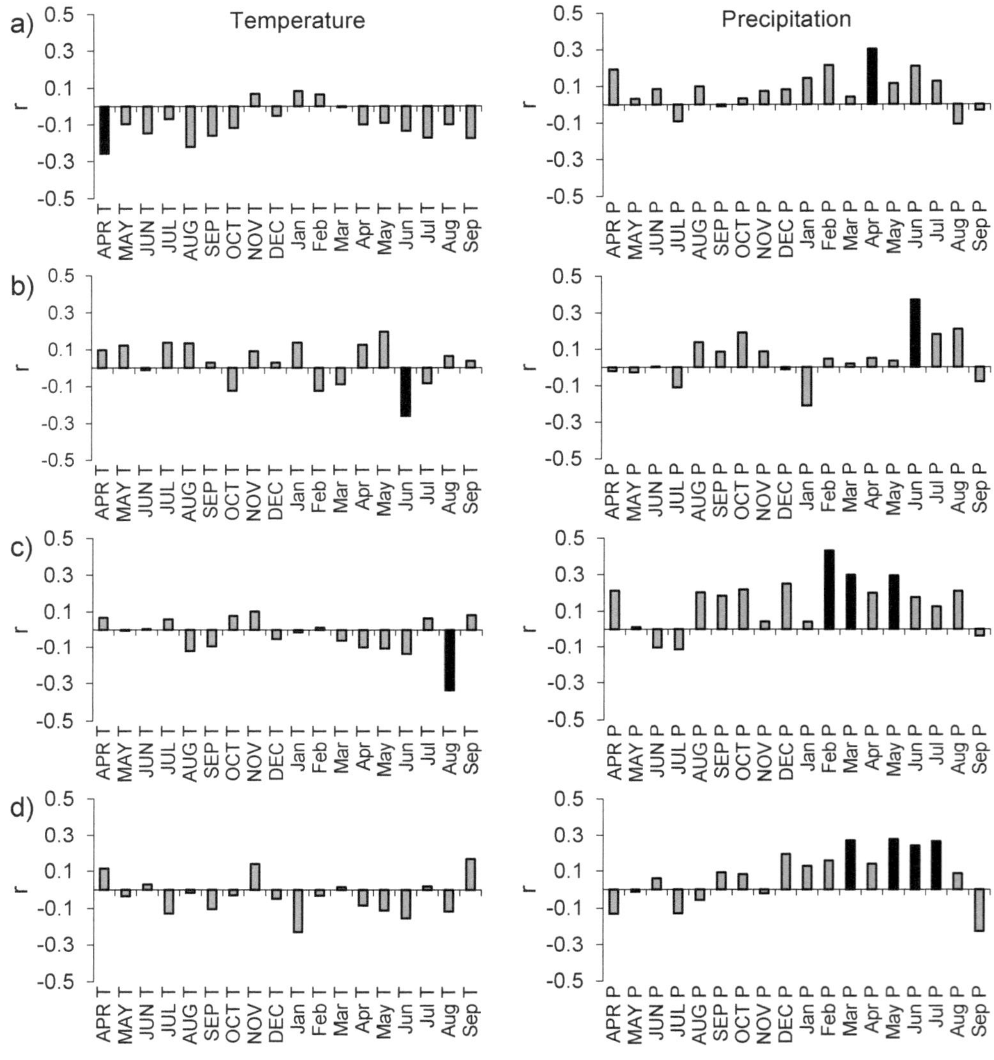

Figure 5. Correlation coefficients of the site chronology of the Turkey oak ring-width index with the average monthly air temperature (left) and sum of precipitation (right) from April to December of the previous year (uppercase letters) and from January to September of the current year (lowercase letters) in the relative year derived from the period 1968–2022: (**a**) IT_1 (Italy); (**b**) BG_2 (Bulgaria); (**c**) SK_3 (Slovakia); (**d**) CZ_4 (Czechia). Significant months ($p < 0.05$) are highlighted in black.

3.4. Turkey Oak's RWI Growth Cycles

Spectral analysis shows that the study plots in Italy (IT_1) and Bulgaria (BG_2) undergo longer 6 to 8-year growth cycles in RWI cycles (Figure 6). Research plot IT_1 shows the longest cycles with a frequency of 6.2 years. Research plot BG_2 illustrates the longest growth cycles in RWI with a frequency of 7.7 years. Contrastingly, the research plots in Central Europe (SK_3 and CZ_4) experience significantly shorter cycles ranging from 2.4 to 4.8 years. The RWI growth cycles of Turkey oak in the Central European sites are shorter than in Southern Europe, so it can be determined that Central Europe has the most frequent RWI cycles of 2 to 5 years. Southern European sites of Turkey oak have more pronounced

6 to 8-year RWI cycles. Overall, on the more northerly plots in SK_3 and CZ_4, the growth cycles of RWI are shorter and much more intense, as indicated by spectral analysis, which reveals predominantly 2 to 5-year cycles. In contrast, in Southern Europe research plots, the growth cycles of RWI are longer and much more intense over a longer time span, ranging from 6 to 7 years.

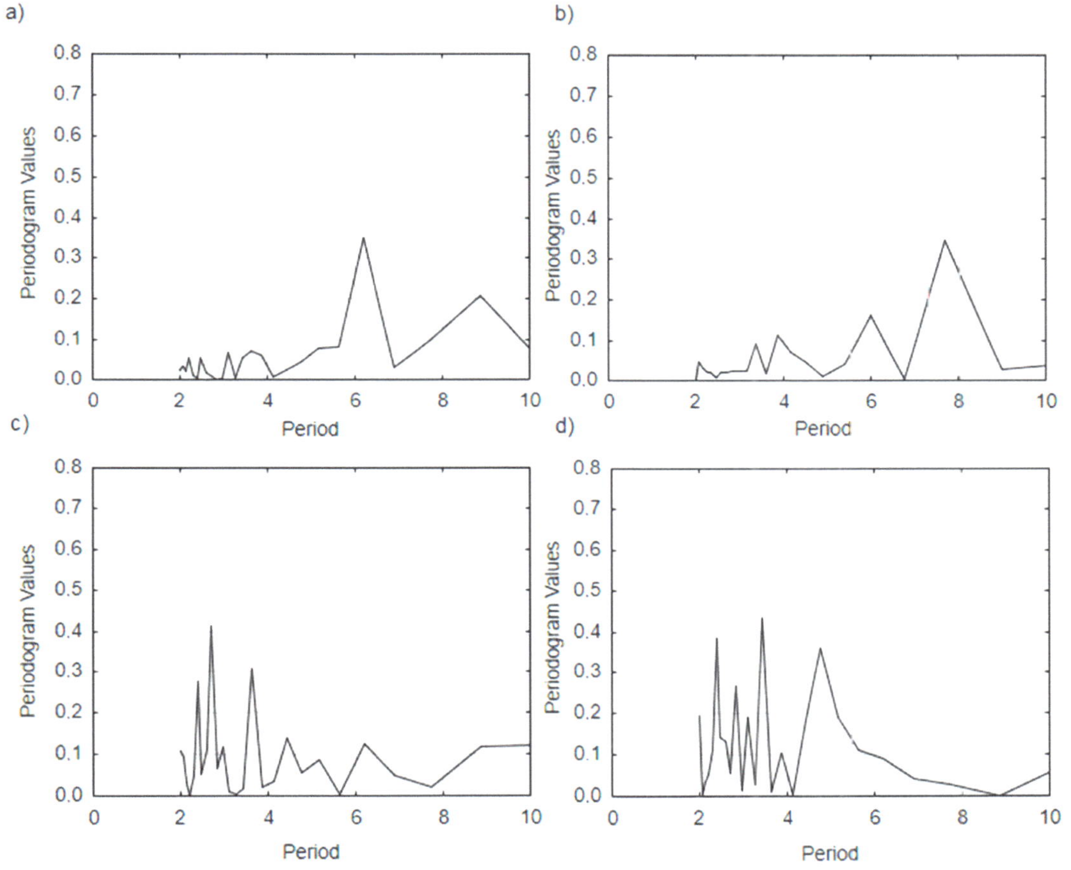

Figure 6. Single spectral analysis of ring-width index (RWI) of Turkey oak from 1968 to 2022: (**a**) IT_1 (Italy); (**b**) BG_2 (Bulgaria); (**c**) SK_3 (Slovakia); (**d**) CZ_4 (Czechia).

3.5. Interaction between Stand Structure, Production Parameters, and Growth

The results of PCA are presented in an ordination diagram in Figure 7. The first ordination axis explains 62.2% of data variability, and the first two axes together explain 83.6%. The x-axis illustrates the stand volume and basal area, and the y-axis represents the vertical structure (A index) combined with the diameter structure (TM_d index). The total diversity was positively correlated with diameter differentiation, while these indices were negatively correlated with horizontal structure (tendency to regular spatial pattern). Production parameters, such as tree height, stand volume, basal area, DBH, and tree volume, were positively correlated to each other. The radial growth of oak increased with a decreasing number of trees in stands and the slenderness coefficient. The vertical structure (A index) was the lowest explanatory variable in the ordination diagram.

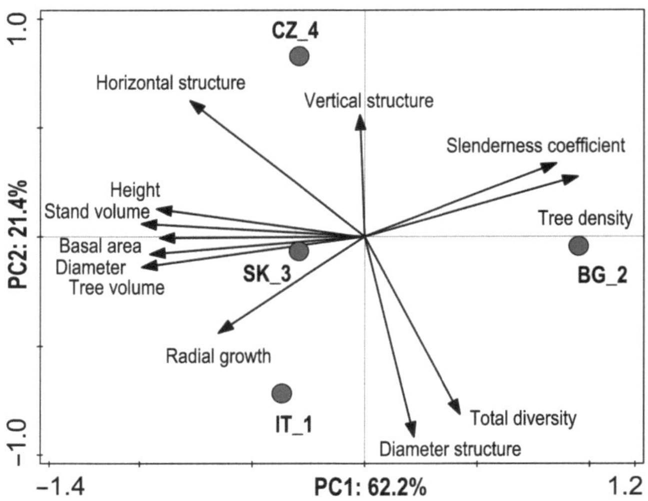

Figure 7. Ordination diagram showing the results of the PCA analysis of relationships between stand characteristics (Stand volume, Stem volume, Basal area, Diameter, Slenderness coefficient, Height, Tree density), diversity (Total diversity—B index, Diameter structure—TM_d index, Vertical structure—A index, Horizontal structure—R index), and radial growth. Grey symbols indicate research plots.

4. Discussion

4.1. Production Potential and Stand Density

Regarding basic production parameters, the average number of trees in the study plots ranged from 475 to 775 trees ha^{-1}, the stand basal area ranged from 22 to 42 m^2 ha^{-1}, and the volume of roundwood (timber to the top of 7 cm o.b.) ranged from 141 to 407 m^3 ha^{-1}. With regard to the distribution of Turkey oak, most of the comparable data on similar stands came from Italy. For example, stand characteristics of Turkey oak coppices at 55 years of age in the Italian Mediterranean region corresponded to an average observed number of 578 to 1018 trees ha^{-1}, with a stand basal area ranging from 29 to 35 m^2 ha^{-1}, and the stand volume was similar, ranging from 260 to 308 m^3 ha^{-1} [94]. Another study from Central Italy reported that in stands of over-aged coppices dominated by Turkey oak at the age of 55 years, the mean number of trees was higher again, ranging from 794 to 891 trees ha^{-1}, and the stand basal area reached values of 29 to 30 m^2 ha^{-1}, while the stand volume reached lower values in a very narrow range of 230 to 231 m^3 ha^{-1} [95]. Šrámek et al. [96] cite that the number of trees in over-aged coppices in Türkiye aged 70–75 years was around 577 trees ha^{-1}, and the stand basal area reached 38 m^2 ha^{-1}. However, other tree species such as sessile oak (*Quercus petraea*), Italian oak (*Quercus frainetto*), European hornbeam (*Carpinus betulus*), or sweet chestnut (*Castanea sativa*) were also significantly represented in these Turkish oak stands. In this case, it is necessary to consider the species composition of the tree layer, which significantly influences the production characteristics of the coppice forest. A significantly higher number of trees (1870 trees ha^{-1}) in oak coppice stands at 71 years of age was reported by [97]. Again, the significant number of trees is a result of the diverse species composition of the coppice forest, with the presence of not only oaks but also small-leaved lime (*Tilia cordata*) or European hornbeam (*Carpinus betulus*).

With regard to stand density, the number of oak trees was distinctly correlated with tree-ring growth. The radial growth of oak increased with a decreasing number of trees in stands and a closely related slenderness coefficient. Similarly, mean tree growth was significantly higher under low density compared with maximum stand density in the case of sessile oak (*Quercus petraea* [Matt.] Liebl.) [98]. A significant response of radial growth to different stand densities was also confirmed for other tree species [99]. The effects of competition on tree radial growth were found to be much higher for shade-intolerant species, such as oaks [100]. Moreover, the reduction in tree density increases subsequent growth for remaining trees and decreases sensitivity to climate change, especially drought stress [101–103]. The resilience of trees after reducing the stand density was confirmed, e.g., for Scots pine stands due to the lower competition of remaining trees to available water sources with growing resistance during a drought event [104].

4.2. Stand Biodiversity and Structure

Stand diversity (and structure) significantly influences growth processes and tree resilience to climate change [4,105]. Šimůnek et al. [66] describe the disparate response of radial tree growth in homogeneous vs. heterogeneous stands to climatic factors. Therefore, a detailed analysis of the structure and diversity of the studied stands is an essential starting point for further dendrochronological analyses. In the case of the evaluated stands of Turkey oak coppice forests aged 50 to 60 years, the overall stand diversity was very low, or rather, the stands reached a monotonous to uniform structure (B index 3.92–5.08). A significantly higher overall diversity (B index 9.73–10.46) of oak coppice forests in Czechia is described by [97], chiefly due to the addition of other broadleaved species and leaving the stands to develop spontaneously. In terms of the horizontal structure of the tree layer, the spatial pattern of the studied stand was predominantly random, with a tendency to aggregate or cluster in Bulgaria. Another study from Czechia describes a similarly aggregated structure of oak coppice forests, especially in the initial distances between trees of up to 2 m [39]. Of all the parameters studied, crown differentiation reached the highest values of structural diversity [39].

4.3. Turkey Oak Coppices and Climate

Radial growth of Turkey oak research plots in this study were clearly more influenced by climatic factors (monthly temperatures and precipitation) in Central European countries than in Italy and Bulgaria. Overall, RWI growth significantly correlated positively to the course of monthly precipitation ($p < 0.05$), especially from March to June, and negatively to the course of monthly temperatures, especially from June to August. Amorini et al. [106] show minimal increments in years with substantial drought on Turkey oak in Central Italy. In addition to the positive effect of May and June precipitation on radial growth, this study also shows a positive correlation for May and June minimum temperatures and the March and April maximum temperatures of the current year. In contrast, no statistically significant correlation was found between the previous year's climate data and radial growth. Similar results were also found in a study by [107] carried out in Turkey oak stands in Slovakia, where it was confirmed that the most significant positive effect on the magnitude of growth is due to the amount of precipitation during the growing season, especially in May and June. Another study from Central Italy confirms the importance of the May and June rainfall of the current year [95]. Conversely, the effect of temperature on the radial growth of Turkey oak is not as significant as the effect of precipitation, as in the case of this study or the study in Bulgaria [108].

In Central Europe, specifically in the plots SK_3 and CZ_4, a higher number of observed positive correlations between RWI and precipitation were identified when compared to the Southern regions represented by IT_1 and BG_2. In the research plots of Central Europe, seven significant correlation coefficients were found, while only three significant correlation coefficients were observed in Southern Europe. For instance, in Northern Europe, the English oak (*Quercus robur* L.) exhibits strong positive associations with precipitation, particularly during the summer, where growth anomalies are linked to oak growth, especially in combination with higher temperatures [109]. The lower positive correlations with precipitation are linked to the lower available precipitation during the vegetation season in Southern Europe, while this general statement is also well documented [110].

4.4. Tree-Ring Growth Cycles across the Studied Plots

The Turkey oak has an advantage over other Central European oaks in that it is known to be significantly more drought tolerant, but at the same time, its growth is not as aggressive as, for example, sessile oak or English oak (*Quercus robur* L.) [64,111,112]. The results in the investigated plots in Southern Europe show that Turkey oak grows steadily, and the largest fluctuations in growth are recorded specifically concerning monthly temperatures and lack of precipitation during the summer months. This is accompanied by the results of the spectral analyses, which show that the tree-ring chronologies of Turkey oak in Italy and Bulgaria fluctuate between six and eight years. These cycles of around seven years are most often associated with the 7-year temperature cycle, which is typical of the European continent [113,114]. The tree-ring series of Turkey oak may be most closely associated with the temperature cycle. Contrastingly, the study plots in Central Europe (SK_3 and CZ_4) follow a 2 to 5-year cycle, which, again, corresponds to a significantly shorter precipitation cycle in Europe [115–117]. Based on these findings, a greater influence of precipitation on tree-ring increment is evident in Central Europe research plots, as indicated by the stronger correlations. Thus, the drier climate of Southern Europe produces a longer cycle of tree-ring increment compared to plots located in Central Europe. The cycles studied may also be closely related to fructification. Masting of Turkey oak in Southern European countries is around five to seven years but can be shorter based on sufficient moisture [118,119]. This correlates with the results found in this study or with the tree-ring series from Italy and Bulgaria (6 to 8-year cycle). On the contrary, as mentioned in Central Europe, due to abundant precipitation, the situation is different and seed years repeat significantly more frequently in 2 to 5-year cycles. It is due to higher precipitation during the growing season in Central Europe, where acorn fruiting occurs more frequently in Turkey oak, which is close to the natural 2-year period [120]. Tree seed production is associated with tree-ring growth and the influence of weather in both the previous and current growing seasons [121–123]. The frequency of tree-ring growth can serve as an indicator of the theoretical fertility of trees, where we observed 2 to 5-year cycles.

4.5. Potential for Coppice Forests

Due to climate change, rising temperatures, and more frequent long-term droughts, the species composition and stand structure will change in the coming decades [124]. The Turkey oak may become a valuable alternative tree species with great potential for adaptation to changing environmental conditions, especially in Central Europe [4]. This tree species generally needs lower amounts of air and soil moisture for its growth [125]. The resilience of Turkey oak to climatic extremes is also confirmed by [112], comparing the effect of the climate on the growth of Turkey oak and sessile oak in northern Hungary, proving that Turkey oak can better recover from prolonged periods of drought. Similar characteristics are described in this study, indicating the high resilience of this tree species, which is documented by the low number of negative years in the RWI and the relatively small influence of climatic factors. This is because after the second rotation of the coppice, surface roots are being formed as the old ones are already dying [108]. Generally, high temperatures in the growing season can induce increased water stress and a subsequent reduction in radial growth due to increased water loss through evapotranspiration and soil moisture evaporation. Stafasani & Toromani [126] reported that most of the coppice Turkey oak mixture stands in Albania showed extreme drought in June of the current year as a limiting factor for growth. The negative effect of June and July temperatures on the growth of young oaks has previously been observed at several sites in continental Europe [127,128], in northern Spain [129] and also in the Mediterranean [130,131]. Moreover, the positive relationship of radial growth with June precipitation shows that water balance in this month is critical for phenology [132,133].

4.6. Study Limitations and Ideas

The number of samples collected for dendrochronological analysis was sufficient, which was confirmed by EPS analysis. The gathered data exhibit satisfactory EPS ranging from 0.90 to 0.96. It is crucial to note that the minimum EPS threshold for data utilization is 0.85, as stipulated by this indicator [88,134]. Furthermore, the collected data, supported by EPS, demonstrate that an ample number of samples has been collected to describe the chronology relating to climate, indicating a robust data series [135].

Before the end of this discussion, it is necessary to mention that this study was constrained by the limited number of research plots in each of the evaluated countries. Consequently, the study was not focused on a detailed description of regional climatic conditions but rather on depicting conditions within the research plots. For this reason, information from the nearest meteorological stations in the studied locations was also utilized. The repetition of research plots is ensured by two robust chronologies from Central Europe and two from Southern Europe. This study presents information on tree-ring growth frequency in Central and Southern Europe.

On the other hand, similar future studies, in addition to the limitations mentioned above (such as the small number of plots), should also focus on other important factors influencing the structure, production, and response of trees to climate change, such as genetics. The provenance of tree species significantly affects the production potential of wood, carbon sequestration, and resistance to climatic extremes [136,137]. Significant genetic diversity was also found in Turkey oak [138,139]. In the future, further research should also focus on the influence of various silviculture regimes or the admixture effect of other tree species in oak stands in the context of adaptation to climate change.

5. Conclusions

In conclusion, it could be argued that coppice forests represent a suitable alternative to standard forest management practices. The suitability can be affected by tree species composition, especially during ongoing climate change. This was confirmed in the presented study where the evaluated Turkey oak coppice forests in Italy, Bulgaria, Slovakia, and Czechia showed, on the one hand, relatively high resistance to unfavorable climatic factors, including climatic extremes, and on the other, adequate values of timber production.

The lowest influence of climatic factors on growth was found in Italy and Bulgaria compared to the tree's climate sensitivity in Central Europe—on the northern edge of its natural distribution range. The spectral analysis also showed that the research areas in Southern Europe go through longer 6 to 8-year growth cycles in radial growth compared to Central Europe (shorter cycles of 2.4 to 4.8 years). It was found that the main limiting factor for growth was the lack of precipitation during the growing season, whereas temperatures played almost no role in the radial growth processes. Regarding the fact that in recent years there has been significant warming in Europe, therefore Turkey oak can be identified as a crucial tree species in terms of adaptation strategies to climate change. In general, this study is the basis for understanding and predicting the growth responses of Turkey oak coppice stands to the climate under conditions of ongoing global climate change in Europe. For future long-term research, however, it is necessary to further focus on other factors, such as genetic origin or different silviculture practices in the context of climate change.

Author Contributions: M.Š. and S.V. designed the research. M.Š., Z.V., V.H. and L.B. collected samples in the field. V.Š., V.H. and A.P. analyzed the data. I.Š., Z.S. and L.B. sourced meteorological data. M.Š., S.V., I.L., Z.V., V.Š. and J.C. prepared first draft of the manuscript, All authors have read and agreed to the published version of the manuscript.

Funding: This research was funded by the Internal Grant Agency, Faculty of Forestry and Wood Sciences, Czech University of Life Sciences Prague (IGA FFWS 2021, Project No A_21_17). This research received funding from the LIFE Climate Action sub-programme of the European Union—project CLIMAFORCEELIFE (LIFE19 CCA/SK/001276).

Data Availability Statement: Monthly temperature and precipitation data for the Czech Republic are available from the Czech Hydrometeorological Institute (www.chmi.cz, accessed on 25 January 2023). Monthly temperature and precipitation data for Slovakia are available from the Slovak Hydrometeorological Institute (www.shmu.sk, accessed on 25 January 2023). Monthly climatic data for Italy are available from the Italian Civil Protection Authority, Basilicata Region (www.centrofunzionalebasilicata.it, accessed on 16 February 2023). Monthly climatic data for Bulgaria are available from the National Institute of Meteorology and Hydrology (www.weather.bg, accessed on 16 February 2023). The tree-ring data presented in this study are available on request from the corresponding author.

Acknowledgments: Acknowledgement goes to the Czech Hydrometeorological Institute, Slovak Hydrometeorological Institute, Italian Civil Protection Authority, and National Institute of Meteorology and Hydrology in Bulgaria for providing the datasets. We would also like to thank both Richard Lee Manore, a native speaker, and Jitka Šišáková, an expert in the field, for checking the English of this paper.

Conflicts of Interest: The authors declare no conflict of interest.

References

1. Canadell, J.G.; Raupach, M.R. Managing Forests for Climate Change Mitigation. *Science* **2008**, *320*, 1456–1457. [CrossRef] [PubMed]
2. Lindner, M.; Fitzgerald, J.; Zimmerman, N.; Reyer, C.; Delzon, S.; van der Maaten, E.; Schelhaas, M.-J.; Lasch, P.; Eggers, J.; van der Maaten-Thunissen, M.; et al. Climate Change and European Forests: What Do We Know, What Are the Uncertainties, and What Are the Implications for Forest Management? *J. Environ. Manag.* **2014**, *146*, 69–83. [CrossRef] [PubMed]
3. Vacek, Z.; Vacek, S.; Cukor, J.; Bulušek, D.; Slávik, M.; Lukáčik, I.; Štefančík, I.; Sitková, Z.; Eşen, D.; Ripullone, F.; et al Dendrochronological Data from Twelve Countries Proved Definite Growth Response of Black Alder ([L.] Gaertn.) to Climate Courses across Its Distribution Range. *Cent. Eur. For. J.* **2022**, *68*, 139–153. [CrossRef]
4. Vacek, Z.; Vacek, S.; Cukor, J. European Forests under Global Climate Change: Review of Tree Growth Processes, Crises and Management Strategies. *J. Environ. Manag.* **2023**, *332*, 117353. [CrossRef] [PubMed]
5. Lal, R. Soil Carbon Sequestration to Mitigate Climate Change. *Geoderma* **2004**, *123*, 1–22. [CrossRef]
6. Wiesmeier, M.; Prietzel, J.; Barthold, F.; Spörlein, P.; Geuß, U.; Hangen, E.; Reischl, A.; Schilling, B.; von Lützow, M.; Kögel-Knabner, I. Storage and Drivers of Organic Carbon in Forest Soils of Southeast Germany (Bavaria)—Implications for Carbon Sequestration. *For. Ecol. Manag.* **2013**, *295*, 162–172. [CrossRef]
7. Cukor, J.; Vacek, Z.; Vacek, S.; Linda, R.; Podrázský, V. Biomass Productivity, Forest Stability, Carbon Balance, and Soil Transformation of Agricultural Land Afforestation: A Case Study of Suitability of Native Tree Species in the Submontane Zone in Czechia. *Catena* **2022**, *210*, 105893. [CrossRef]
8. Lévesque, M.; Saurer, M.; Siegwolf, R.; Eilmann, B.; Brang, P.; Bugmann, H.; Rigling, A. Drought Response of Five Conifer Species under Contrasting Water Availability Suggests High Vulnerability of Norway Spruce and European Larch. *Glob. Chang. Biol.* **2013**, *19*, 3184–3199. [CrossRef]
9. González de Andrés, E.; Seely, B.; Blanco, J.A.; Imbert, J.B.; Lo, Y.-H.; Castillo, F.J. Increased Complementarity in Water-Limited Environments in Scots Pine and European Beech Mixtures under Climate Change. *Ecohydrology* **2017**, *10*, e1810. [CrossRef]
10. Niemelä, J.; Young, J.; Alard, D.; Askasibar, M.; Henle, K.; Johnson, R.; Kurttila, M.; Larsson, T.-B.; Matouch, S.; Nowicki, P.; et al. Identifying, Managing and Monitoring Conflicts between Forest Biodiversity Conservation and Other Human Interests in Europe. *For. Policy Econ.* **2005**, *7*, 877–890. [CrossRef]
11. Agnoletti, M. *The Conservation of Cultural Landscapes*; CAB International: Chatham, UK, 2006; ISBN 1845931548.
12. Agnoletti, M. The Degradation of Traditional Landscape in a Mountain Area of Tuscany during the 19th and 20th Centuries: Implications for Biodiversity and Sustainable Management. *For. Ecol. Manag.* **2007**, *249*, 5–17. [CrossRef]
13. McGrath, M.J.; Luyssaert, S.; Meyfroidt, P.; Kaplan, J.O.; Bürgi, M.; Chen, Y.; Erb, K.; Gimmi, U.; McInerney, D.; Naudts, K.; et al. Reconstructing European Forest Management from 1600 to 2010. *Biogeosciences* **2015**, *12*, 4291–4316. [CrossRef]
14. Poleno, Z.; Vacek, S.; Podrázský, V.; Remeš, J.; Štefančík, I.; Mikeska, M.; Kobliha, J.; Kupka, I.; Malík, V.; Turčáni, M.; et al. *Pěstování Lesů III—Praktické Postupy Pěstování Lesů [Silviculture III. Practical Methods in Silviculture]*; Lesnická Práce: Kostelec nad Černými lesy, Czech Republic, 2009; ISBN 978-80-87154-34-2.
15. Corcuera, L.; Camarero, J.J.; Sisó, S.; Gil-Pelegrín, E. Radial-Growth and Wood-Anatomical Changes in Overaged Quercus Pyrenaica Coppice Stands: Functional Responses in a New Mediterranean Landscape. *Trees—Struct. Funct.* **2006**, *20*, 91–98. [CrossRef]
16. Fabbio, G.; Cutini, A. Il Ceduo Oggi: Quale Gestione Oltre Le Definizioni? *Forest* **2017**, *14*, 257–274. [CrossRef]
17. Cutini, A.; Ferretti, M.; Bertini, G.; Brunialti, G.; Bagella, S.; Chianucci, F.; Fabbio, G.; Fratini, R.; Riccioli, F.; Caddeo, C.; et al. Testing an Expanded Set of Sustainable Forest Management Indicators in Mediterranean Coppice Area. *Ecol. Indic.* **2021**, *130*, 108040. [CrossRef]

18. Camponi, L.; Cardelli, V.; Cocco, S.; Serrani, D.; Salvucci, A.; Cutini, A.; Agnelli, A.; Fabbio, G.; Bertini, G.; Roggero, P.P.; et al. Effect of Coppice Conversion into High Forest on Soil Organic C and Nutrients Stock in a Turkey Oak (*Quercus cerris* L.) Forest in Italy. *J. Environ. Manag.* **2022**, *312*, 114935. [CrossRef]
19. Bohensky, E.L.; Maru, Y. Indigenous Knowledge, Science, and Resilience: What Have We Learned from a Decade of International Literature on "Integration"? *Ecol. Soc.* **2011**, *16*, 19. [CrossRef]
20. Johnson, J.T.; Howitt, R.; Cajete, G.; Berkes, F.; Louis, R.P.; Kliskey, A. Weaving Indigenous and Sustainability Sciences to Diversify Our Methods. *Sustain. Sci.* **2016**, *11*, 1–11. [CrossRef]
21. Scullion, J.J.; Vogt, K.A.; Winkler-Schor, S.; Sienkiewicz, A.; Peña, C.; Hajek, F. Designing Conservation-Development Policies for the Forest Frontier. *Sustain. Sci.* **2016**, *11*, 295–306. [CrossRef]
22. Suchomel, C.; Becker, G.; Pyttel, P. Fully Mechanized Harvesting in Aged Oak Coppice Stands. *For. Prod. J.* **2011**, *61*, 290–296. [CrossRef]
23. Kull, K. Growth Form Parameters of Clonal Herbs. In *Consortium Masingii: A Festschrift for Viktor Masing*; Aaviksoo, K., Kull, K., Paal, J., Trass, H., Eds.; Tartu University: Tartu, Estonia, 1995; pp. 106–115.
24. Buckley, P. Coppice Restoration and Conservation: A European Perspective. *J. For. Res.* **2020**, *25*, 125–133. [CrossRef]
25. Cervellini, M.; Fiorini, S.; Cavicchi, A.; Campetella, G.; Simonetti, E.; Chelli, S.; Canullo, R.; Gimona, A. Relationships between Understory Specialist Species and Local Management Practices in Coppiced Forests—Evidence from the Italian Apennines. *For. Ecol. Manag.* **2017**, *385*, 35–45. [CrossRef]
26. Vrška, T.; Janík, D.; Pálková, M.; Adam, D.; Trochta, J. Below-and above-Ground Biomass, Structure and Patterns in Ancient Lowland Coppices. *IForest* **2017**, *10*, 23–31. [CrossRef]
27. Rackham, O. *Woodlands*; Collins: London, UK, 2006.
28. Peterken, G.F. *Natural Woodland: Ecology and Conservation in Northern Temperate Regions*; Cambridge University Press: Cambridge, UK, 1996; ISBN 0521367921.
29. Verheyen, K.; Bossuyt, B.; Hermy, M.; Tack, G. The Land Use History (1278-1990) of a Mixed Hardwood Forest in Western Belgium and Its Relationship with Chemical Soil Characteristics. *J. Biogeogr.* **1999**, *26*, 1115–1128. [CrossRef]
30. Schweingruber, F.H. *Wood Structure and Environment (Springer Series in Wood Science)*; Springer: Berlin/Heidelberg, Germany, 2007; ISBN 978-3-540-48299-4.
31. Campetella, G.; Canullo, R.; Gimona, A.; Garadnai, J.; Chiarucci, A.; Giorgini, D.; Angelini, E.; Cervellini, M.; Chelli, S.; Bartha, S. Scale-Dependent Effects of Coppicing on the Species Pool of Late Successional Beech Forests in the Central Apennines, Italy. *Appl. Veg. Sci.* **2016**, *19*, 474–485. [CrossRef]
32. Bartha, S.; Merolli, A.; Campetella, G.; Canullo, R. Changes of Vascular Plant Diversity along a Chronosequence of Beech Coppice Stands, Central Apennines, Italy. *Plant Biosyst.* **2008**, *142*, 572–583. [CrossRef]
33. Vild, O.; Roleček, J.; Hédl, R.; Kopecký, M.; Utinek, D. Experimental Restoration of Coppice-with-Standards: Response of Understorey Vegetation from the Conservation Perspective. *For. Ecol. Manag.* **2013**, *310*, 234–241. [CrossRef]
34. Vacek, S.; Vacek, Z.; Ulbrichová, I.; Bulušek, D.; Prokupková, A.; Král, J.; Vančura, K. Biodiversity Dynamics of Differently Managed Lowland Forests Left to Spontaneous Development in Central Europe. *Austrian J. For. Sci.* **2019**, *136*, 249–282.
35. Unrau, A.; Becker, G.; Spinelli, R.; Lazdina, D.; Magagnotti, N.; Nicolescu, V.N.; Buckley, P.; Bartlett, D.; Kofman, P.D. *Coppice Forests in Europe*; Unrau, A., Becker, G., Spinelli, R., Lazdina, D., Magagnotti, N., Nicolescu, V.-N., Buckley, P., Barlett, D., Kofman, P.D., Eds.; Albert Ludwig University Freiburg: Freiburg im Breisgau, Germany, 2018; ISBN 9783981734027.
36. Matula, R.; Šrámek, M.; Kvasnica, J.; Uherková, B.; Slepička, J.; Matoušková, M.; Kutchartt, E.; Svátek, M. Pre-Disturbance Tree Size, Sprouting Vigour and Competition Drive the Survival and Growth of Resprouting Trees. *For. Ecol. Manag.* **2019**, *446*, 71–79. [CrossRef]
37. Dickmann, D.I. Silviculture and Biology of Short-Rotation Woody Crops in Temperate Regions: Then and Now. *Biomass Bioenergy* **2006**, *30*, 696–705. [CrossRef]
38. Del Tredici, P. Sprouting in Temperate Trees: A Morphological and Ecological Review. *Bot. Rev.* **2001**, *67*, 121–140. [CrossRef]
39. Vacek, Z.; Vacek, S.; Bílek, L.; Král, J.; Ulbrichová, I.; Simon, J.; Bulušek, D. Impact of Applied Silvicultural Systems on Spatial Pattern of Hornbeam-Oak Forests. *Cent. Eur. For. J.* **2018**, *64*, 33–45. [CrossRef]
40. Čížek, L.; Šebek, P.; Bače, R.; Beneš, J.; Doležal, J.; Dvorský, M.; Miklín, J.; Svoboda, M. *Metodika Péče o Druhově Bohaté (Světlé) Lesy*; Entomologický ústav, Biologické centrum AV ČR, v.v.i.: České Budějovice, Czech Republic, 2016; p. 126.
41. Müllerová, J.; Hédl, R.; Szabó, P. Coppice Abandonment and Its Implications for Species Diversity in Forest Vegetation. *For. Ecol. Manag.* **2015**, *343*, 88–100. [CrossRef]
42. Volařík, D.; Svátek, M.; Šenfeldr, M.; Kučera, A.; Šrámek, M.; Dreslerová, J.; Matula, R. Variation in Canopy Openness among Main Structural Types of Woody Vegetation in a Traditionally Managed Landscape. *Folia Geobot.* **2017**, *52*, 15–32. [CrossRef]
43. Franklin, J.F.; Spies, T.A.; Van Pelt, R.; Carey, A.B.; Thornburgh, D.A.; Berg, D.R.; Lindenmayer, D.B.; Harmon, M.E.; Keeton, W.S.; Shaw, D.C. Disturbances and Structural Development of Natural Forest Ecosystems with Silvicultural Implications, Using Douglas-Fir Forests as an Example. *For. Ecol. Manag.* **2002**, *155*, 399–423. [CrossRef]
44. Madera, P.; Slach, T.; Úradnícek, L.; Lacina, J.; Cernušáková, L.; Friedl, M.; Repka, R.; Bucek, A. Tree Shape and Form in Ancient Coppice Woodlands. *J. Landsc. Ecol. Republic)* **2017**, *10*, 49–62. [CrossRef]
45. Bouvet, A.; Paillet, Y.; Archaux, F.; Tillon, L.; Denis, P.; Gilg, O.; Gosselin, F. Effects of Forest Structure, Management and Landscape on Bird and Bat Communities. *Environ. Conserv.* **2016**, *43*, 148–160. [CrossRef]

46. UN-ECE; FAO. *Forest Resources of Europe, CIS, North America, Australia, Japan and New Zealand (Industrialized Temperate/Boreal Countries): UN-ECE/FAO Contribution to the Global Forest Resources Assessment 2000: Main Report*; United Nations: Geneva Switzerland, 2000; ISBN 9211167353.
47. Cutini, A.; Brunialti, G.; Amici, V.; Bagella, S.; Bertini, G.; Caddeo, C.; Calderisi, M.; Chianucci, F.; Ciucchi, B.; Corradini, S Report: Integrated Scientific Synthesis and Evaluation of Project Results–LIFE FutureForCoppiceS–Shaping Future Forestry for Sustainable Coppices in Southern Europe: The Legacy of Past Management Trials (with Synthesis for Resource Managers and Policy Mak. *Deliv. LIFE Futur. Proj. Action B* **2019**, *9*, 108.
48. MPRV. *Správa o Lesnom Hospodárstve v Slovenskej Republike Za Rok 2021*; Zelená Správa: Bratislava, Slovakia, 2022.
49. Maděra, P.; Machala, M.; Slach, T.; Friedl, M.; Černušakova, L.; Volařik, D.; Buček, A. Predicted Occurrence of Ancient Coppice Woodlands in the Czech Republic. *IForest* **2017**, *10*, 788–795. [CrossRef]
50. Kadavý, J.; Kneifl, M.; Servus, M.; Knott, R.; Hurt, V.; Flora, M. *Nízký a Střední Les—Plnohodnotná Alternativa Hospodaření Malých a Středních Vlastníků Lesa—Obecná Východiska*; Lesnická práce, s. r. o. nakladatelství a vydavatelství: Kostelec nad Černými lesy Czech Republic, 2011; ISBN 978-80-87154-96-0.
51. Mairota, P.; Buckley, P.; Suchomel, C.; Heinsoo, K.; Verheyen, K.; Hédl, R.; Terzuolo, P.G.; Sindaco, R.; Carpanelli, A. Integrating Conservation Objectives into Forest Management: Coppice Management and Forest Habitats in Natura 2000 Sites *iForest—Biogeosci. For.* **2016**, *9*, 560–568. [CrossRef]
52. Spiecker, H.; Hein, S.; Makkonen-Spiecker, K.; Thies, M. (Eds.) *Valuable Broadleaved Forests in Europe (European Forest Institute Research Report; 22)*; Brill: Leiden, The Netherlands, 2009; ISBN 978 90 04 16795 7.
53. Šplíchalová, M. Aspects of Oak (*Quercus* Sp.) Management in Spain and Its Application. Coppice Forests: Past, Present and Future. In Proceedings of the Coppice Forests: Past, Present and Future, Brno, Czech Republic, 9–11 April 2015; Vild, O., Ed.; pp 9–11.
54. Nicolescu, V.-N.; Carvalho, J.; Hochbichler, E.; Bruckman, V.; Piqué-Nicolau, M.; Hernea, C.; Viana, H.; Štochlová, P.; Ertekin, M. Tijardovic, M.; et al. *Silvicultural Guidelines for European Coppice Forests. COST Action FP1301 Reports*; Albert Ludwig University of Freiburg: Freiburg, Germany, 2017.
55. Coppini, M.; Hermanin, L. Restoration of Selective Beech Coppices: A Case Study in the Apennines (Italy). *For. Ecol. Manag.* **2007** *249*, 18–27. [CrossRef]
56. Johann, E. Coppice Forests in Austria: The Re-Introduction of Traditional Management Systems in Coppice Forests in Response to the Decline of Species and Landscape and under the Aspect of Climate Change. *For. Ecol. Manag.* **2021**, *490*, 119129. [CrossRef]
57. Valente, A.M.; Pelayo, A.; Figueiredo, A.M.; Fonseca, C.; Torres, R.T. Overabundant Wild Ungulate Populations in Europe Management with Consideration of Socio-Ecological Consequences. *Mamm. Rev.* **2020**, *50*, 353–366. [CrossRef]
58. Carpio, A.J.; Apollonio, M.; Acevedo, P. Wild Ungulate Overabundance in Europe: Contexts, Causes, Monitoring and Management Recommendations. *Mamm. Rev.* **2021**, *51*, 95–108. [CrossRef]
59. Bran, D.; Lobréaux, O.; Maistre, M.; Perret, P.; Romane, F. Germination of Quercus Ilex and Q. Pubescens in a Q. Ilex Coppice—Long-Term Consequences. *Vegetatio* **1990**, *87*, 45–50. [CrossRef]
60. Viscosi, V.; Fortini, P.; D'Imperio, M. A Statistical Approach to Species Identification on Morphological Traits of European White Oaks: Evidence of Morphological Structure in Italian Populations of Quercus Pubescens Sensu Lato. *Acta Bot. Gall.* **2011**, *158* 175–188. [CrossRef]
61. Gasparini, P. *Italian National Forest Inventory—Methods and Results of the Third Survey*; Gasparini, P., Di Cosmo, L., Floris, A., De Laurentis, D., Eds.; Springer: Cham, Switzerland, 2022; ISBN 978-3-030-98677-3.
62. Rossnev, B.; Petkov, P.; Mirchev, P.; Georgiev, G.; Georgieva, M.; Matova, M. System Approach for Determination and Improvement of *Quercus cerris* L. Forests Status in Bulgaria. In Proceedings of the Integral Protection of Forests; Scientific-Technological Platform Belgrade, Serbia, 2007; pp. 186–191.
63. Rossnev, B. *Quercus cerris Forests Status in Bulgaria and Measures for Improvement*; Bulgarian Academy of Sciences, Forest Research Institute: Sofia, Bulgaria, 2006; p. 120.
64. Kostić, S.; Levanič, T.; Orlović, S.; Matović, B.; Stojanović, D.B. Turkey Oak (*Quercus cerris* L.) Is More Drought Tolerant and Better Reflects Climate Variations Compared to Pedunculate Oak (*Quercus Robur* L.) in Lowland Mixed Forests in Northwestern Serbia A Stable Carbon Isotope Ratio (Δ13C) and Radial Growth Approach. *Ecol. Indic.* **2022**, *142*, 109242. [CrossRef]
65. Toth, D.; Maitah, M.; Maitah, K.; Jarolínová, V. The Impacts of Calamity Logging on the Development of Spruce Wood Prices in Czech Forestry. *Forests* **2020**, *11*, 283. [CrossRef]
66. Šimůnek, V.; Sharma, R.P.; Vacek, Z.; Vacek, S.; Hůnová, I. Sunspot Area as Unexplored Trend inside Radial Growth of European Beech in Krkonoše Mountains: A Forest Science from Different Perspective. *Eur. J. For. Res.* **2020**, *139*, 999–1013. [CrossRef]
67. Sharma, R.P.; Vacek, Z.; Vacek, S. Individual Tree Crown Width Models for Norway Spruce and European Beech in Czech Republic. *For. Ecol. Manag.* **2016**, *366*, 208–220. [CrossRef]
68. Kraft, G. *Beiträge Zur Lehre von Den Durchforstungen, Schlagstellungen Und Lichtungshieben*; Klindworth's Verlag: Hannover Germany, 1884.
69. Remeš, J.; Bílek, L.; Novák, J.; Vacek, Z.; Vacek, S.; Putalová, T.; Koubek, L. Diameter Increment of Beech in Relation to Social Position of Trees, Climate Characteristics and Thinning Intensity. *J. For. Sci.* **2015**, *61*, 456–464. [CrossRef]
70. RRinntech. *TSAP-WINTM: Time Series Analysis and Presentation for Dendrochronology and Related Applications*; Rinntech: Heidelberg Switzerland, 2010.

71. Fabrika, M.; Ďurský, J. Algorithms and Software Solution of Thinning Models for SIBYLA Growth Simulator. *J. For. Sci.* **2005**, *51*, 431–445. [CrossRef]
72. Clark, P.J.; Evans, F.C. Distance to Nearest Neighbor as a Measure of Spatial Relationships in Populations. *Ecology* **1954**, *35*, 445–453. [CrossRef]
73. Pretzsch, H. Wissen Nutzbar Machen Für Das Management von Waldökosystemen. *Allg. Forstz./Der Wald.* **2006**, *61*, 1158–1159.
74. Füldner, K. Strukturbeschreibung in Mischbeständen. *Forstarchiv* **1995**, *66*, 235–606.
75. Jaehne, S.; Dohrenbusch, A. A Method to Evaluate Forest Stand Diversity. *Forstwiss* **1997**, *116*, 1–6.
76. Seifert, T.; Schuck, J.; Block, J.; Pretzsch, H. Simulation von Biomasse-Und Nährstoffgehalt von Waldbäumen. In Proceedings of the Deutscher Verband Forstlicher Forschungsanstalten Sektion Ertragskunde: Jahrestagung; Nagel, J., Ed.: Nordwestdeutsche Forstliche Versuchsanstalt, Abteilung Waldwachstum: Göttingen, Germany, 2006; Volume 29, pp. 208–223.
77. Vacek, Z.; Vacek, S.; Esen, D.; Yildiz, O.; Král, J.; Gallo, J. Effect of Invasive Rhododendron Ponticum l. on Natural Regeneration and Structure of Fagus Orientalis Lipsky Forests in the Black Sea Region. *Forests* **2020**, *11*, 603. [CrossRef]
78. Petráš, R.; Pajtík, J. Sústava Česko-Slovenských Objemových Tabuliek Drevín. *Lesn. časopis* **1991**, *37*, 49–56.
79. Reineke, L.H. Prefecting a Stand-Density Index for Evenaged Forests. *J. Agric. Res.* **1933**, *46*, 627–638.
80. Crookston, N.L.; Stage, A.R. *Percent Canopy Cover and Stand Structure Statistics from the Forest Vegetation Simulator*; US Department of Agriculture, Forest Service, Rocky Mountain Research Station: Fort Collins, CO, USA, 1999.
81. R Core Team. *R: A Language and Environment for Statistical Computing 2014*; R Foundation Statistical Computing: Vienna, Austria, 2018.
82. Bunn, A.G. A Dendrochronology Program Library in R (DplR). *Dendrochronologia* **2008**, *26*, 115–124. [CrossRef]
83. Bunn, A.G. Statistical and Visual Crossdating in R Using the DplR Library. *Dendrochronologia* **2010**, *28*, 251–258. [CrossRef]
84. van der Maaten-Theunissen, M.; van der Maaten, E.; Bouriaud, O. PointRes: An R Package to Analyze Pointer Years and Components of Resilience. *Dendrochronologia* **2015**, *35*, 34–38. [CrossRef]
85. Bunn, A.; Mikko, K. *Chronology Building in Dplr*; R Foundation for Statistical Computing: Vienna, Austria, 2018.
86. Schweingruber, F.H.; Eckstein, D.; Serre-Bachet, F.; Braker, O.U. Identification, Presentation and Interpretation of Event Years and Pointer Years in Dendrochronology. *Dendrochronologia* **1990**, *8*, 9–38.
87. Neuwirth, B.; Schweingruber, F.H.; Winiger, M. Spatial Patterns of Central European Pointer Years from 1901 to 1971. *Dendrochronologia* **2007**, *24*, 79–89. [CrossRef]
88. Fritts, H.C. *Tree Rings and Climate.*; Academic Press Inc.: New York, NY, USA, 1976.
89. Speer, J.H. *Fundamentals of Tree-Ring Research*; University of Arizona Press: Tucson, Arizona, 2010; ISBN 978-0-816-52684-0.
90. Biondi, F.; Waikul, K. DENDROCLIM2002: A C++ Program for Statistical Calibration of Climate Signals in Tree-Ring Chronologies. *Comput. Geosci.* **2004**, *30*, 303–311. [CrossRef]
91. StatSoft Power Solutions, Inc. *Statistica Electronic Manual 2013*; StatSoft Power Solutions, Inc.: Tulsa, OK, USA, 2013.
92. Šmilauer, P.; Lepš, J. *Multivariate Analysis of Ecological Data Using CANOCO 5*; Cambridge University Press: New York, NY, USA, 2014.
93. Karamizadeh, S.; Abdullah, S.M.; Manaf, A.A.; Zamani, M.; Hooman, A. An Overview of Principal Component Analysis. *J. Signal Inf. Process.* **2013**, *4*, 173–175. [CrossRef]
94. Notarangelo, M.; La Marca, O.; Moretti, N. Long-Term Effects of Experimental Cutting to Convert an Abandoned Oak Coppice into Transitional High Forest in a Protected Area of the Italian Mediterranean Region. *For. Ecol. Manag.* **2018**, *430*, 241–249. [CrossRef]
95. Di Filippo, A.; Alessandrini, A.; Biondi, F.; Blasi, S.; Portoghesi, L.; Piovesan, G. Climate Change and Oak Growth Decline: Dendroecology and Stand Productivity of a Turkey Oak (*Quercus cerris* L.) Old Stored Coppice in Central Italy. *Ann. For. Sci.* **2010**, *67*, 706. [CrossRef]
96. Šrámek, M.; Volařík, D.; Ertas, A.; Matula, R. The Effect of Coppice Management on the Structure, Tree Growth and Soil Nutrients in Temperate Turkey. *J. For. Sci.* **2015**, *61*, 27–34. [CrossRef]
97. Vancura, K.; Simkova, M.; Vacek, Z.; Vacek, S.; Gallo, J.; Simunek, V.; Podrazsky, V.; Stefancik, I.; Hájek, V.; Prokupkova, A.; et al. Effects of Environmental Factors and Management on Dynamics of Mixed Calcareous Forests under Climate Change in Central European Lowlands. *Dendrobiology* **2022**, *87*, 79–100. [CrossRef]
98. Longuetaud, F.; Seifert, T.; Leban, J.-M.; Pretzsch, H. Analysis of Long-Term Dynamics of Crowns of Sessile Oaks at the Stand Level by Means of Spatial Statistics. *For. Ecol. Manag.* **2008**, *255*, 2007–2019. [CrossRef]
99. Plauborg, K.U. Analysis of Radial Growth Responses to Changes in Stand Density for Four Tree Species. *For. Ecol. Manag.* **2004**, *188*, 65–75. [CrossRef]
100. Kunstler, G.; Albert, C.H.; Courbaud, B.; Lavergne, S.; Thuiller, W.; Vieilledent, G.; Zimmermann, N.E.; Coomes, D.A. Effects of Competition on Tree Radial-Growth Vary in Importance but Not in Intensity along Climatic Gradients. *J. Ecol.* **2011**, *99*, 300–312. [CrossRef]
101. Bravo-Oviedo, A.; Condés, S.; Del Río, M.; Pretzsch, H.; Ducey, M.J. Maximum Stand Density Strongly Depends on Species-Specific Wood Stability, Shade and Drought Tolerance. *Forestry* **2018**, *91*, 459–469. [CrossRef]
102. Giuggiola, A.; Bugmann, H.; Zingg, A.; Dobbertin, M.; Rigling, A. Reduction of Stand Density Increases Drought Resistance in Xeric Scots Pine Forests. *For. Ecol. Manag.* **2013**, *310*, 827–835. [CrossRef]

103. Sterck, F.; Vos, M.; Hannula, S.E.; de Goede, S.; de Vries, W.; den Ouden, J.; Nabuurs, G.J.; van der Putten, W.; Veen, C. Optimizing Stand Density for Climate-Smart Forestry: A Way Forward towards Resilient Forests with Enhanced Carbon Storage under Extreme Climate Events. *Soil Biol. Biochem.* **2021**, *162*, 108396. [CrossRef]
104. Sohn, J.A.; Hartig, F.; Kohler, M.; Huss, J.; Bauhus, J. Heavy and Frequent Thinning Promotes Drought Adaptation in Pinus Sylvestris Forests. *Ecol. Appl.* **2016**, *26*, 2190–2205. [CrossRef]
105. Seidl, R.; Rammer, W.; Lexer, M.J. Adaptation Options to Reduce Climate Change Vulnerability of Sustainable Forest Management in the Austrian Alps. *Can. J. For. Res.* **2011**, *41*, 694–706. [CrossRef]
106. Amorini, E.; Biocca, M.; Manetti, M.C.; Motta, E. Dendroecological Study in a Declining Coppice Stand. *Ann. Des Sci. For.* **1996**, *53*, 731–742. [CrossRef]
107. Priwitzer, T.; Pajtík, J.; Ištoňa, J.; Pavlenda, P. Vplyv Zrážok Na Dynamiku Rastu, Fenológiu a Opad Lesných Drevín. In Proceedings of the International Scientific Conference Bioclimatology and Natural Hazards: Proceedings, Zvolen-Polana Nad Detvou, Slovakia, 17–20 September; Střelcová, K., Škvarenina, J., Blažec, M., Eds.; Slovenská Bioklimatologická Spoločnost: Poľana nad Detvou, Slovakia, 2007.
108. Zafirov, N.; Kostov, G. Main Stress Factors in Coppice Oak Forests in Western Bulgaria. *Silva Balc.* **2019**, *20*, 37–52. [CrossRef]
109. Drobyshev, I.; Niklasson, M.; Eggertsson, O.; Linderson, H.; Sonesson, K. Influence of Annual Weather on Growth of Pedunculate Oak in Southern Sweden. *Ann. For. Sci.* **2008**, *65*, 512. [CrossRef]
110. St. George, S. An Overview of Tree-Ring Width Records across the Northern Hemisphere. *Quat. Sci. Rev.* **2014**, *95*, 132–150. [CrossRef]
111. Deligöz, A.; Bayar, E. Drought Stress Responses of Seedlings of Two Oak Species (Quercus cerris and Quercus Robur). *Turkish J. Agric. For.* **2018**, *42*, 114–123. [CrossRef]
112. Mészáros, I.; Adorján, B.; Nyitrai, B.; Kanalas, P.; Oláh, V.; Levanič, T. Long-Term Radial Growth and Climate-Growth Relationships of Quercus Petraea (Matt.) Liebl. and Quercus cerris L. in a Xeric Low Elevation Site from Hungary. *Dendrochronologia* **2022**, *76*, 126014. [CrossRef]
113. Jajcay, N.; Hlinka, J.; Kravtsov, S.; Tsonis, A.A.; Paluš, M. Time Scales of the European Surface Air Temperature Variability: The Role of the 7—8 Year Cycle. *Geophys. Res. Lett.* **2016**, *43*, 902–909. [CrossRef]
114. Sen, A.K.; Ogrin, D. Analysis of Monthly, Winter, and Annual Temperatures in Zagreb, Croatia, from 1864 to 2010: The 7.7-Year Cycle and the North Atlantic Oscillation. *Theor. Appl. Climatol.* **2016**, *123*, 733–739. [CrossRef]
115. Qian, B.; Xu, H.; Corte-Real, J. Spatial-temporal structures of quasi-periodic oscillations in precipitation over europe. *Int. J. Climatol.* **2000**, *20*, 1583–1598. [CrossRef]
116. Domonkos, P.; Tar, K. Long-Term Changes in Observed Temperature and Precipitation Series 1901–1998 from Hungary and Their Relations to Larger Scale Changes. *Theor. Appl. Climatol.* **2003**, *75*, 131–147. [CrossRef]
117. Ilyés, C.; Turai, E.; Szucs, P. Examination of Rainfall Data for 110 Years Using Spectral and Wavelet Analysis. *Cent. Eur. Geol.* **2018**, *61*, 1–15. [CrossRef]
118. Cutini, A.; Chianucci, F.; Chirichella, R.; Donaggio, E.; Mattioli, L.; Apollonio, M. Mast Seeding in Deciduous Forests of the Northern Apennines (Italy) and Its Influence on Wild Boar Population Dynamics. *Ann. For. Sci.* **2013**, *70*, 493–502. [CrossRef]
119. Bisi, F.; Chirichella, R.; Chianucci, F.; Von Hardenberg, J.; Cutini, A.; Martinoli, A.; Apollonio, M. Climate, Tree Masting and Spatial Behaviour in Wild Boar (Sus Scrofa L.): Insight from a Long-Term Study. *Ann. For. Sci.* **2018**, *75*, 9. [CrossRef]
120. Simeone, M.C.; Zhelev Stojanov, P.; Kandemir, G. *EUFORGEN Technical Guidelines for Genetic Conservation and Use of Turkey Oak (Quercus cerris)*; European Forest Genetic Resources Programme (EUFORGEN), European Forest Institute: Bonn, Germany, 2019. ISBN 978-952-5980-43-1.
121. Garcia-Barreda, S.; Sangüesa-Barreda, G.; Madrigal-González, J.; Seijo, F.; González de Andrés, E.; Camarero, J.J. Reproductive Phenology Determines the Linkages between Radial Growth, Fruit Production and Climate in Four Mediterranean Tree Species. *Agric. For. Meteorol.* **2021**, *307*, 108493. [CrossRef]
122. Drobyshev, I.; Övergaard, R.; Saygin, I.; Niklasson, M.; Hickler, T.; Karlsson, M.; Sykes, M.T. Masting Behaviour and Dendrochronology of European Beech (*Fagus Sylvatica* L.) in Southern Sweden. *For. Ecol. Manag.* **2010**, *259*, 2160–2171. [CrossRef]
123. Drobyshev, I.; Niklasson, M.; Mazerolle, M.J.; Bergeron, Y. Agricultural and Forest Meteorology Reconstruction of a 253-Year Long Mast Record of European Beech Reveals Its Association with Large Scale Temperature Variability and No Long-Term Trend in Mast Frequencies. *Agric. For. Meteorol.* **2014**, *192–193*, 9–17. [CrossRef]
124. Bertini, G.; Amoriello, T.; Fabbio, G.; Piovosi, M. Forest Growth and Climate Change: Evidences from the ICP-Forests Intensive Monitoring in Italy. *iForest—Biogeosci. For.* **2011**, *4*, 262–267. [CrossRef]
125. Móricz, N.; Illés, G.; Mészáros, I.; Garamszegi, B.; Berki, I.; Bakacsi, Z.; Kámpel, J.; Szabó, O.; Rasztovits, E.; Cseke, K.; et al. Different Drought Sensitivity Traits of Young Sessile Oak (Quercus Petraea (Matt.) Liebl.) and Turkey Oak (*Quercus cerris* L.) Stands along a Precipitation Gradient in Hungary. *For. Ecol. Manag.* **2021**, *492*, 119165. [CrossRef]
126. Stafasani, M.; Toromani, E. Growth-Climate Response of Young Turkey Oak (*Quercus cerris* L.) Coppice Forest Stands along Longitudinal Gradient in Albania. *South-East Eur. For.* **2015**, *6*, 25–38. [CrossRef]
127. Gray, B.M.; Wigley, T.M.L.; Pilcher, J.R. Statistical Significance of Reproducibility of Tree-Ring Response Functions. *Tree-Ring Bull.* **1981**, *41*, 21–35.
128. Gray, B.M.; Pilcher, J.R. Testing the Significance of Summary Response Functions. *Tree Ring Bull.* **1983**, *43*, 31–38.

129. Pérez, A.; Fernández, A. Dendroclimatic Reconstruction of the Last of the XVIII Century in Galicia(Spain) (in Spanish). *Inv Agrar. Rec. F* **1997**, *6*, 17–37.
130. Santini, A.; Bottacci, A.; Gellini, R. Preliminary Dendroecological Survey on Pedunculate Oak (*Quercus Robur* L.) Stands in Tuscany (Italy). *Ann. des Sci. For.* **1994**, *51*, 1–10. [CrossRef]
131. Tessier, L.; Nola, P.; Serre-Bachet, F. Deciduous Quercus in the Mediterranean Region: Tree-ring/Climate Relationships. *New Phytol.* **1994**, *126*, 355–367. [CrossRef]
132. Corcuera, L.; Camarero, J.J.; Gil-Pelegrín, E. Effects of a Severe Drought on Growth and Wood Anatomical Properties of Quercus Faginea. *IAWA J.* **2004**, *25*, 185–204. [CrossRef]
133. Zweifel, R.; Zimmermann, L.; Zeugin, F.; Newbery, D.M. Intra-Annual Radial Growth and Water Relations of Trees: Implications towards a Growth Mechanism. *J. Exp. Bot.* **2006**, *57*, 1445–1459. [CrossRef] [PubMed]
134. Fritts, H.C. Growth-Rings of Trees: Their Correlation with Climate. *Science* **1966**, *154*, 973–979. [CrossRef] [PubMed]
135. Fowler, A.; Boswijk, G. Chronology Stripping as a Tool for Enhancing the Statistical Quality of Tree-Ring Chronologies. *Tree-Ring Res.* **2003**, *59*, 53–62.
136. Taeger, S.; Zang, C.; Liesebach, M.; Schneck, V.; Menzel, A. Impact of Climate and Drought Events on the Growth of Scots Pine (*Pinus Sylvestris* L.) Provenances. *For. Ecol. Manag.* **2013**, *307*, 30–42. [CrossRef]
137. Fulín, M.; Dostál, J.; Čáp, J.; Novotný, P. Evaluation of Silver Fir Provenances at 51 Years of Age in Provenance Trials in the Předhoří Hrubý Jeseník and Nízký Jeseník Mts. Regions, Czech Republic. *J. For. Sci.* **2023**, *69*, 44–59. [CrossRef]
138. Bertolasi, B.; Zago, L.; Gui, L.; Cossu, P.; Vanetti, I.; Rizzi, S.; Cavallini, M.; Lombardo, G.; Binelli, G. Genetic Variability and Admixture Zones in the Italian Populations of Turkey Oak (*Quercus cerris* L.). *Life* **2023**, *13*, 18. [CrossRef] [PubMed]
139. Berg, E.E.; Hamrick, J.L. Fine-scale genetic structure of a turkey oak forest. *Evolution (N.Y.).* **1995**, *49*, 110–120. [CrossRef]

forests

MDPI

Article

Study of the Genetic Adaptation Mechanisms of Siberian Larch (*Larix sibirica* Ledeb.) Regarding Climatic Stresses Based on Dendrogenomic Analysis

Serafima V. Novikova [1,2], Natalia V. Oreshkova [1,2,3,4], Vadim V. Sharov [1,5,6], Dina F. Zhirnova [7,8], Liliana V. Belokopytova [7,8], Elena A. Babushkina [7,8] and Konstantin V. Krutovsky [2,4,9,10,11,12,*]

1 Laboratory of Genomic Research and Biotechnology, Federal Research Center, Krasnoyarsk Science Center of the Siberian Branch of the Russian Academy of Sciences, 660036 Krasnoyarsk, Russia; serafima_novikova_11@mail.ru (S.V.N.); oreshkova@ksc.krasn.ru (N.V.O.); vsharov@sfu-kras.ru (V.V.S.)

2 Laboratory of Forest Genomics, Genome Research and Education Center, Institute of Fundamental Biology and Biotechnology, Siberian Federal University, 660041 Krasnoyarsk, Russia

3 Laboratory of Forest Genetics and Selection, V. N. Sukachev Institute of Forest, Siberian Branch of Russian Academy of Sciences, 660036 Krasnoyarsk, Russia

4 Department of Genomics and Bioinformatics, Institute of Fundamental Biology and Biotechnology, Siberian Federal University, 660041 Krasnoyarsk, Russia

5 Department of High-Performance Computing, Institute of Space and Information Technologies, Siberian Federal University, 660074 Krasnoyarsk, Russia

6 Tauber Bioinformatics Research Center, University of Haifa, Haifa 3498838, Israel

7 Laboratory of Dendroecology and Ecological Monitoring, Khakass Technical Institute, Siberian Federal University, 655017 Abakan, Russia; dina-zhirnova@mail.ru (D.F.Z.); white_lili@mail.ru (L.V.B.); babushkina70@mail.ru (E.A.B.)

8 Institute of Ecology and Geography, Siberian Federal University, 660041 Krasnoyarsk, Russia

9 Department of Forest Genetics and Forest Tree Breeding, Georg-August University of Göttingen, 37077 Göttingen, Germany

10 Center for Integrated Breeding Research, George-August University of Göttingen, 37075 Göttingen, Germany

11 Laboratory of Population Genetics, N. I. Vavilov Institute of General Genetics, Russian Academy of Sciences, 119333 Moscow, Russia

12 Scientific and Methodological Center, G. F. Morozov Voronezh State University of Forestry and Technologies, 394087 Voronezh, Russia

* Correspondence: konstantin.krutovsky@forst.uni-goettingen.de; Tel.: +49-551-339-3537

Citation: Novikova, S.V.; Oreshkova, N.V.; Sharov, V.V.; Zhirnova, D.F.; Belokopytova, L.V.; Babushkina, E.A.; Krutovsky, K.V. Study of the Genetic Adaptation Mechanisms of Siberian Larch (*Larix sibirica* Ledeb.) Regarding Climatic Stresses Based on Dendrogenomic Analysis. *Forests* **2023**, *14*, 2358. https://doi.org/10.3390/f14122358

Academic Editors: Yassine Messaoud, Jan Světlík and Giorgio Alberti

Received: 23 October 2023
Revised: 27 November 2023
Accepted: 28 November 2023
Published: 30 November 2023

Abstract: Dendrogenomics is a new interdisciplinary approach that allows joint analysis of dendrological and genomic data and opens up new ways to study the temporal dynamics of forest treelines, delineate spatial and temporal population structures, decipher individual tree responses to abiotic and biotic stresses, and evaluate the adaptive genetic potential of forest tree populations. These data are needed for the prediction of climate change effects and mitigation of the negative effects. We present here an association analysis of the variation of 27 individual tree traits, including adaptive dendrophenotypes reflecting the individual responses of trees to drought stress, such as the resistance (Rt), recovery (Rc), resilience (Rs), and relative resilience (RRs) indexes measured in 136 Siberian larch trees in 5 populations in the foothills of the Batenevsky Ridge (Kuznetsk Alatau, Republic of Khakassia, Russia), with variation of 9742 SNPs genotyped using ddRADseq in the same trees. The population structure of five closely located Siberian larch populations was relatively weak (F_{ST} = 0.018). We found that the level of individual heterozygosity positively correlated with the Rc and RR indices for the five studied drought periods and partly with the Rs indices for three drought periods. It seems that higher individual heterozygosity improves the adaptive capabilities of the trees. We also discovered a significant negative relationship between individual heterozygosity and the Rt index in four out of five periods, which means that growth slows down during droughts more in trees with higher individual heterozygosity and is likely associated with energy and internal resource reallocation toward more efficient water and energy usage and optimization of larch growth during drought years. We found 371 SNPs with potentially adaptive variations significantly associated with the variation of adaptive dendrophenotypes based on all three different methods of association analysis. Among them, 26 SNPs were located in genomic regions carrying functional genes: 21 in intergenic regions and 5 in gene-coding regions. Based on the obtained results, it can be assumed that these populations of Siberian larch have relatively high standing adaptive genetic variation and adaptive potential underlying the adaptations of larch to various climatic conditions.

Keywords: adaptation; climate change; ddRADseq; dendrophenotypes; drought; *Larix sibirica* Ledeb.; Siberian larch; SNPs; tree growth

1. Introduction

Droughts have become more frequent and drastic due to climate change [1,2], especially in boreal forests [3,4]. Therefore, it is very important to find whether tree populations possess sufficient standing genetic variation that may help trees to adapt to environmental stresses such as droughts. One of the most efficient approaches to find such variation is to search for genes whose variation is associated with the variation of adaptive dendrophenotypes using genome-wide genetic markers [5]. Our main hypothesis was that there are genes whose variation is significantly associated with individual responses of trees to the droughts, and therefore, our main objective was to study the association between the phenotypic variation of important adaptive traits related to the drought response and the genetic variation of individual trees and populations of Siberian larch (*Larix sibirica* Ledeb.) in the south area mostly affected by climate change [6–11].

Siberian larch is a coniferous tree common in Russia, Kazakhstan, Mongolia, and China. This is one of the key species of Siberian boreal forests, whose wood is superior in performance to many other coniferous species and has high ecological and economic significance. The wide range of Siberian larch includes both tundra zones in the north and forest and forest-steppe zones of Altai and Sayan Mountains in southern East Siberia [12–14].

The constant moisture deficit typical of the forest-steppe zone of southern East Siberia is gradually increasing with an increase in average annual temperatures due to global climate change, which, along with periodic droughts, affects the structure and dynamics of forest ecosystems. *L. sibirica* is a relatively drought-resistant species because of its high level of phenotypic plasticity and genetic variation. However, under the influence of rapidly increasing environmental stress, such as a decrease in the amount of summer precipitation, intense and recurring droughts, fires, and pest infestations, there is a concern that larch will not be able to effectively adapt to the new conditions associated with global warming [10,15,16].

Dendrochronological data based on measurements of annual tree wood growth rings contain a lot of information about the individual responses of trees to biotic and abiotic environmental factors [7,8,17–23]. In turn, the ability to individually genotype a large number of trees in populations for thousands of genes and genetic markers makes it possible to link genetic variation with the variation of specific dendrophenotypes [5,24–28]. This interdisciplinary approach, allowing the integration of dendrochronology, dendroecology, and genomics, has been gaining popularity in recent years [5,29–31].

According to the results of dendroclimatic studies conducted in the region of southern East Siberia, Siberian larch has successfully adapted to the modern conditions of the forest-steppe zone [20]. To understand the genetic mechanisms of this adaptation, we searched for DNA markers whose variation is associated with the variation of individual dendrophenotypes, reflecting the individual response of a tree to abiotic stress such as drought. For this purpose, genome-wide genotyping of a large number of trees in five populations was performed, and the structure of Siberian larch populations, genetic diversity, and genetic basis of adaptation to growing conditions within the forest-steppe zone of the foothills of the Batenevsky Ridge (Kuznetsk Alatau, southern East Siberia, Russia) were analyzed.

2. Materials and Methods

2.1. Sampling and Dendrophenotypes

The samples were collected from 150 trees in five populations in the foothills of the Batenevsky Ridge in the eastern part of the Kuznetsk Alatau, Republic of Khakassia (southern East Siberia, Russia) (Figure 1). This mountain system is characterized by a slight difference in elevation (500–1200 m above sea level), and most of its area is covered with mixed forest consisting of Siberian larch, Scots pine (*Pinus sylvestris* L.), and silver birch (*Betula pendula* Roth.). On the drier southern and southeastern slopes, open forest is interspersed with steppe areas. The forests of the study area are part of the state forest; forest districts are located on the ridge. The climate of the region is sharply continental, with large seasonal and daily temperature differences, hot summers, and frosty winters with little snow lasting from November to March. Despite the significant size of the study area, the high and similar variation in larch growth allows us to consider the forest-steppe ecotone of the Batenevsky Range as an integral geographical object, which is characterized by a unity of fluctuations in environmental conditions and the responses of vegetation to them [14]. Samples of larch wood and needles were collected at five population sites located 30–50 km apart in the forest-steppe zone along the foothills of the Batenevsky Ridge (Figure 1). The sites were selected on well-lit slopes oriented from southwest to southeast (Table 1).

Table 1. The five population sampling sites of Siberian larch used in the study.

Population (Abbreviation Used in the Study)	Description	Coordinates	Altitude above Sea Level, m
Tuim (TUI)	Individuals and groups of larch trees along the steppe vegetation	54°24′20″ N 89°57′27″ E	550–600
Son (SON)	Mixed larch and birch forest	54°21′55″ N 90°22′04″ E	530–600
Bograd (BOG)	Mixed larch and birch forest	54°15′58″ N 90°41′30″ E	550–620
Bidja (BID)	Mixed larch, pine, and birch forest, with individuals and groups of larch trees in the steppe	54°00′20″ N 90°58′52″ E	640–670
Kamyziak (KAM)	Mixed larch and birch forest	53°55′52″ N 90°37′30″ E	700–770

Wood core samples were extracted from 150 mature, undamaged, live larch trees at chest level. The samples were collected and processed using standard dendrochronological techniques [32]. Individual tree ring widths (TRWs) were measured using the LINTAB device (Rinntech-Metriwerk GmbH & Co. KG, Heidelberg, Germany; https://rinntech.info/products/lintab, accessed on 14 October 2023) using the TSAP-Win program [33]. The TRWs were cross-dated, and the dating was checked using the COFECHA program [34]. The average TRW, generalized raw chronologies (avTRW) and their variance (varTRW) were used as dendrophenotypes.

Five drought periods (1951, 1963–1965, 1974–1976, 1999, and 2015, respectively) that occurred more than 6 years apart were selected from climatic extremes that significantly suppressed larch growth on a regional scale. To select drought periods, years with low values (<mean − SD) for the standardized precipitation–evapotranspiration index (SPEI) and/or precipitation averaged for the current April–July and/or previous June–September, as well as high temperatures (>mean + SD) over the current May–July and/or previous July-September, were identified in the study area (53.5–54.5° N, 89.5–91.5° E). Then, those periods with the simultaneous occurrence of several of these climatic deviations that also were accompanied by the formation of narrow tree rings (<mean − SD for averaged local tree-ring chronology) across all or most of the sampling

sites were selected as drought periods. More detailed explanations of the selection of the drought periods exactly in these sampling sites, as well as the reasoning for the used seasonal timeframe of the climatic factors impacting significantly larch growth, can be found in [35]. To analyze the effect of this stress on larch growth, we used the resistance (Rt = Gd/Gprev), recovery (Rc = Gpost/Gd), and resilience (Rs = Gpost/Gprev) indices proposed by Lloret et al. (2011) [36], where Gprev is the average growth during the 3 years before the drought, Gd is the average growth during the drought, and Gpost is the average growth during the 3 years after the drought, as calculated from the tree-ring series of each tree after their standardization (omission of age-related trends fitted by cubic smoothing splines) using the ARSTAN program [37]. A more detailed description can also be found in [35]. The relative resilience indices (RRs) were also used, calculated as RRs = Rs − Rt. Additionally, the age of the tree at the time of collection (Age), the average length of 15 needles (avLn), and the variance of these 15 needle length measurements (varLn) were used as phenotypic traits. The radial growth trends (trendTRW) for each tree and the mean chronology for each population were calculated as the slope of the linear regression of the respective indexed time series after standardization (i.e., after elimination of the long-term trends related to tree size, age, and forest stand dynamics) for the 30-year period of the years 1990–2019. Phenotypic data that were not normally distributed according to the Shapiro–Wilk test were normalized using the rank-based inverse normal transformation (INT) algorithm. A total of 26 dendrophenotypic traits were considered in this study (Table 2).

Table 2. Description of the dendrophenotypes.

Dendrophenotype	Abbreviation				
Age of the tree at the time of collection	Age				
Average needle length	avLn				
Variance of needle length	varLn				
Average tree ring width	avTRW				
Variation of tree ring width	varTRW				
Radial growth trends (1990–2019)	trendTRW				
Index	Drought years				
	1951	1963–1965	1974–1976	1999	2015
Resistance	Rt1	Rt2	Rt3	Rt4	Rt5
Recovery	Rc1	Rc2	Rc3	Rc4	Rc5
Resilience	Rs1	Rs2	Rs3	Rs4	Rs5
Relative resilience	RRs1	RRs2	RRs3	RRs4	RRs5

A pairwise Wilcoxon rank-sum test was performed to compare the dendrophenotypes and individual heterozygosity data between populations using the pairwise.wilcox.test command from the basic R package stats [38]. The Benjamini–Hochberg procedure was used as a method of multiple testing correction.

2.2. DNA Extraction and ddRAD Sequencing

DNA was isolated from the needles of the collected larch samples using the CTAB method [39]. The DNA concentration was assessed using a Qubit 2.0 fluorimeter (Thermo Fisher Scientific, Waltham, MA, USA). The purity and quality of the isolated DNA were also assessed using an Implen NanoPhotometer P330 (Implen, München, Germany). High-quality DNA samples with an A260/230 of approximately 1.8 and a concentration of 20–150 ng/μL were selected for this study.

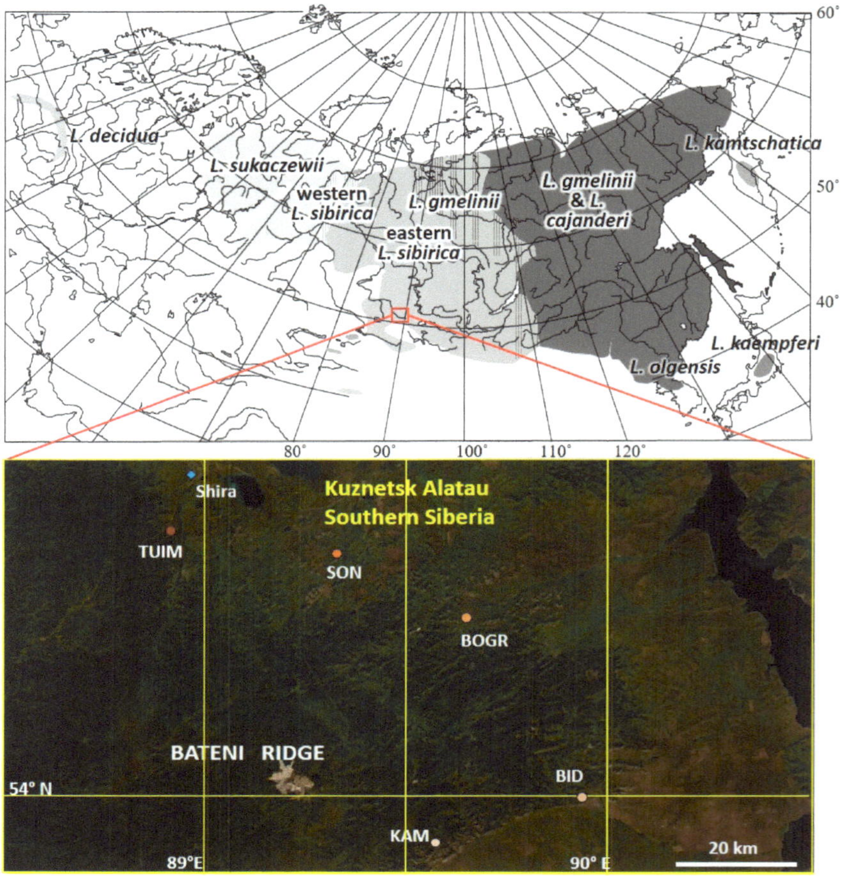

Figure 1. Larch distribution area in Eurasia according to [14] (upper part, adapted with permission from Figure 1 in [13]) and location of the five population sampling sites of Siberian larch used in this study (lower part, adapted with permission from [35]): Bidja (BID), Bograd (BOGR), Kamyziak (KAM), Son (SON), and Tuim (TUIM). See also Table 1 for details.

The preparation of the ddRAD-seq libraries was based on a modified version of the protocol [40]. The DNA samples were treated with two restriction enzymes, *Eco*RI and *Mse*I [41]. Validation of the prepared ddRAD-seq library pools before sequencing was performed on an Agilent 2200 TapeStation System (Agilent Technologies, Santa Clara, CA, USA). Single-end sequencing of the ddRAD-seq libraries was performed in three lanes using 100 cycles on a NovaSeq 6000 sequencer (Illumina, San Diego, CA, USA).

2.3. SNP Calling

The obtained sequencing data underwent several stages of initial processing: filtering, trimming according to quality indicators, and demultiplexing using the process_radtags utility included in the Stacks software version 2.65 [42]. Reads from each sample were then aligned to the Siberian larch reference genome [43] using Bowtie2 version 2.3 [44].

Variant calling was performed using the Gstacks utility from the Stacks software. The resulting set of alignment-covered loci was subjected to several filtering steps using the Populations utility to retain only loci that were present in at least 75% of all samples and in 75% of samples within each population. The maximum level of observed heterozygosity of the site should not exceed 0.85, and the frequency of the minor allele should not be less than 0.05. SNPs that were in linkage disequilibrium (LD) ($r^2 > 0.8$) were removed. Imputation of missing values was performed using the LD-kNNi method in TASSEL v.5.0 [45].

2.4. Genetic Structure of Populations

General indicators of genetic diversity, such as the mean observed (H_O) and expected (H_E) heterozygosities, nucleotide diversity (π), mean individual fixation index (F_{IS}), and their associated standard errors (s.e.), were estimated using the Stacks v2.5 software. The pairwise F_{ST} values among all the populations tested and their confidence intervals (using 10,000 bootstrap samples) were estimated using the R package StAMPP [46]. To determine the population structure, principal component analysis (PCA) was performed based on allele frequency data using the R ade4 package [47]. Sparse non-negative matrix factorization (sNMF) analysis was performed using LEA [48], and the estimation of individual ancestries was performed using the Admixture algorithm implemented in the AdmixPipe program [49]. Hierarchical analysis of molecular variance (AMOVA) was performed using Arlequin v. 3.5.1.2 [50]. The Mantel test to identify correlations between genetic and geographic distances was performed on two datasets: for 84 individual trees of the KAM, BID, and SON populations, for which the exact coordinates of the trees were recorded and the IBS genetic distance matrix was calculated, and for five populations using the average coordinates and the matrix of Nei's genetic distances using the R package vegan v. 2.6-2 [51].

2.5. Associations between Dendrophenotypes and Individual Heterozygosity

For each sample tree, the individual level of heterozygosity was calculated as the number of polymorphic sites in the heterozygous state divided by the total number of polymorphic sites genotyped for a given sample. The correlations between individual heterozygosity and dendrophenotypes were analyzed using Spearman's rank correlation and Pearson's correlation coefficient using the R stats package [38].

2.6. Genotype–Phenotype Associations

The genotype–dendrophenotype associations were analyzed using the general linear model (GLM) and mixed linear model (MLM) in the TASSEL software version 5.0, where each site is an independent variable and each dendrophenotype is a dependent (response) variable. MLM analysis considers the confounding factors such as the pairwise kinship matrix K (reflecting relatedness between individuals) and population structure (the Q-matrix coefficients reflecting ancestry contributions into individuals), whereas GLM analysis accounts only for the population structure. The K-matrix of relatedness was calculated using the centered IBS method. The Q-matrix was obtained using the sNMF algorithm for the most probable number of population clusters (K), which equaled two for our samples. To search for associations whose weak signals were systematically observed in several different periods of drought, common overlapping SNPs were found for the SNP–dendrophenotype associations that passed the filtering threshold of p-value < 0.01.

The genotype–phenotype associations were also assessed using the Bayesian sparse linear mixed model (BSLMM) implemented in the GEMMA software package v.0.97 [52]. BSLMM is a polygenic model that accounts for the contribution of single SNPs with large effects, as well as the simultaneous contribution of multiple SNPs with smaller effects, on phenotypic variation. To achieve this, BLSMM includes the main effects of individual SNPs as predictors of the phenotype and the polygenic effect resulting from the combination of multiple SNPs with small effects. To identify significant genotype–phenotype associations for each trait, a posterior inclusion probability (PIP) threshold of >0.25 was used.

2.7. SNP Annotation

To analyze the genomic regions where SNPs associated with dendrophenotypes were located, annotation of the Siberian larch reference genome assembly was used [53].

Annotation of selected markers was performed using the SNPdat program [54]. Additionally, for SNPs associated with dendrophenotypes, the ±1000 bp long nucleotide sequences flanking the SNPs were extracted from the genome assembly using SNP coordinates, and a search for homologs of these extracted sequences was performed in the NCBI BLAST "nt" database [55].

3. Results

3.1. Dendrophenotypes

In order to understand how the values of dendrophenotypic traits differ between populations and among individuals, the pairwise Wilcoxon rank-sum test was performed (Table S1) and box-plots for each trait were generated (Figure 2), respectively. They demonstrated that some population samples were different from each other for some traits, and there was significant phenotypic variation between individuals (Table S1).

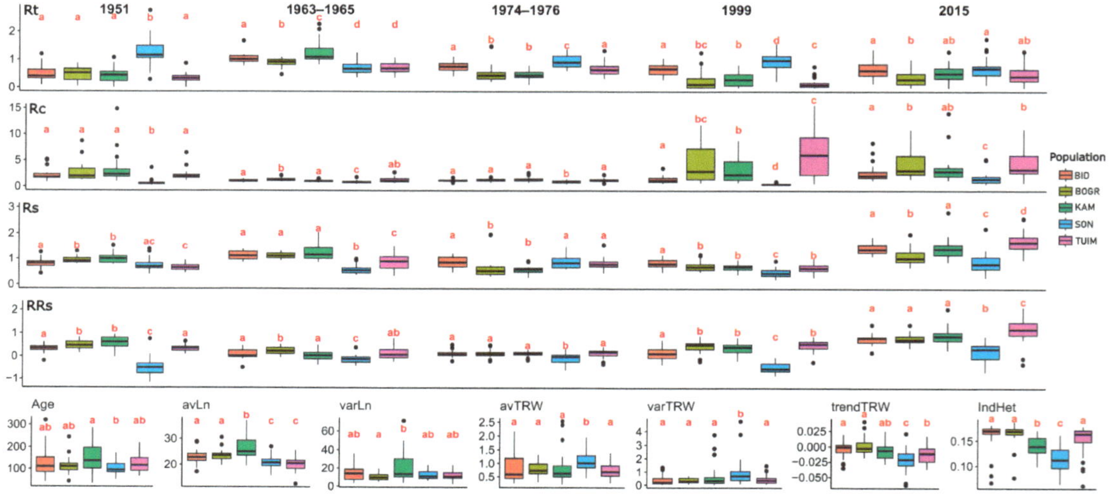

Figure 2. Box-plot of 27 dendrophenotypic traits, including age of the tree (Age), average needle length (avLn), variance of needle length (varLn), average tree ring width (avTRW), variation of tree ring width (varTRW), radial growth trends (1990–2019) (trendTRW), individual heterozygosity (IndHet) and indexes of the tree response to the drought in different years (1951, 1963–1965, 1974–1976, 1999 and 2015): resistance (Rt), recovery (Rc), resilience (Rs), and relative resilience (RRs1) in five population samples BID, BOGR, KAM, SON, and TUIM (see Figure 1 and Table 1). Letters a, b and c under boxes depict whether differences between population samples are statistically significant, at least at $p < 0.05$, based on the pairwise Wilcoxon rank-sum test (see also Table S1).

The average age (Age) of the studied trees was 124 years, ranging from 34 (kam2 tree) to 319 (bid24 tree) (Table S1). The highest average age was in the KAM sample, 153.8 ± 13.9 years; the lowest average age was in the SON sample, 102.8 ± 5.6 years (Table S1). Only between these two samples was there a significant difference in age based on the Wilcoxon rank-sum test ($p = 0.019$; Table S1).

The average needle length (avLn) for all the trees was 22.8 ± 0.3 cm, varying widely from 12.5 cm in the kam28 tree to 36.9 cm in the kam7 tree, and on average, from 20.1 ± 0.5 in the TUIM sample to 21.1 ± 0.5 in SON, 22.9 ± 0.5 in BID, 23.5 ± 0.5 in BOGR, and 26.5 ± 0.9 in KAM (Table S1). The samples differed significantly between each other ($p < 0.05$; Table S1), except for the BID–BOGR and SON–TUIM pairs.

The average variance of the needle length (varLn) across all the trees was 14.4 ± 0.8 cm (Table S1), varying insignificantly between samples from 10.5 ± 0.9 in BOGR to 12.2 ± 0.9

in TUIM, 12.7 ± 1.0 in SON, 14.3 ± 1.6 in BID, and 21.5 ± 3.0 in KAM, except only for the BOGR and KAM pair ($p = 0.048$; Table S1).

The average tree ring width (avTRW) for all the trees was 0.87 ± 0.08, varying widely among individual trees from 0.21 ± 0.12 in the kam28 tree to 2.53 ± 1.93 in the kam3 tree, and on average, in the samples from 0.77 ± 0.06 in TUIM to 0.78 ± 0.06 in BOGR, 0.83 ± 0.11 in KAM, 0.85 ± 0.10 in BID, and 1.10 ± 0.07 in SON (Table S1). Among all the sample pairs, only the SON sample was significantly different from all the other samples ($p < 0.05$; Table S1).

The variance of the tree ring width (varTRW) was varying also widely among individual trees, from 0.01 in the kam28 tree to 4.74 in the son21 tree (Table S1). On average, the largest varTRW was observed in the SON sample (0.97 ± 0.19), while in the other samples, it varied from 0.33 ± 0.04 in BOGR, 0.36 ± 0.07 in BID, 0.43 ± 0.06 in TUIM, and 0.63 ± 0.16 in KAM population samples (Table S1). The only SON sample was significantly different from the all the other samples ($p < 0.05$; Table S1).

The mean tree radial growth trend (trendTRW) values were negative in 99 out of 135 trees, varying also widely among individual trees, from −0.063 in the son4 tree to 0.040 in the bog9 tree (Table S1). On average, they were slightly negative in all the samples except BOGR (0.001 ± 0.003), varying from −0.022 ± 0.003 in SON to −0.012 ± 0.002 in TUIM, −0.008 ± 0.002 in KAM, and −0.004 ± 0.002 in BID (Table S1). The SON sample was significantly different from all the other samples, and the TUIM sample was also significantly different from the BOGR and BID samples (Table S1).

The growth practically completely ceased (Rt = 0) in 21 trees: bogr1, bogr3, bogr7, bogr9, bogr12, bogr14, kam17, and kam18 during both drought periods 1999 and 2015; tuim2 and tuim26 during the drought period of 1951; kam3, kam20, kam23, kam25, kam26, tuim8, tuim13, tuim15, tuim16, tuim26 during the drought period of 1999; and tuim14 during the drought period of 2015. The resistance (Rt) index very widely varied among individual trees, from Rt4 = 0.00173 in the bogr10 tree during the drought period of 1999 to Rt1 = 2.75 in the son30 tree during the drought period of 1951 (Table S1). Among all the drought periods studied, the highest average resistance index was observed for the drought period of 1963–1965 (Rt2 = 0.91 ± 0.03). Among all the population samples, the highest mean resistance index values were observed in the SON population sample for all the drought periods (Rt1 = 1.26 ± 0.10, Rt3 = 0.95 ± 0.04, Rt4 = 0.94 ± 0.06, Rt5 = 0.70 ± 0.07), except for the 1963–1965 period (Rt2 = 0.71 ± 0.03).

The recovery (Rc) index varied extremely widely among individual trees, from Rc1 = 0.26 in the son18 tree during the drought period of 1951 to as large a value as Rc4 = 290.04 in the bogr10 tree during the drought period of 1999 (Table S1). The highest average value of the Rc index was observed for the 1999 drought (8.31 ± 3.09). For the droughts of 1951 and 1974–1976, the highest Rc values were observed in the KAM population (Rc1 = 4.48 ± 1.31 and Rc3 = 1.40 ± 0.08, respectively); for the droughts of 1963–1965 and 1999, the highest Rc values were observed in the BOGR population (Rc2 = 1.33 ± 0.06 and Rc4 = 24.74 ± 15.38, respectively); and for the 2015 drought, the highest average value recovery index was observed in the TUIM population (Rc5 = 6.29 ± 1.34). The SON population showed the lowest Rc values (significantly different from the other populations) for all the drought periods.

The resilience (Rs) index varied very widely among individual, from Rs4 = 0.21 in the son23 tree during the drought period of 1999 to Rs5 = 2.83 in the kam26 tree during the drought period of 2015 (Table S1). The average value of the Rs index for 136 trees varied from Rs4 = 0.68 ± 0.02 for the 1999 drought to Rs5 = 1.33 ± 0.04 for the 2015 drought. Among the population samples, the highest average values of the Rs index were observed for trees of the KAM population for the droughts of 1951 and 1963–1965 (Rs1 = 1.01 ± 0.04 and Rs2 = 1.30 ± 0.07, respectively, Table S1) and for trees of the SON, BID, and TUIM populations for the droughts of 1974–1976, 1999, and 2015 (Rs3 = 0.89 ± 0.05, Rs4 = 0.85 ± 0.03, and 1.69 ± 0.07, respectively, Table S1). For three of the five drought periods, 1963–1965, 1999, and 2015, the lowest values of the Rs index were observed in the SON population

(Rs2 = 0.60 ± 0.04, Rs4 = 0.48 ± 0.03, Rs5 = 0.90 ± 0.07, Table S1), which was significantly different from the other populations (Table S1).

The relative resilience (RRs) index varied also very widely among individual trees from RRs1 = −1.75 in the son30 tree during the drought period of 1951 to RRs5 = 2.15 in the tuim1 tree during the drought period of 1951 (Table S1). The average value of the RRs index for all the trees varied from RRs5 = 0.78 ± 0.04 for the 2015 drought to RRs2 = 0.09 ± 0.02 for the 1963–1965 drought. The highest values of RRs were observed for the 1951 drought for trees of the KAM population (RRs1 = 0.59 ± 0.05), for the 1963–1965 drought for trees of the BOGR population (RRs2 = 0.25 ± 0.04), and for the last three droughts for trees of the TUIM population (RRs3 = 0.16 ± 0.03, RRs4 = 0.50 ± 0.04, and RRs5 = 1.20 ± 0.10, respectively, Table S1). The lowest values of the RRs index were observed in the SON population for all five drought periods (RRs1 = −0.50 ± 0.10, RRs2 = −0.11 ± 0.04, RRs3 = −0.06 ± 0.05, RRs4 = −0.46 ± 0.04, RRs5 = 0.20 ± 0.08, respectively, Table S1).

In general, the obtained dendrophenotypic data demonstrate that individual trees had very different individual responses to climatic stresses, such as drought, which well justifies the main objective of this study to associate the variation of individual dendrophenotypes that reflect individual tree responses to drought with the genetic variation of the large number of genetic markers genotyped in the same trees.

3.2. SNP Calling

For 150 larch tree samples, 1.8 billion 100 bp long reads were obtained via ddRADseq sequencing. Fourteen trees were excluded from further analysis because of insufficient sequencing quality. After the initial processing and quality filtering, approximately 1.4 billion reads with lengths from 32 to 92 bp were selected for further analysis, with an average of 10,496,058 ± 391,691 reads per sample. On average, 97% of the reads for each sample were successfully mapped to the Siberian larch reference genome. In total, 11,550 SNPs that met the filtering criteria were selected. For pairs of SNPs with LD (r^2 > 0.8), only one SNP was selected for further analysis. The final dataset contained 9742 biallelic SNPs genotyped in 136 Siberian larch trees (Data S1).

3.3. Genetic Structure of Populations

For each sample, general parameters of genetic variation, such as the nucleotide diversity (π), observed (H_O) and expected (H_E) heterozygosities, and fixation index (F_{IS}) were calculated and are presented in Table 3. The levels of heterozygosity based on all 9742 SNPs was almost the same as those based on the 371 SNPs significantly associated with the variation of adaptive dendrophenotypes variation, although the latter ones were slightly higher if they were calculated for all the trees together as a single population.

The degree of genetic differentiation was assessed using the F_{ST} parameter (Table 4). The maximum F_{ST} value was between the SON and BOG samples (0.022), and the minimum value between the TUI and BOG (0.007). The low pairwise F_{ST} values obtained indicate that the samples were relatively poorly differentiated and belong to genetically similar populations.

PCA was performed using allele frequency data to identify also the population structure. The analysis revealed that the studied samples do not form clearly defined unambiguous clusters, although two tentative clusters can presumably be distinguished: BID, BOG and TUI form one cluster, and the SON and KAM samples form another (Figure 3).

In addition, for a better understanding of the population structure and further association analysis, a search for the most probable number of clusters or subpopulations K for K from 1 to 20 was conducted, with 20 replicates for each number K, and estimates of the contribution of each cluster to individual admixtures of trees (Q-values) were obtained. The ΔK method for selecting the most probable number K showed that the most probable numbers are K = 2 or K = 3. The contributions ("admixture") of each cluster to individual trees (Q-values) according to the results of the Admixture program are shown in Figure 4. The sNMF algorithm, which was also used to calculate the Q-value matrix, showed similar results. When the number of clusters is $K = 2$, the trees belonging to the SON and KAM samples represent a separate cluster and are genetically different from the trees collected at other sites.

Table 3. Values of mean nucleotide diversity (π), observed (H_O) and expected (H_E) heterozygosities, individual fixation index (F_{IS}), and their associated standard errors (s.e.) calculated for five population samples of Siberian larch based on all 9742 SNPs and 371 SNPs significantly associated with the variation of adaptive dendrophenotypes.

Population	$\pi \pm$ s.e.	$H_O \pm$ s.e.	$H_E \pm$ s.e.	$F_{IS} \pm$ s.e.
		9742 SNPs		
TUI	0.00033 ± 0.00000	0.162 ± 0.001	0.162 ± 0.001	0.036 ± 0.017
SON	0.00030 ± 0.00000	0.118 ± 0.001	0.146 ± 0.001	0.163 ± 0.016
KAM	0.00033 ± 0.00000	0.147 ± 0.001	0.161 ± 0.001	0.099 ± 0.019
BOG	0.00034 ± 0.00000	0.173 ± 0.001	0.166 ± 0.001	0.012 ± 0.014 *
BID	0.00034 ± 0.00000	0.170 ± 0.001	0.170 ± 0.001	0.033 ± 0.015
All	0.00289 ± 0.00003	0.154 ± 0.001	0.166 ± 0.001	0.095 ± 0.049
		371 SNPs		
TUI	0.00189 ± 0.00012	0.171 ± 0.009	0.166 ± 0.007	0.029 ± 0.100 *
SON	0.00160 ± 0.00011	0.125 ± 0.008	0.141 ± 0.007	0.117 ± 0.020
KAM	0.00215 ± 0.00013	0.184 ± 0.009	0.189 ± 0.007	0.071 ± 0.101 *
BOG	0.00191 ± 0.00013	0.182 ± 0.010	0.168 ± 0.007	−0.010 ± 0.075 *
BID	0.00180 ± 0.00012	0.171 ± 0.009	0.159 ± 0.006	−0.005 ± 0.091 *
All	0.00192 ± 0.00011	0.166 ± 0.007	0.173 ± 0.005	0.073 ± 0.292 *

Note. All the values were significantly different from zero except the F_{IS} values marked by *.

Table 4. Pairwise F_{ST} fixation indexes with 95%confidence intervals in brackets calculated between 5 population samples.

Population	TUI	SON	KAM	BOG
SON	0.016 (0.014–0.017)			
KAM	0.014 (0.012–0.016)	0.008 (0.007–0.009)		
BOG	0.007 (0.005–0.008)	0.022 (0.019–0.024)	0.018 (0.015–0.020)	
BID	0.007 (0.005–0.008)	0.020 (0.018–0.022)	0.017 (0.015–0.018)	0.007 (0.006–0.008)

According to the Admixture analysis, all the genotypes were divided into two groups: the first group included the TUI, BOG, and BID populations, and the second group was SON–KAM.

AMOVA was conducted and Wright's fixation indices (F-indices) were calculated for each hierarchical level (Table 5). The highest level of genetic variation was within populations (98.2%).

The correlation of the matrices of the pairwise geographic distances with the matrices of the pairwise genetic differences was calculated using the Mantel test: for 84 trees of the KAM, BID, and TUI samples and for 5 populations using average coordinates and Nei's genetic distance matrix. The resulting test p-values indicate that there is no relationship between the geographic and genetic distances of populations ($p = 0.171$ and $p = 0.558$, respectively).

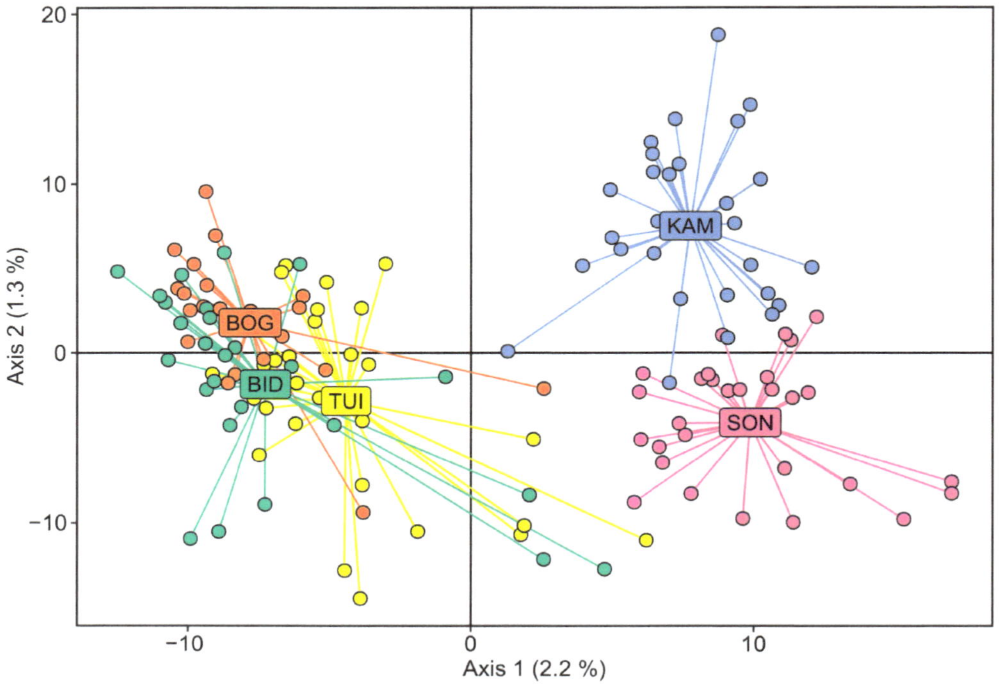

Figure 3. Principal component analysis (PCA) results for five Siberian larch population samples.

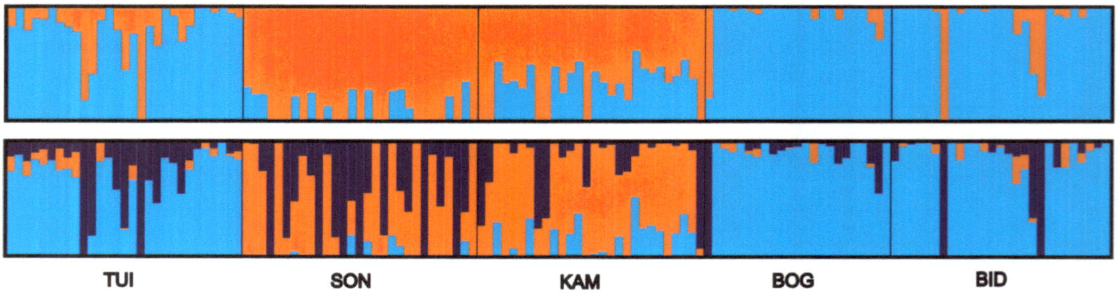

Figure 4. Admixture of each of the two ($K = 2$, above) and three ($K = 3$, below) clusters in individual Siberian larch trees representing five populations (TUI, SON, KAM, BOG, and BID) highlighted by two (for $K = 2$) or three (for $K = 3$) different colors, respectively.

Table 5. AMOVA results for five population samples of Siberian larch.

Source of Variation	Sum of Squares	Variance Components	Percentage Variation	*F*-Statistics over All Loci
Among groups	2079.3	7.7	1.0	$F_{ST} = 0.018$ $F_{SC} = 0.008$ $F_{CT} = 0.010$
Among populations within groups	3139.3	5.8	0.8	
Within populations	195,233.5	731.7	98.2	
Total	200,452.1	745.3	100.0	

3.4. Associations between Dendrophenotypes and Individual Heterozygosity

For each sample, the individual level of heterozygosity was calculated, with an average 0.144 for all the samples. The lowest average values of individual heterozygosity were obtained for the SON population (0.112 ± 0.005) and the highest for the BOG population (0.164 ± 0.005). The Pearson's and Spearman's correlation coefficients of individual heterozygosity and each dendrophenotype were calculated (Table 6). These coefficients were calculated both for data not standardized by chronology and for indices calculated after the standardization procedure. The results were very similar; therefore, Table 6 presents coefficients only based on the unstandardized data.

Table 6. Pearson's product-moment (r) and Spearman's rank (r_S) correlation coefficients between dendrophenotypes and individual heterozygosity.

Dendrophenotype	r	p	r_S	p
Age	−0.006	0.9413	0.011	0.8979
avLn	0.047	0.5858	0.030	0.7325
varLnr	−0.055	0.5262	−0.150	0.0808
avTRW	−0.049	0.5718	−0.046	0.5911
varTRW	−0.118	0.1701	−0.164	0.0560
trendTRW	0.372 *	<0.0001	0.371 *	<0.0001
Rc1	0.322 *	0.0004	0.366 *	<0.0001
Rc2	0.345 *	<0.0001	0.351 *	<0.0001
Rc3	0.339 *	<0.0001	0.437 *	<0.0001
Rc4	0.307 *	0.0007	0.435 *	<0.0001
Rc5	0.318 *	0.0003	0.331 *	0.0002
Rs1	0.064	0.4918	−0.028	0.7615
Rs2	0.228 *	0.0111	0.278 *	0.0019
Rs3	0.074	0.4098	0.062	0.4898
Rs4	0.180 *	0.0370	0.311 *	0.0002
Rs5	0.142	0.1097	0.200	0.0237
Rt1	−0.432 *	<0.0001	−0.378 *	<0.0001
Rt2	−0.076	0.4014	0.021	0.8144
Rt3	−0.286 *	0.0011	−0.258 *	0.0034
Rt4	−0.342 *	<0.0001	−0.280 *	0.0010
Rt5	−0.203 *	0.0216	−0.215 *	0.0146
RRs1	0.471 *	<0.0001	0.316 *	0.0005
RRs2	0.378 *	<0.0001	0.353 *	<0.0001
RRs3	0.405 *	<0.0001	0.493 *	<0.0001
RRs4	0.423 *	<0.0001	0.401 *	<0.0001
RRs5	0.216 *	0.0144	0.258 *	0.0033

* Significant values.

There was a significant positive relationship between the recovery index (Rc), relative resilience index (RRs), and individual heterozygosity for all five drought periods (p-value < 0.05). The resistance index (Rt) showed a significant negative correlation with individual heterozygosity for four droughts, excluding the drought period of 1963–1965 (Rt2). The resilience index (Rs) correlated according to both the Pearson's and Spearman's correlation coefficients with individual heterozygosity based on the 1963–1965 and 1999 droughts, but only based on Spearman's rank correlation coefficient for the 2015 drought data.

Dot-plot graphs of individual heterozygosity (IndHet) and dendrophenotypes (Rt, Rc, Rs, RRs, varTRW, and trendTRW), together with the Pearson's (r) and Spearman's rank (r_S) correlation coefficients, regression coefficients (R^2) and trendlines based on linear, quadratic, cubic and local polynomial regression fitting, are presented for significant cases in Figure S1.

The mean radial growth and mean radial growth trend (trendTRW) are presented in Table 7. The positive trend values mean that the growth rate was increasing over the past 30 years during the period of 1990–2019, taking into account the size–age trend, while negative values mean opposite—that it was decreasing. If the mean value of the growth index is more than one, then the trees have grown on average faster over the past 30 years than in their entire life; if less than one, then slower. The growth rate was slightly decreasing in four out of five populations, despite the fact that the trees have grown on average faster over the past 30 years than in their entire life in three out of five populations. A significant positive correlation between the radial growth trend (trendTRW) and individual heterozygosity was discovered. That is, the trees with higher heterozygosity demonstrate less of a decrease in the growth rate.

Table 7. The mean radial growth and mean radial growth trend (trendTRW) over the past 30 years during the period of 1990–2019 in five population sites of Siberian larch.

Population	BID	BGD	KAM	SON	TUIM
trendTRW	−0.002	0.002	−0.009	−0.017	−0.009
Mean radial growth	0.973	1.017	0.982	1.042	1.058

3.5. Genotype–Dendrophenotype Associations

According to the BSLMM, the proportion of variation in dendrophenotypes explained by the available genotypes (PVE) averaged 0.77, varying from the minimum value of 0.39 for the Rt5 index to the maximum value of 0.98 for avTRW. At the same time, the proportion of genetic variation explained by the variants with a relatively large effect (PGE) was 0.44 on average and varied from 0.03 in Rc4 to 0.96 in avTRW. For 21 out of 26 dendrophenotypes, the BSLMM method found associated SNPs with PIP values >25% (the estimated frequency by which an SNP has a sparse effect in the MCMC). In total, 2523 SNPs with presumably adaptive variation were selected using the BSLMM method, 453 of which were associated with more than 1 dendrophenotype.

According to the MLM, 1440 SNPs in total were associated with at least 1 of 26 dendrophenotypes, among which 565 were associated with 2 to 8 dendrophenotypes simultaneously.

According to the GLM, 2102 SNPs in total were associated with at least 1 of 26 dendrophenotypes, among which 1179 were associated with more than 1 dendrophenotype and 1 SNP (LS.4109781.252) was associated with 10 different dendrophenotypes.

A total of 4109 significant SNPs were detected using any of the 3 methods. Among them, 965 SNPs were detected by both GLM and MLM, and 371 SNPs by all 3 methods (Figure 5).

3.6. SNP Annotation

For the 371 SNPs found by all 3 association analysis methods, we analyzed the genomic regions in which these SNPs were localized (Table S2). According to the genomic annotation data of Siberian larch, only 26 SNPs were located in scaffolds with annotated genome features, whereas the rest were located in unannotated scaffolds. The majority of the successfully annotated adaptive markers were located in intergenic regions; only five were located in the coding regions of genes, and three of them were nonsynonymous and resulted in amino acid substitutions (Table 8).

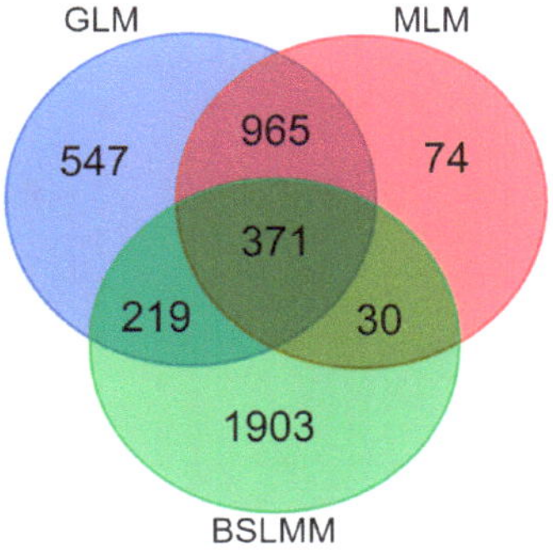

Figure 5. Venn diagram summarizing the results of the search for SNPs associated with at least one of 26 dendrophenotypes using three methods: BSLMM, GLM and MLM.

Table 8. Annotation of 26 potentially adaptive SNPs.

SNP	Dendrophenotype (GWAS Method)	Location *	Gene	Function	Reference
LS.3651056.309	Rt1 (GLM, MLM), avTRW (BSLMM)	Exon	Cytochrome P450 89A2	Oxidoreductase activity	[56,57]
LS.3905950.5973	avTRW, varTRW (GLM, MLM), Rs4 (BSLMM)	Exon	Xyloglucan endotransglucosylase/hydrolase protein B	Cell wall construction of growing tissues	[58]
LS.4033175.6261	Rs1 (GLM, MLM), RRs1 (BSLMM)	Exon	Cellulose synthase-like protein E6	Cellulose biosynthetic process	[59]
LS.4447596.15042	Age (GLM, MLM), Rc1 (BSLMM)	Exon	Uncharacterized protein LOC116133164	-	[53]
LS.4817075.929	varLn (GLM, MLM), Rc5 (GLM, BSLMM), RRs5 (GLM)	Exon	Protein Brevis radix-like 4	Modulator of root growth	[60]
LS.3903843.2180	Rc2, Rc3, RRs2, RRs3 (MLM, GLM) Rs2 (GLM), Rs3 (BSLMM)	78	RING-H2 finger protein ATL3	Ubiquitin ligase	[61,62]
LS.4185574.3908	varLn (GLM, MLM), Rs3 (BSLMM)	325	Alpha-mannosidase 2	Production of complex-type glycans	[63]
LS.7774.12431	RRs4, Rs2 (GLM, MLM), RRs5 (GLM, BSLMM) Rc5, Rt5 (GLM), Age (BSLMM)	3951	O-fucosyltransferase 19-like	Component of the gibberellin signaling pathway	[64]
LS.10638.27578	avTRW (GLM, MLM, BSLMM), Age, Rc4 (GLM)	5283	GPALPP motifs-containing protein 1 (lipopolysaccharide-specific response protein)	Lipopolysaccharide-specific response	[53]
LS.56935.13023	Rs4 (GLM, MLM, BSLMM)	5957	Probable aldo-keto reductase 4	Glyphosate degradation	[65]
LS.21791.50056	RRs1, Rc5 (GLM, MLM), Rs3 (BSLMM)	6066	Unknown	-	[53]
LS.61616.10689	Rc1, Rt1 (GLM, MLM), Rs5 (BSLMM)	6109	Hypothetical protein	-	[53]
LS.24961.61616	RRs5 (GLM, MLM), Rt4 (BSLMM)	6200	Unknown	-	[53]
LS.7196.6722	RRs5, Rs5 (GLM, MLM), RRs2 (BSLMM)	8888	Hypothetical heat shock protein	-	[53]
LS.105004.9682	avLn, Rt2 (GLM, MLM), Rc3 (GLM, BSLMM), RRs1, Rc1, Rt3 (GLM)	12,733	Phospholipase A1-Igamma3, chloroplastic	Catalyzes the hydrolysis of phosphatidylcholine	[66]
LS.7184.3035	Rs1 (GLM, MLM), RRs2 (BSLMM)	12,939	Unknown	-	[53]
LS.7252.104584	Rc3 (GLM, MLM, BSLMM), RRs1, Rc1, (MLM, GLM), Rs1 (GLM)	15,386	PLAT/LH2 domain-containing lipoxygenase family protein isoform 2	Defense mechanisms against pathogens	[67,68]

Table 8. *Cont.*

SNP	Dendrophenotype (GWAS Method)	Location *	Gene	Function	Reference
LS.14907.46110	Rt2 (GLM, MLM, BSLMM), Rs1 (GLM)	15,426	E3 ubiquitin ligase BIG BROTHER-related-like protein	It may limit the duration of organ growth and ultimately organ size by actively degrading critical growth stimulators	[69,70]
LS.15117.58377	RRs2, Rc2 (GLM, MLM), Age (GLM), Rs4 (BSLMM)	34,970	Cysteine and histidine-rich domain-containing protein RAR1 (CHRD1)	Gene-mediated disease resistance	[71]
LS.30352.56788	Rs4 (GLM, MLM), Rs3 (BSLMM)	54,635	Unknown	-	[53]
LS.16524.116919	Rs1 (GLM, MLM), Rs5 (BSLMM)	58,871	Uncharacterized FCP1 homology domain-containing protein C1271.03c	-	[53]
LS.12450.3129	Rs2, avTRW, varTRW (GLM, MLM), avLn (BSLMM)	64,166	RING-H2 finger protein	Ubiquitin ligase	[61,62]
LS.1920.81278	Rc1 (GLM, MLM), Rt1 (GLM), Rt3 (BSLMM)	65,622	U-box domain-containing protein 52-like	Functions as an E3 ubiquitin ligase.	[72]
LS.5783.87994	Rt1 (GLM, MLM), Rc1 (GLM), Rs3 (BSLMM)	69,130	ATP-dependent DNA helicase RECG-like	Recombination and DNA repair	[73]
LS.2549.42867	Rc3 (GLM, MLM), RRs2 (BSLMM)	69,752	Hypothetical protein GW17_00026793	-	[53]
LS.16630.144091	varLn (GLM, MLM), avLn (BSLMM)	83,423	Adenylosuccinate lyase	Synthesis of purine nucleotides	[74]

* Location in gene or distance to the nearest gene (bp).

For all the intervals of the nucleotide sequences (±1000 bp from the SNP or the maximum possible interval when the range of 2000 bp exceeded the scaffold size limits) containing SNPs associated with dendrophenotypes, homologs were searched in the NCBI BLAST database. Highly homologous sequences were found for 57 intervals containing SNPs, and most matches were found with genomic sequences and mostly unknown mRNAs and hypothetical proteins of white spruce (*Picea glauca*), Sitka spruce (*Picea sitchensis*), and Loblolly pine (*Pinus taeda*). Only for eight intervals were homologous sequences containing annotated genes or other structural units of the plant genome found in the NCBI GenBank database. For example, for the SNP locus LS.4643215.2651 which is associated with the dendrophenotype avLn according to GLM and MLM, as well as with the Rc3 index according to BSLMM, a homology was found with the mRNA encoding multiprotein-bridging factor 1c (MBF1, XM_028068828) of cowpea (*Vigna unguiculata*). It is a transcription cofactor whose molecular function is to form a bridge between transcription factors and the basal transcription machinery. They are involved in developmental processes and stress responses [75].

SNP LS.3970711.165 associated with the dendrophenotypes Rc4 and RRs4 according to GLM and MLM and with the RRs4 index according to BSLMM is located in a locus homologous to the mRNA encoding copper methylamine oxidase-like (XM_022162594) of oilseed sunflower (*Helianthus annuus*), which is involved in the mechanisms of protection against pathogens [76].

SNP LS.890128.210, associated with Rs3 according to BSLMM and Rs2 according to MLM and GLM is located in a locus homologous to the gene encoding the MYB6 transcription factor of Monterey pine (*Pinus radiata*, KY196189), which is involved in the regulation of the biosynthesis of secondary metabolites [77].

SNP LS.3903843.2180, which is associated with five dendrophenotypes (Rc2, Rc3, Rs2, RRs2 and RRs3) according to GLM, Rc2, Rc3, RRs2 and RRs3 according to MLM, and with Rs3 according to BSLMM, is located in a locus homologous to the mRNA of RING-H2 finger protein ATL3-like of wild soybean (*G. soja*, XM_028350422). There are also large annotation data for this SNP (Table 8).

SNP LS.4382060.16010, which is associated with dendrophenotypes RRs1, Rs1, Rs3, Rs5, and Rt3 according to GLM, Rs3 according to MLM, and Rc1 according to BSLMM, is located in a locus homologous to the mitochondrial gene encoding ribosomal protein S4 of eastern white pine (*Pinus strobus*, MN9653 08), which is involved in the formation of plant disease resistance [78].

SNP LS.29920.18803, which is associated with avLn, Rs1, and Rt4 according to GLM and MLM, Rs2 and Rt2 according to GLM, and Rc1 according to BSLMM, is located in a locus homologous to the gene encoding polygalacturonase of breadseed poppy (*Papaver somniferum*, XM_026530000), which is responsible for the degradation of cell wall pectin and is involved in the processes of organ aging and the response to biotic stress in plants [79].

Homologues related to retrotransposons were identified for two sequences. Thus, the interval containing SNP LS.9095140.133, which is associated with the dendrophenotype Rs4 according to GLM and MLM and with avLn according to BSLMM, was aligned to the sequence encoding the RNA-directed DNA polymerase of cabbage (*Brassica oleracea*, XM_013761928). The sequence containing SNP LS.4042787.3127, which is associated with dendrophenotypes RRS1, RC1, RS2, RT1 and RT3 according to GLM, with RRS1, RC1 and RT1 according to MLM, and with RS5 according to BSLMM, was homologous to the sequence of retrotransposons of maritime pine (*Pinus pinaster*, DQ394069) and Monterey pine (AJ004945).

4. Discussion

The results of the presented genome-wide analysis of the structure and genetic variation of natural populations of Siberian larch in the foothills of the Batenevsky Range are generally consistent with previous conclusions about the relatively weak population structure of closely located Siberian larch populations: the value of $F_{ST} = 0.018$ observed in this study is very close to the value of $F_{ST} = 0.017$ based on a larger dataset of 25,143 SNPs obtained earlier for Siberian larch in the Altai-Sayan region [80].

The genetic differentiations between two groups identified on the basis of population structure analysis (F_{CT}), between populations within groups (F_{SC}), and between all the populations (F_{ST}) are consistent with the results previously obtained for larch and based on a larger dataset of SNPs [80,81]. The largest proportion of the genetic variance, 98%, was within the studied samples, and only 1% of the variance was due to differences between groups.

The results of the PCA showed that the samples were weakly genetically differentiated, which was expected because of the relatively short distances between sampling sites (about 50 km). Some trees from some populations demonstrated higher genetic similarity to trees from other populations according to the results of the cluster analysis (Figure 3). This is not surprising and can be expected, since the distance between populations is only tens of kilometers, there are no clear gaps between them such as treeless areas or forests that do not include larch, and there are no pronounced geographical barriers. Thus, some gene flow and genetic exchange can be expected between these populations, especially considering that larch is a wind-pollinated and wind-propagated species with small and light winged seeds. It is possible that the observed distribution reflects the predominance of southwestern and western winds in the basin of this region in combination with landscape restrictions. In particular, the SON and KAM populations are located approximately upwind of the remaining populations. Minimal genetic differentiation was observed between the TUI and BOG populations. There was no correlation between the geographic distance and the relatedness of trees in the study area.

The lack of a significant correlation between individual heterozygosity and the avTRW and varTRW dendrophenotypes is consistent with those previously obtained by Babushkina et al. [24] for two populations of Siberian larch based on the eight most polymorphic microsatellite loci (SSRs), although in the case of varTRW, it was negative according to Spearman's rank correlation coefficient ($r_S = -0.164$) and almost significant ($p = 0.056$), which may partly indicate that increased heterozygosity stabilizes individual development, reducing the variance of wood growth. There was also no correlation between individual heterozygosity and age, average needle length, or variance in needle length.

According to recently obtained preliminary data for 124 Siberian pine (*Pinus sibirica* Du Tour) trees growing in the same region 150–250 km to the south of study area there was no significant correlation between individual heterozygosity based on 13,914 biallelic SNPs and avTRW, although a significant negative correlation of the varTRW with individual heterozygosity was also found ($r_S = -0.303$) (data not published).

The presence of a significant positive correlation between individual heterozygosity and the recovery (Rc) and relative resilience (RRs) indices for all the studied drought periods and partly the resilience (Rs) indices for three drought periods suggests that differences in the drought resistance of individual Siberian larch trees are associated with the level of individual heterozygosity, the increase in which improves the adaptive capabilities of the organism.

However, we also discovered a significant negative relationship between the level of heterozygosity and the resistance index (Rt), which directly reflects the efficiency of larch growth during the drought years in four out of five periods. Perhaps one of the adaptive mechanisms of larch involves a sharp reduction in growth during periods of drought as part of a survival strategy, which is more pronounced in trees with a higher level of individual heterozygosity. Whether this is a consequence of increased homeostasis or heterosis and overdominance is difficult to determine and requires additional research.

It is interesting to note that the confidence intervals for the correlation trends are narrower in the upper-to-middle intervals (0.14–0.15) of the heterozygosity values, and he trendlines based on the quadratic, cubic and local polynomial regression fitting better approximate the correlation between individual heterozygosity and dendrophenotypes with higher regression coefficients (Figure S1). It can be a signature of the "optimal" heterozygosity reflecting the stable individual growth and response to the stress factors and providing the maximum adaptation of an individual or a population to the environment in which the individual or population was formed and in which it now exists (in sensu [82–84]).

The search for genotype–phenotype associations using three different methods revealed 371 potentially adaptive SNPs. Only 26 SNPs were located in genomic regions carrying functional genes: 21 in intergenic regions and 5 in gene-coding regions. The genetic diversity summary parameters based on 371 potentially adaptive SNPs were only slightly higher than those that were based on all 9742 SNPs (Table 3). It seems that a limited number of SNPs are needed to appropriately assess genetic diversity.

It is important to note that the growth rate was slightly decreasing in four out of five populations over the past 30 years during the period of 1990–2019. This is likely related to the climate change, particularly reflecting the increased water deficit caused by warming. These observations are corroborated by recent climate-driven decreasing trends in the grain crop yields observed in the steppe territories of the study region [85]. However, creation of large water reservoirs has so far mitigated most of the drying climatic trends in the region [86] comparing to most of continental Asia, where severely declining tree growth and forest die-off were observed recently in many places on southern and lower fringes of forested areas [6,87].

5. Conclusions

Several regions of the genome and genes have been identified, the variation of which is associated with the variation of important adaptive traits, dendrophenotypes. We certainly realize the limits of the RADseq method and that the obtained genomic data could be insufficient to uncover all the genes and genetic mechanism that are responsible for adaptation to drought, although those that are found could be of interest for the scientific community. We identified several SNPs in candidate genes whose variation was associated with important adaptive dendrophenotypes. Based on these results it can be assumed that Siberian larch has relatively high adaptive genetic variation and potential in the studied area. It seems well adapted to abiotic stresses such as droughts. However, the negative growth trend observed in four out of five populations could be a sign of the beginning negative effect of climate change. The results of this study will allow for a deeper understanding of the genetic mechanisms underlying the adaptations of larch to various climatic conditions. This study allowed us to detect important genes, and we obtained SNPs with significant adaptive variation that can be potentially used as genetic markers in the marker-aided breeding for stress tolerance as well as for creating a SNP genotyping chip for monitoring adaptive genetic variation in other larch populations, although more SNP genotyping data, preferably based on the whole-genome sequencing, are required for developing a comprehensive high-density SNP genotyping assay to study larch. The presented approach, which includes both dendrophenotypic and genetic data, can serve as a scientific basis for optimizing nature management, developing methods for the rational use of the studied species, identifying populations with good genetic potential and conducting environmental monitoring.

Supplementary Materials: The following supporting information can be downloaded at https://www.mdpi.com/article/10.3390/f14122358/s1, Figure S1: Dot-plot graphs of individual heterozygosity (IndHet) and dendrophenotypes (Rt, Rc, Rs, RRs, varTRW, and trendTRW) for significant cases with the Pearson's (r) and Spearman's rank (r_S) correlation coefficients, regression coefficients (R^2) and trendlines based on linear (L), quadratic (Q), cubic (C) and local polynomial (LP) regression fitting. The gray area highlights the 95% confidence interval; Table S1: Individual and mean values of 27 traits in 136 trees and 5 populations and pairwise Wilcoxon rank-sum test p-values for 27 traits adjusted according to the Benjamini–Hochberg procedure; Table S2: Annotation summary of 371 SNPs; Data S1: Genotypes of 9742 SNPs in the vcf format.

Author Contributions: Conceptualization, K.V.K.; methodology, S.V.N., D.F.Z., L.V.B., E.A.B. and K.V.K.; software, S.V.N., V.V.S., D.F.Z. and L.V.B.; validation, S.V.N.; formal analysis, S.V.N., N.V.O., V.V.S., D.F.Z., L.V.B., E.A.B. and K.V.K.; investigation, S.V.N., N.V.O., V.V.S., D.F.Z., L.V.B., E.A.B. and K.V.K.; resources, N.V.O., D.F.Z., L.V.B., E.A.B. and K.V.K.; data curation, S.V.N., V.V.S., D.F.Z., L.V.B. and E.A.B.; writing—original draft preparation, S.V.N. and K.V.K.; writing—review and editing, S.V.N., N.V.O., V.V.S., D.F.Z., L.V.B., E.A.B. and K.V.K.; visualization, S.V.N. and K.V.K.; supervision, N.V.O., D.F.Z., E.A.B. and K.V.K.; project administration, K.V.K., N.V.O. and E.A.B.; funding acquisition, K.V.K. All authors have read and agreed to the published version of the manuscript.

Funding: The Siberian larch and Siberian stone pine parts used in this study were supported by the two Russian Science Foundation grants 19-14-00120 and 22-14-00083, respectively.

Data Availability Statement: The data presented in this study are all available in the article and the files with the supplementary material.

Acknowledgments: The authors are grateful to the Department of High-Performance Computing of the Siberian Federal University and to its head, Dmitry A. Kuzmin, for providing access to the high-performance computer cluster.

Conflicts of Interest: The authors declare no conflict of interest.

References

1. Mukherjee, S.; Mishra, A.; Trenberth, K.E. Climate Change and Drought: A Perspective on Drought Indices. *Curr. Clim. Chang. Rep.* **2018**, *4*, 145–163. [CrossRef]
2. Cook, B.I.; Mankin, J.S.; Anchukaitis, K.J. Climate Change and Drought: From Past to Future. *Curr. Clim. Chang. Rep.* **2018**, *4*, 164–179. [CrossRef]
3. Gauthier, S.; Bernier, P.; Kuuluvainen, T.; Shvidenko, A.Z.; Schepaschenko, D.G. Boreal forest health and global change. *Science* **2015**, *349*, 819–822. [CrossRef] [PubMed]
4. Brecka, A.F.J.; Shahi, C.; Chen, H.Y.H. Climate change impacts on boreal forest timber supply. *For. Policy Econ.* **2018**, *92*, 11–21. [CrossRef]
5. Krutovsky, K.V. Dendrogenomics is a new interdisciplinary field of research of the adaptive genetic potential of forest tree populations integrating dendrochronology, dendroecology, dendroclimatology, and genomics. *Russ. J. Genet.* **2022**, *58*, 1273–1286. [CrossRef]
6. Liu, H.; Park Williams, A.; Allen, C.D.; Guo, D.; Wu, X.; Anenkhonov, O.A.; Liang, E.; Sandanov, D.V.; Yin, Y.; Qi, Z.; et al. Rapid warming accelerates tree growth decline in semi-arid forests of inner Asia. *Global Chang. Biol.* **2013**, *19*, 2500–2510. [CrossRef] [PubMed]

7. Zhirnova, D.F.; Belokopytova, L.V.; Krutovsky, K.V.; Kholdaenko, Y.A.; Babushkina, E.A.; Vaganov, E.A. Spatial-coherent dynamics and climatic signals in the radial growth of Siberian stone pine (*Pinus sibirica* Du Tour) in subalpine stands along the Western Sayan Mountains. *Forests* **2022**, *13*, 1994. [CrossRef]
8. Belokopytova, L.V.; Zhirnova, D.F.; Krutovsky, K.V.; Mapitov, N.B.; Vaganov, E.A.; Babushkina, E.A. Species- and age-specific growth reactions to extreme droughts of the keystone tree species across forest-steppe and sub-taiga habitats of South Siberia *Forests* **2022**, *13*, 1027. [CrossRef]
9. Kharuk, V.I.; Petrov, I.A.; Dvinskaya, M.L.; Im, S.T.; Shushpanov, A.S. Comparative reaction of larch (*Larix sibirica* Ledeb.) radial increment on climate change in the forest steppe and highlands of Southern Siberia. *Contemp. Probl. Ecol.* **2018**, *11*, 388–395. [CrossRef]
10. Dulamsuren, C.; Hauck, M.; Leuschner, C. Recent drought stress leads to growth reductions in *Larix sibirica* in the western Khentey, Mongolia. *Glob. Chang. Biol.* **2010**, *16*, 3024–3035. [CrossRef]
11. Barchenkov, A.P.; Petrov, I.A.; Shushpanov, A.S.; Golyukov, A.S. Climatic response of larch (*Larix* sp.) radial increment in provenances on the Krasnoyarsk forest steppe. *Contemp. Probl. Ecol.* **2023**, *16*, 620–630. [CrossRef]
12. Abaimov, A.P. Geographical distribution and genetics of Siberian larch species. In *Permafrost Ecosystems. Ecological Studies*; Osawa A., Zyryanova, O., Matsuura, Y., Kajimoto, T., Wein, R., Eds.; Springer: Dordrecht, The Netherlands, 2010; Volume 209, pp. 41–58 [CrossRef]
13. Semerikov, V.L.; Iroshnikov, A.I.; Lascoux, M. Mitochondrial DNA variation pattern and postglacial history of the Siberian larch (*Larix sibirica* Ledeb.). *Russ. J. Ecol.* **2007**, *38*, 147–154. [CrossRef]
14. Bobrov, E.G. *Forest-Forming Conifers of the USSR*; Nauka Publishing: Leningrad, Russia, 1978; 188p.
15. Urban, J.; Rubtsov, A.V.; Urban, A.V.; Shashkin, A.V.; Benkova, V.E. Canopy transpiration of a *Larix sibirica* and *Pinus sylvestris* forest in Central Siberia. *Agric. For. Meteorol.* **2019**, *271*, 64–72. [CrossRef]
16. Dulamsuren, C.; Hauck, M.; Bader, M.; Oyungerel, S.; Osokhjargal, D.; Nyambayar, S.; Leuschner, C. The different strategies of *Pinus sylvestris* and *Larix sibirica* to deal with summer drought in a northern Mongolian forest–steppe ecotone suggest a future superiority of pine in a warming climate. *Can. J. For. Res.* **2009**, *39*, 2520–2528. [CrossRef]
17. Schweingruber, F.H. *Tree Rings and Environment Dendroecology*; Haupt: Berne, Switzerland; Stuttgart, Germany; Vienna, Austria 1996; ISBN 978-3-258-05458-2.
18. Fritts, H.C. *Tree Rings and Climate*; Blackburn Press: Caldwell, NJ, USA, 2001; ISBN 978-1-930665-39-2.
19. Vaganov, E.A.; Hughes, M.K.; Shashkin, A.V. *Growth Dynamics of Conifer Tree Rings: Images of Past and Future Environments* Ecological Studies; Springer: Berlin/Heidelberg, Germany, 2006; ISBN 978-3-540-26086-8.
20. Babushkina, E.A.; Zhirnova, D.F.; Belokopytova, L.V.; Tychkov, I.I.; Vaganov, E.A.; Krutovsky, K.V. Response of four tree species to changing climate in a moisture-limited area of South Siberia. *Forests* **2019**, *10*, 999. [CrossRef]
21. Belokopytova, L.V.; Babushkina, E.A.; Zhirnova, D.F.; Panyushkina, I.P.; Vaganov, E.A. Climatic response of conifer radial growth in forest-steppes of South Siberia: Comparison of three approaches. *Contemp. Probl. Ecol.* **2018**, *11*, 366–376. [CrossRef]
22. Belokopytova, L.V.; Zhirnova, D.F.; Meko, D.M.; Babushkina, E.A.; Vaganov, E.A.; Krutovsky, K.V. Tree rings reveal the impact of soil temperature on larch growth in the forest-steppe of Siberia. *Forests* **2021**, *12*, 1765. [CrossRef]
23. Rozenberg, P.; Pâques, L.; Huard, F.; Roques, A. Direct and Indirect analysis of the elevational shift of larch budmoth outbreaks along an elevation gradient. *Front. For. Glob. Chang.* **2020**, *3*, 86. [CrossRef]
24. Babushkina, E.A.; Vaganov, E.A.; Grachev, A.M.; Oreshkova, N.V.; Belokopytova, L.V.; Kostyakova, T.V.; Krutovsky, K.V. The effect of individual genetic heterozygosity on general homeostasis, heterosis and resilience in Siberian larch (*Larix sibirica* Ledeb. using dendrochronology and microsatellite loci genotyping. *Dendrochronologia* **2016**, *38*, 26–37. [CrossRef]
25. Johnson, J.S.; Gaddis, K.D.; Cairns, D.M.; Konganti, K.; Krutovsky, K.V. Landscape genomic insights into the historic migration of mountain hemlock in response to Holocene climate change. *Am. J. Bot.* **2017**, *104*, 439–450. [CrossRef]
26. Johnson, J.S.; Chhetri, P.; Krutovsky, K.V.; Cairns, D.M. Growth and its relationship to individual genetic diversity of mountain hemlock (*Tsuga mertensiana*) at alpine treeline in Alaska: Combining dendrochronology and genomics. *Forests* **2017**, *8*, 418. [CrossRef
27. Johnson, J.S.; Gaddis, K.D.; Cairns, D.M.; Krutovsky, K.V. Seed dispersal at alpine treeline: An assessment of seed movement within the alpine treeline ecotone. *Ecosphere* **2017**, *8*, e01649. [CrossRef]
28. Depardieu, C.; Gérardi, S.; Nadeau, S.; Parent, G.J.; Mackay, J.; Lenz, P.; Lamothe, M.; Girardin, M.P.; Bousquet, J.; Isabel, N Connecting tree-ring phenotypes, genetic associations and transcriptomics to decipher the genomic architecture of drought adaptation in a widespread conifer. *Mol. Ecol.* **2021**, *30*, 3898–3917. [CrossRef] [PubMed]
29. Cappa, E.P.; Klutsch, J.G.; Sebastian-Azcona, J.; Ratcliffe, B.; Wei, X.; Da Ros, L.; Liu, Y.; Chen, C.; Benowicz, A.; Sadoway, S.; et al Integrating genomic information and productivity and climate-adaptability traits into a regional white spruce breeding program *PLoS ONE* **2022**, *17*, e0264549. [CrossRef] [PubMed]
30. Fasanella, M.; Suarez, M.L.; Hasbún, R.; Premoli, A.C. Individual-based dendrogenomic analysis of forest dieback driven by extreme droughts. *Can. J. For. Res.* **2021**, *51*, 420–432. [CrossRef]
31. Heer, K.; Behringer, D.; Piermattei, A.; Bässler, C.; Brandl, R.; Fady, B.; Jehl, H.; Liepelt, S.; Lorch, S.; Piotti, A.; et al. Linking Dendroecology and association genetics in natural populations: Stress responses archived in tree rings associate with SNP genotypes in silver fir (*Abies alba* Mill.). *Mol. Ecol.* **2018**, *27*, 1428–1438. [CrossRef]
32. Cook, E.R.; Kairiukstis, L.A. (Eds.) *Methods of Dendrochronology*; Springer: Dordrecht, The Netherlands, 1990; ISBN 978-90-481-4060-2
33. Rinn, F. TSAP-Win: Time Series Analysis and Presentation for Dendrochronology and Related Applications. Version 0.55 User Reference. Heidelberg, Germany. 2003. Available online: https://software.rinntech.com/tsap (accessed on 14 October 2023).

34. Holmes, R.L. Computer-assisted quality control in tree-ring dating and measurement. *Tree-Ring Bull.* **1983**, *43*, 69–78.
35. Zhirnova, D.F.; Babushkina, E.A.; Belokopytova, L.V.; Vaganov, E.A. To which side are the scales swinging? Growth stability of Siberian larch under permanent moisture deficit with periodic droughts. *For. Ecol. Manag.* **2020**, *459*, 117841. [CrossRef]
36. Lloret, F.; Keeling, E.G.; Sala, A. Components of tree resilience: Effects of successive low-growth episodes in old ponderosa pine forests. *Oikos* **2011**, *120*, 1909–1920. [CrossRef]
37. Cook, E.R.; Krusic, P.J. *Program ARSTAN: A Tree-Ring Standardization Program Based on Detrending and Autoregressive Time Series Modeling, with Interactive Graphics*; Lamont-Doherty Earth Observatory, Columbia University: Palisades, NY, USA, 2005; 14p.
38. R Core Team. *R: A Language and Environment for Statistical Computing*; R Foundation for Statistical Computing: Vienna, Austria, 2013. Available online: https://www.r-project.org (accessed on 14 October 2023).
39. Porebski, S.; Bailey, L.G.; Baum, B.R. Modification of a CTAB DNA extraction protocol for plants containing high polysaccharide and polyphenol components. *Plant Mol. Biol. Rep.* **1997**, *15*, 8–15. [CrossRef]
40. Peterson, B.K.; Weber, J.N.; Kay, E.H.; Fisher, H.S.; Hoekstra, H.E. Double digest RADseq: An inexpensive method for de novo SNP discovery and genotyping in model and non-model species. *PLoS ONE* **2012**, *7*, e37135. [CrossRef] [PubMed]
41. Parchman, T.L.; Gompert, Z.; Mudge, J.; Schilkey, F.D.; Benkman, C.W.; Buerkle, C.A. Genome-wide association genetics of an adaptive trait in lodgepole pine: Association mapping of serotiny. *Mol. Ecol.* **2012**, *21*, 2991–3005. [CrossRef] [PubMed]
42. Catchen, J.; Hohenlohe, P.A.; Bassham, S.; Amores, A.; Cresko, W.A. Stacks: An analysis tool set for population genomics. *Mol. Ecol.* **2013**, *22*, 3124–3140. [CrossRef] [PubMed]
43. Kuzmin, D.A.; Feranchuk, S.I.; Sharov, V.V.; Cybin, A.N.; Makolov, S.V.; Putintseva, Y.A.; Oreshkova, N.V.; Krutovsky, K.V. Stepwise large genome assembly approach: A case of Siberian larch (*Larix sibirica* Ledeb.). *BMC Bioinf.* **2019**, *20*, 37. [CrossRef] [PubMed]
44. Langmead, B.; Salzberg, S.L. Fast gapped-read alignment with Bowtie 2. *Nat. Methods* **2012**, *9*, 357–359. [CrossRef]
45. Bradbury, P.J.; Zhang, Z.; Kroon, D.E.; Casstevens, T.M.; Ramdoss, Y.; Buckler, E.S. TASSEL: Software for association mapping of complex traits in diverse samples. *Bioinformatics* **2007**, *23*, 2633–2635. [CrossRef] [PubMed]
46. Pembleton, L.W.; Cogan, N.O.I.; Forster, J.W. St AMPP: An R package for calculation of genetic differentiation and structure of mixed-ploidy level populations. *Mol. Ecol. Resour.* **2013**, *13*, 946–952. [CrossRef] [PubMed]
47. Dray, S.; Dufour, A.-B. The Ade4 package: Implementing the duality diagram for ecologists. *J. Stat. Soft.* **2007**, *22*, 1–20. [CrossRef]
48. Frichot, E.; François, O. LEA: An R package for landscape and ecological association studies. *Methods Ecol. Evol.* **2015**, *6*, 925–929. [CrossRef]
49. Mussmann, S.M.; Douglas, M.R.; Chafin, T.K.; Douglas, M.E. AdmixPipe: Population analyses in admixture for non-model organisms. *BMC Bioinf.* **2020**, *21*, 337. [CrossRef]
50. Excoffier, L.; Lischer, H.E.L. Arlequin suite ver 3.5: A new series of programs to perform population genetics analyses under Linux and Windows. *Mol. Ecol. Resour.* **2010**, *10*, 564–567. [CrossRef] [PubMed]
51. Oksanen, J.; Simpson, G.; Blanchet, F.; Kindt, R.; Legendre, P.; Minchin, P.; O'Hara, R.; Solymos, P.; Stevens, M.; Szoecs, E.; et al. Vegan: Community Ecology Package. R Package Version 2.6-5. 2023. Available online: https://github.com/vegandevs/vegan (accessed on 14 October 2023).
52. Zhou, X.; Carbonetto, P.; Stephens, M. Polygenic Modeling with Bayesian Sparse Linear Mixed Models. *PLoS Genet.* **2013**, *9*, e1003264. [CrossRef] [PubMed]
53. Bondar, E.I.; Feranchuk, S.I.; Miroshnikova, K.A.; Sharov, V.V.; Kuzmin, D.A.; Oreshkova, N.V.; Krutovsky, K.V. Annotation of Siberian larch (*Larix sibirica* Ledeb.) nuclear genome—One of the most cold-resistant tree species in the only deciduous genus in Pinaceae. *Plants* **2022**, *11*, 2062. [CrossRef] [PubMed]
54. Doran, A.G.; Creevey, C.J. Snpdat: Easy and rapid annotation of results from de novo SNP discovery projects for model and non-model organisms. *BMC Bioinform.* **2013**, *14*, 45. [CrossRef] [PubMed]
55. Clark, K.; Karsch-Mizrachi, I.; Lipman, D.J.; Ostell, J.; Sayers, E.W. GenBank. *Nucleic Acids Res.* **2016**, *44*, D67–D72. [CrossRef] [PubMed]
56. Duan, F.; Ding, J.; Lee, D.; Lu, X.; Feng, Y.; Song, W. Overexpression of SoCYP85A1, a spinach cytochrome P450 gene in transgenic tobacco enhances root development and drought stress tolerance. *Front. Plant Sci.* **2017**, *8*, 1909. [CrossRef] [PubMed]
57. Rao, M.J.; Xu, Y.; Tang, X.; Huang, Y.; Liu, J.; Deng, X.; Xu, Q. CsCYT75B1, a citrus CYTOCHROME P450 gene, is involved in accumulation of antioxidant flavonoids and induces drought tolerance in transgenic *Arabidopsis*. *Antioxidants* **2020**, *9*, 161. [CrossRef]
58. Xu, W.; Purugganan, M.M.; Polisensky, D.H.; Antosiewicz, D.M.; Fry, S.C.; Braam, J. *Arabidopsis* TCH4, regulated by hormones and the environment, encodes a xyloglucan endotransglycosylase. *Plant Cell* **1995**, *7*, 1555–1567. [CrossRef]
59. Zhu, J.; Lee, B.-H.; Dellinger, M.; Cui, X.; Zhang, C.; Wu, S.; Nothnagel, E.A.; Zhu, J.-K. A Cellulose synthase-like protein is required for osmotic stress tolerance in *Arabidopsis*: SOS6 is important for osmotic stress tolerance in plants. *Plant J.* **2010**, *63*, 128–140. [CrossRef]
60. Jinu, J.; Visarada, K.B.R.S.; Kanti, M.; Malathi, V.M. Dehydration stress influences the expression of *brevis radix* gene family members in sorghum (*Sorghum bicolor*). *Proc. Indian Natl. Sci. Acad.* **2022**, *88*, 324–335. [CrossRef]
61. Kim, S.J.; Kim, W.T. Suppression of *Arabidopsis* RING E3 ubiquitin ligase *AtATL78* increases tolerance to cold stress and decreases tolerance to drought stress. *FEBS Lett.* **2013**, *587*, 2584–2590. [CrossRef] [PubMed]

62. Suh, J.Y.; Kim, S.J.; Oh, T.R.; Cho, S.K.; Yang, S.W.; Kim, W.T. Arabidopsis Tóxicos En Levadura 78 (AtATL78) Mediates ABA-dependent ROS signaling in response to drought stress. *Biochem. Biophys. Res. Commun.* **2016**, *469*, 8–14. [CrossRef]
63. Wang, W.; Wu, Y.; Shi, R.; Sun, M.; Li, Q.; Zhang, G.; Wu, J.; Wang, Y.; Wang, W. Overexpression of wheat α-mannosidase gene *TaMP* impairs salt tolerance in transgenic *Brachypodium distachyon*. *Plant Cell Rep.* **2020**, *39*, 653–667. [CrossRef]
64. Liang, L.; Wang, Q.; Song, Z.; Wu, Y.; Liang, Q.; Wang, Q.; Yang, J.; Bi, Y.; Zhou, W.; Fan, L.-M. *O*-fucosylation of CPN20 by SPINDLY derepresses abscisic acid signaling during seed germination and seedling development. *Front. Plant Sci.* **2021**, *12*, 724144. [CrossRef] [PubMed]
65. Simpson, P.J.; Tantitadapitak, C.; Reed, A.M.; Mather, O.C.; Bunce, C.M.; White, S.A.; Ride, J.P. Characterization of two novel aldo–keto reductases from *Arabidopsis*: Expression patterns, broad substrate specificity, and an open active-site structure suggest a role in toxicant metabolism following stress. *J. Mol. Biol.* **2009**, *392*, 465–480. [CrossRef]
66. Seo, Y.S.; Kim, E.Y.; Kim, J.H.; Kim, W.T. Enzymatic characterization of Class I DAD1-like acylhydrolase members targeted to chloroplast in *Arabidopsis*. *FEBS Lett.* **2009**, *583*, 2301–2307. [CrossRef] [PubMed]
67. Oliw, E.H. Plant and Fungal Lipoxygenases. *Prostaglandins Other Lipid Mediat.* **2002**, *68*, 313–323. [CrossRef] [PubMed]
68. Bateman, A.; Sandford, R. The PLAT domain: A new piece in the PKD1 puzzle. *Curr. Biol.* **1999**, *9*, R588–R590. [CrossRef]
69. Disch, S.; Anastasiou, E.; Sharma, V.K.; Laux, T.; Fletcher, J.C.; Lenhard, M. The E3 ubiquitin ligase BIG BROTHER controls *Arabidopsis* organ size in a dosage-dependent manner. *Curr. Biol.* **2006**, *16*, 272–279. [CrossRef]
70. Al-Saharin, R.; Hellmann, H.; Mooney, S. Plant E3 ligases and their role in abiotic stress response. *Cells* **2022**, *11*, 890. [CrossRef]
71. Yuan, C.; Li, C.; Zhao, X.; Yan, C.; Wang, J.; Mou, Y.; Sun, Q.; Shan, S. Genome-wide identification and characterization of HSP90-RAR1-SGT1-complex members from *Arachis* genomes and their responses to biotic and abiotic stresses. *Front. Genet.* **2021**, *12*, 689669. [CrossRef] [PubMed]
72. Song, J.; Mo, X.; Yang, H.; Yue, L.; Song, J.; Mo, B. The U-box family genes in *Medicago truncatula*: Key elements in response to salt, cold, and drought stresses. *PLoS ONE* **2017**, *12*, e0182402. [CrossRef] [PubMed]
73. Mohapatra, M.D.; Poosapati, S.; Sahoo, R.K.; Swain, D.M. Helicase: A genetic tool for providing stress tolerance in plants. *Plant Stress* **2023**, *9*, 100171. [CrossRef]
74. Zhang, Z.; Hao, Z.; Chai, R.; Qiu, H.; Wang, Y.; Wang, J.; Sun, G. Adenylsuccinate synthetase MoADE12 plays important roles in the development and pathogenicity of the rice blast fungus. *J. Fungi* **2022**, *8*, 780. [CrossRef]
75. Jaimes-Miranda, F.; Chávez Montes, R.A. The Plant MBF1 protein family: A bridge between stress and transcription. *J. Exp. Bot.* **2020**, *71*, 1782–1791. [CrossRef]
76. Rea, G.; Metoui, O.; Infantino, A.; Federico, R.; Angelini, R. Copper amine oxidase expression in defense responses to wounding and *Ascochyta rabiei* invasion. *Plant Physiol.* **2002**, *128*, 865–875. [CrossRef]
77. Ambawat, S.; Sharma, P.; Yadav, N.R.; Yadav, R.C. MYB transcription factor genes as regulators for plant responses: An overview. *Physiol. Mol. Biol. Plants* **2013**, *19*, 307–321. [CrossRef]
78. Gassmann, W.; Hinsch, M.E.; Staskawicz, B.J. The *Arabidopsis* RPS4 bacterial-resistance gene is a member of the TIR-NBS-LRR family of disease-resistance genes. *Plant J.* **1999**, *20*, 265–277. [CrossRef]
79. Liu, H.; Ma, Y.; Chen, N.; Guo, S.; Liu, H.; Guo, X.; Chong, K.; Xu, Y. Overexpression of stress-inducible OsBURP16, the β subunit of polygalacturonase 1, decreases pectin content and cell adhesion and increases abiotic stress sensitivity in rice. *Plant Cell Environ.* **2014**, *37*, 1144–1158. [CrossRef]
80. Novikova, S.V.; Sharov, V.V.; Oreshkova, N.V.; Simonov, E.P.; Krutovsky, K.V. Genetic adaptation of Siberian larch (*Larix Sibirica* Ledeb.) to high altitudes. *Int. J. Mol. Sci.* **2023**, *24*, 4530. [CrossRef]
81. Novikova, S.V.; Oreshkova, N.V.; Sharov, V.V.; Semerikov, V.L.; Krutovsky, K.V. Genetic structure and geographical differentiation of Siberian larch (*Larix sibirica* Ledeb.) populations based on genome genotyping by sequencing. *Contemp. Probl. Ecol.* **2023**, *16*, 631–644. [CrossRef]
82. Altukhov, Y.P.; Gafarov, N.I.; Krutovskii, K.V.; Dukharev, V.A. Allozyme variability in a natural population of Norway spruce (*Picea abies* [L.] Karst.). III. Correlation between levels of individual heterozygosity and relative number of inviable seeds. *Genetika* **1986**, *22*, 2825–2830, (In Russian, translated in English as *Soviet Genetics* **1987**, *22*, 1580–1585).
83. Altukhov, Y.P. The role of balancing selection and overdominance in maintaining allozyme polymorphism. *Genetica* **1991**, *85*, 79–90. [CrossRef] [PubMed]
84. Dubrova, Y.E.; Salmenkova, E.A.; Altukhov, Y.P.; Kartavtsev, Y.F.; Kalkova, E.V.; Omel'chenko, V.T. Family heterozygosity and progeny body length in pink salmon *Oncorhynchus gorbuscha* (Walbaum). *Heredity* **1995**, *75*, 281–289. [CrossRef]
85. Babushkina, E.A.; Belokopytova, L.V.; Zhirnova, D.F.; Shah, S.K.; Kostyakova, T.V. Climatically driven yield variability of major crops in Khakassia (South Siberia). *Int. J. Biometeorol.* **2018**, *62*, 939–948. [CrossRef]
86. Zhirnova, D.F.; Belokopytova, L.V.; Meko, D.M.; Babushkina, E.A.; Vaganov, E.A. Climate change and tree growth in the Khakass-Minusinsk Depression (South Siberia) impacted by large water reservoirs. *Sci. Rep.* **2021**, *11*, 14266. [CrossRef]
87. Xu, H.J.; Wang, X.P.; Zhang, X.X. Decreased vegetation growth in response to summer drought in Central Asia from 2000 to 2012. *Int. J. Appl. Earth Observ. Geoinf.* **2016**, *52*, 390–402. [CrossRef]

forests

MDPI

Article

Dendrochronological Analysis of One-Seeded and Intermediate Hawthorn Response to Climate in Poland

Anna Cedro * and Bernard Cedro

Institute of Marine & Environmental Sciences, University of Szczecin, Adama Mickiewicza 16, 70-383 Szczecin, Poland; bernard.cedro@usz.edu.pl
* Correspondence: anna.cedro@usz.edu.pl

Abstract: Although the hawthorn is not a forest-forming species, and it has no high economic significance, it is a very valuable component of forests, mid-field woodlots or roadside avenues. The literature, however, lacks information on the growth rate, growth phases, or growth–climate–habitat relationship for trees of this genus. This work aimed to establish the rate of growth of *Craraegus monogyna* and *C. xmedia* Bechst growing in various parts of Poland, in various habitats; analyze the growth–climate relationship; and distinguish dendrochronological regions for these species. Samples were taken using a Pressler borer from nine populations growing in different parts of Poland, from a total of 192 trees (359 samples). The tree-ring width was measured down to 0.01 mm. The average tree-ring width in the studied hawthorn populations ranged from 1.42 to 3.25 mm/year. Using well-established cross-dating methods, nine local chronologies were compiled with tree ages between 45 and 72 years. Dendroclimatic analyses (pointer year analysis, correlation and response function analysis) were performed for a 33-year period from 1988 to 2020, for which all local chronologies displayed EPS > 0.85. The tree-ring width in the hawthorn populations depended mostly on temperature and rainfall through the May–August period. High rainfall and the lack of heat waves through these months cause an increase in cambial activity and the formation of wide tree rings. Conversely, rainfall shortages through this period, in conjunction with high air temperatures, caused growth depressions. Cluster analysis enabled the identification of two dendrochronological regions among the hawthorn in Poland: a western and eastern region, and a single site (CI), whose separation was most likely caused by contrasting habitat and genetic conditions. The obtained results highlight the need for further study of these species in Poland and other countries.

Keywords: one-seeded hawthorn (*Crataegus monogyna*); intermediate hawthorn (*Crataegus media* Bechst.); tree-ring width; dendroclimatology; Poland

Citation: Cedro, A.; Cedro, B. Dendrochronological Analysis of One-Seeded and Intermediate Hawthorn Response to Climate in Poland. *Forests* **2023**, *14*, 2264. https://doi.org/10.3390/f14112264

Academic Editors: Yassine Messaoud, Jan Světlík and Giorgio Alberti

Received: 12 October 2023
Revised: 13 November 2023
Accepted: 14 November 2023
Published: 17 November 2023

1. Introduction

In recent decades, we have witnessed climate change, manifested mostly via air temperature increase (warmer winters, earlier and longer warm seasons, elevated maximum temperatures), and increasingly extreme weather (e.g., heavy rains, droughts, heat waves) [1]. Also, the human impact on the natural environment has been increasing: occupation, transforming and destroying plant and animal habitats, emissions of substances that are harmful to living organisms into soil, water and air, or introducing organisms into geographic regions where they are not indigenous [2–5]. As a result, the ranges of individual species occurrence are shifting, new threats are occurring in areas where they were previously unknown, and phenomena leading to habitat degradation and species extinction are intensifying [6–9]. Trees and forests are just as vulnerable to these changes as other organisms. At the same time, they represent a very valuable element in counteracting global warming as organisms capable of absorbing and storing carbon dioxide, as organisms/habitats creating a microclimate that is favorable to other organisms, retaining water in the habitat, or lowering the surface temperature [10–12]. The list of ecosystem services for trees and forest habitats is very long. For this reason, the protection of each species and each forest habitat is so important. The hawthorn, although neither economically significant nor forest forming, is remarkably valuable ecologically. It most commonly inhabits forest edge zones and grows in clearings, mid-field woodlots, or along roads. The hawthorn trees provide habitat for numerous animal species, and offer both nutrition and shelter [13–17]. By increasing the biodiversity of their occurrence sites, they increase the immunity of such habitats to the ongoing changes, and thus increase their quality and value [18]. For these reasons, each new information on the hawthorn ecology is valuable and may be used for counteracting climate change. Two hawthorn species occur in Poland: the one-seeded hawthorn (*Crataegus monogyna*) and the Midland hawthorn (*C. laevigata*). There is also a natural hybrid of these two, the intermediate hawthorn (*C. xmedia* Bechst.). All these species occur in the study area. In the literature, the hawthorn is described as having low requirements with respect to habitat (resistant to both strong frost and drought) growing mostly in sun-lit places [19,20]. The hawthorn rapidly inhabits deforested sites or abandoned agricultural land [21]. It is noted for its high biocenotic significance, as hawthorn gatherings are a source of nutrition for numerous animals, provide shelter, and protect water and soils [22,23]. On the other hand, however, they are also the habitat of numerous pests that infest both hawthorn individuals and orchard-grown trees (especially apple and pear trees). The hawthorn (both inflorescences and fruits) is also used in folk medicine and herbalism as a remedy for diarrhea and insomnia, and for the treatment of cardiovascular diseases and digestive tract conditions. It is the source of numerous highly biologically active chemical compounds that display, for instance, anti-inflammatory antibacterial and anti-oxidative effects [24–26].

For these reasons, each new information on the hawthorn ecology is valuable and may be used in the fight against climate change. There are few papers on the ecology of the individual hawthorn species, and information regarding tree-ring width, growth rate and growth–climate relationship is virtually absent (apart from Cedro and Cedro [27]).

This paper aims to (i) determine the rate of growth in the hawthorns growing in various parts of Poland, in different habitats; (ii) analyze the growth–climate relationship and (iii) attempt an identification of dendrochronological regionalization for this species.

2. Material and Methods

2.1. Study Area

The fieldwork focused on nine hawthorn populations in Poland. Two populations (LB and ST) are located in northern Poland (Figure 1), within the young glacial relief zone (the last stagnation phases of the ice sheet). Four populations (MA, CI, ZB and DB) are located within the lowlands of central Poland (70–130 m a.s.l.), and three populations are located within mountain areas (above 450 m a.s.l.): WA in the Sudetes, and WG and LE in the Carpathians. The ST study plot is located at the lowest elevation (45 m a.s.l.), and the LE plot is located at the highest elevation (540 m a.s.l.) (Table 1). Two populations are natural hybrids of the one-seeded hawthorn and the Midland hawthorn, i.e., intermediate hawthorn (*C. xmedia*, ST and LB). The remaining seven plots are populations of the one seeded hawthorn (*C. monogyna*).

The intermediate hawthorn plot (*C. xmedia*)—LB—is located about 7 km SE from the city of Lębork, in northern Poland, within the Kaszuby Lake District, in a region characterized by extremely variable surface elevation [28]. The studied trees, however, are growing on an area with less variable elevation, on a gentle slope of a small stream (about 120 m a.s.l.), on a Quaternary, mostly sandy substratum. It is the margin of a former military training site. At present, the training site is disused by the military, and it is undergoing a natural succession. The studied trees are growing on clearings, along the forest edge, and underneath the canopy of older trees. The largest group of trees was probably planted, as an espalier arrangement is preserved.

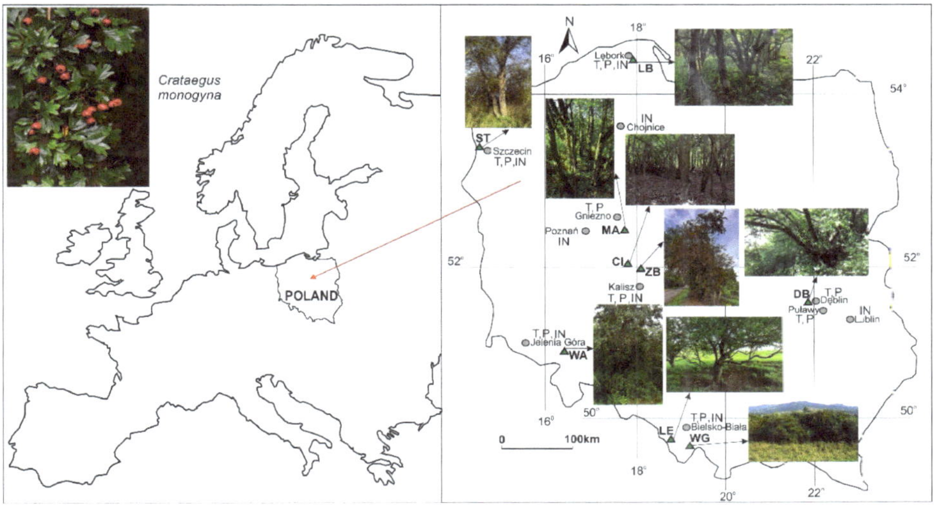

Figure 1. Location of the study plots: triangles—study plots, dots—weather stations, T—temperature, P—precipitation, IN—insolation.

Table 1. List of the study plots along with basic information.

Lab. Code	Name	Species	Geographic Coordinates	Altitude a.s.l. (m)	No. of Trees	No. of Samples	No. of Tree Rings
LB	Lębork	*C.xmedia*	N: 54.5265° N E: 17.8470° E	120	21	41	1584
ST	Stobno	*C.xmedia*	N: 53.4181° N E: 14.4051° E	45	22	30	1990
MA	Malczewo	*C. monogyna*	N: 52.4350° N E: 17.6433° E	123	22	40	1501
CI	Ciemierów	*C. monogyna*	N: 52.1047° N E: 17.7267° E	78	20	38	2080
ZB	Zbiersk	*C. monogyna*	N: 51.9724° N E: 18.1124° E	120	21	41	1508
DB	Dęblin	*C. monogyna*	N: 51.5490° N E: 21.8282° E	115	22	42	1381
WA	Wałbrzych	*C. monogyna*	N: 50.7995° N E: 16.2333° E	481	21	41	1327
WG	Węgierska Górka	*C. monogyna*	N: 49.6016° N E: 19.1010° E	451	22	44	935
LE	Leszna Górna	*C. monogyna*	N: 49.6975° N E: 18.7318° E	540	21	42	1944
				Σ	192	359	14,250

The ST study plot is located in the NW part of Poland, on a morainic plateau composed of Quaternary deposits of the last Ice Age [28], in a typically agricultural landscape. On German topographic maps (e.g., from 1921), a road passes through the center of the study plot. Following World War II, the road was plowed and at present it is annually sown. Only the study plot was excluded from farming. The gathering of the intermediate hawthorns is probably derived from one to several individuals sown by birds next to a dirt road. At present it is a monospecific mid-field woodplot composed of the intermediate hawthorn (*C. xmedia*) trees and shrubs, comprising several hundred individuals of variable age. The woodlot is about 75 m long and 20–25 m wide. It is located on a SE-facing scarp that is up to 4 m high (about 45 m a.s.l.). Cambisols have developed on the till bedrock. The woodlot is surrounded on all sides by agricultural land [27].

The MA study plot, close to Malczewo, is inhabited by the one-seeded hawthorn (*C. monogyna*). It is located in Greater Poland, on the Września Plain (123 m a.s.l.), in a typically agricultural landscape. In between the fields, there is a small pine forest (about 80 years old), with an admixture of ash. A gathering of the hawthorn, the black elderberry and the bird cherry is growing at the edge of this forest, along a dirt road. The forest is growing within a shallow depression, with no streams nearby. The hawthorn occurs mostly at the NE edge of the forest.

The CI study plot is located close to Ciemierów on the South Great Polish Lowland (78 m a.s.l.). The one-seeded hawthorn trees are growing under the canopy of a mixed fresh forest (e.g., with pine, oak and birch) on an abandoned agricultural land. The remains of an old forester's lodge are located nearby. The trees are growing mostly in an avenue layout and were most likely planted by a forester. Gleyed pseudopodzolic soil is developed on clayey sands at this plot.

Also, the ZB study plot, close to Zbiersk, is located on the South Great Polish Lowland (120 m a.s.l.). The one-seeded hawthorns were planted here by foresters in the 1960s–1970s. At present, the trees are growing at the border of a lumber mill and gardens, forming a ~100 m long espalier.

The one-seeded hawthorn trees (*C. monogyna*) from the DB study plot are growing on a flood embankment along the Wisła River (left bank of the river, about 100–300 m to the south of a railroad crossing close to Dęblin, 115 m a.s.l.). Geographically, this area is considered part of the Middle Vistula Valley [28]. A very dense assemblage of the hawthorn is growing at the summit and the slopes of the scarp.

The WA site, located within the city limits of Wałbrzych, is located in the Wałbrzych Mountains, part of the Central Sudetes [28]. In the past, this area hosted numerous pollution emitting industrial plants, including mines and coking plants. The one-seeded hawthorn trees (*C. monogyna*) are growing next to a disused railway leading to Biały Kamień. The hawthorn individuals occur on gentle slopes and scarps (481 m a.s.l.). The bedrock is composed of Carboniferous-aged sandstones. The trees are growing in clearings, in small woodlot patches, or at forest edges. The population is dominated by young and very young individuals (up to 20–30 years old), only a few specimens are older.

The WG study plot, close to Węgierska Górka (451 m a.s.l.), is located in the Żywiec Basin, part of the Western Beskidy range within the Carpathians. A this site, a dynamically developing population of the one-seeded hawthorn (numerous trees and shrubs, mostly young, with few older specimens) is growing on gentle, south-facing, meadowy slopes. Numerous bunkers, part of the Polish defensive system from the 1930s, are located immediately adjacent to the hawthorn population, and during the sampling campaign for this study (August 2021), intense express road construction was in progress in the vicinity, involving a tunnel, a viaduct and the road. The bedrock is composed of marly shales of *Cretaceous* age.

The one-seeded hawthorn from the vicinity of Leszna Górna (LE study plot) is growing on gentle slopes and near the summits of the elevations forming part of the Silesian Beskid Mountains (part of Western Beskidy, Carpathians, 508–540 m a.s.l.). The woodlots are surrounded by pastures used for sheep farming, and the sheep tend to rest underneath the hawthorn trees. Under the oldest hawthorn specimens, there is often no vegetation under the canopy, the root zones are exposed, and there are well-trodden paths. All this indicates that numerous generations of sheep took rest there. Upper *Cretaceous* shales make up the bedrock at this site.

During sampling in northern Poland (LB and ST study plots), we observed numerous individuals of the hawthorn trees infested by the orchard ermine (*Yponomeuta padella*). This insect causes partial damage to the assimilation apparatus [29,30].

2.2. Tree-Ring Data

Samples were taken in September 2020 (the paper by Cedro and Cedro [27] presents residual RES chronology and tree-ring/climate analysis for the period of 1981–2020, 40 years, for the ST site), during the growing season 2021 (MA, CI, ZB, DB, WG and LE), and in late September 2021 (LB and WA). Hawthorn individuals who were dominant and presumably older were selected for sampling. Samples were collected using a Pressler borer, at 1.0–1.3 m height above ground (two cores per individual). A total of 192 trees were sampled (from 20 to 22 per plot), yielding 359 measuring radii (from 30 to 44 per plot). In the laboratory, samples were glued onto boards, dried and sliced with a knife in order to obtain a clear view of the tree rings. In the case of the hawthorn, due to the very weak visibility of annual growth rings, the measurements were carried out using an aqueous filter, and/or the sample surfaces were smeared with chalk. The tree-ring width was measured under a stereomicroscope with an accuracy of 0.01 mm using LDB_Measure software [31]. A total of 14,250 rings were measured. As a next step, local chronologies were compiled using well-established cross-dating methods. Based on the high visual similarity of dendrochronological curves and high values of statistical indicators (Student's *t*-test and correlation coefficient), dendrochronological sequences were selected for building the chronologies. The least visually and statistically correlated sequences were discarded. The quality of the chronologies was tested using COFECHA, part of the DPL software package [32–35]. Student's *t*-test and coherence coefficient (*Gleichläufigkeitswert*, GL) were computed for pairs of chronologies using the TCS 1.0 program [36], in order to determine the similarity between local chronologies. The EPS coefficient was also computed [37]. Age trend and autocorrelation were subsequently removed from the dendrochronological sequences selected for the chronology using an indexing process (a two-phase detrending technique, by fitting either a modified negative exponential curve or a regression line with a negative or zero slope) [33,35]. Standardized (STD) chronologies were selected for dendroclimatic analyses. The period of 1988–2020 (33 years) was adopted for common analyses of chronological similarity and dendroclimatological analyses: pointer years and correlation and response function analyses. Average monthly air temperatures, monthly rainfall totals and monthly insolation values from June of the year preceding growth (pVI) to September of the growing year (IX) were used to analyze correlations and response functions. The analysis was carried out separately for temperature, precipitation and insolation, resulting in r^2 values (coefficient of determination of multiple regression) for each meteorological parameter. The analysis of pointer years was carried out using the TCS program [36]. This was achieved by calculating positive years (+) characterized by an increase in the width of the rings in relation to the previous year and negative years (−) in which the rings decreased in width compared to the previous year [38,39]. Pointer years were calculated based on a minimum of 10 trees, using 90% as the minimum threshold for consistency of the growth trend. As regional pointer years, we considered years in which at least 6 local chronologies (out of 9) had a consistent pointer year in the entire study area.

2.3. Climate Data

Mean monthly air temperature data (T), monthly precipitation sums (P) and monthly insolation data (IN) were retrieved from 16 weather stations of the Institute of Meteorology and Water Management (IMGW). These originated from stations Lębork—12125 (T, P, IN); Chojnice—12235 (IN); Szczecin—12205 (T, P, IN); Gniezno—252170110 (T, P); Poznań—12330 (T, P, IN); Kalisz—12435 (T, P, IN); Dęblin—12490 (T, P); Puławy—12491 (T, P); Lublin—12495 (IN); Jelenia Góra—12500 (T, P, IN); and Bielsko-Biała—12600 (T, P, IN) (Figure 1), located as near as possible to the respective study plots (from 3.5 to 57 km).

3. Results

3.1. Ring Width Chronologies

A local chronology was compiled for each study site (Table 2). The longest chronology was obtained for LE (72 years from 1949 to 2020), and the shortest chronology was obtained for MA (45 years from 1976 to 2020). The average age of the studied hawthorn trees equals just 58 years; the studied populations are therefore young and dynamically developing assemblages of this species. The number of samples included in individual local chronologies varies from 13 (for LE) to 24 (for ST), an average of 17 samples. The average tree ring width is the lowest for the WA population (1.42 mm/year), and the highest for the WG population (3.25 mm/year). The average tree-ring width for all the studied populations equals 2.05 mm/year. The rate of tree growth is well represented by cumulative radial growth: the WG population displays the highest growth rate throughout the entire study period, and the WA population displays the lowest growth rate throughout the study period (Figure 2). EPS values > 0.85 are noted from 1972 to 2020/2021. However, the period with EPS > 0.85 is the shortest for the DB site (1988–2020, 33 years). For this reason, this period was assumed as the time frame for the analyses of chronology convergence and the dendroclimatic analyses: pointer year analysis and correlation and response function analysis, which were performed for all study plots.

The chronology convergence was analyzed using the t coefficient and GL [40,41]. The highest t value was obtained for the DB and MA chronologies (10.99), and t > 9.0 was obtained for the following pairs of chronologies: CI and LB, WA and WG, and LE and WA. The lowest t values were obtained for the CI and WA chronologies (2.16), and the pairs of chronologies: LB and WG, ST and WG, and CI and WA are characterized by t values < 3.0 (Table 3). The highest GL value was noted for the same pair of chronologies as in the case of the t index: DB and MA (96%). GL > 80% values were obtained for the following pairs of chronologies: CI and LB, MA and ST, MA and WA, and LE and ST. The GL value is the lowest for the following pairs of chronologies: DB and LB, CI and DB, CI and ST, and CI and MA (66%). Comparably low GL values (GL < 70%) were noted for LB and WA, CI and WG, and CI and LE (Table 3).

Table 2. Basic statistics of measured and index (standard) hawthorn local chronologies. Abbreviations: TRW—tree-ring width; SD—standard deviation; 1AC—first-order autocorrelation; MS—mean sensitivity; EPS—Expressed Population Signal.

Lab. Code	No. of Years	Time Span	No. of Samples	Mean TRW (Min–Max) (mm)	Measured Chronology			Standard Chronology			EPS >0.85
					SD	1AC	MS	SD	1AC	MS	
LB	63	1959–2021	15	1.79 (0.95–3.39)	0.902	0.571	0.363	0.226	−0.062	0.294	1986–2021
ST	56	1965–2020	24	2.41 (1.48–4.44)	1.587	0.509	0.453	0.320	0.224	0.355	1981–2020
MA	45	1976–2020	16	2.38 (1.37–3.84)	1.491	0.531	0.440	0.290	0.088	0.334	1987–2020
CI	70	1951–2020	15	1.58 (1.11–2.34)	0.967	0.494	0.439	0.306	0.142	0.338	1972–2020
ZB	47	1974–2020	19	1.83 (1.12–3.00)	1.426	0.580	0.509	0.332	0.070	0.411	1985–2020
DB	56	1965–2020	17	2.34 (1.02–4.51)	1.392	0.318	0.507	0.320	−0.017	0.391	1988–2020
WA	62	1960–2021	18	1.42 (0.68–4.30)	0.906	0.435	0.491	0.332	0.007	0.429	1985–2021
WG	50	1971–2020	16	3.25 (1.66–4.19)	1.581	0.556	0.381	0.320	0.368	0.288	1986–2020
LE	72	1949–2020	13	1.46 (0.76–2.39)	0.902	0.571	0.363	0.226	−0.062	0.294	1976–2020

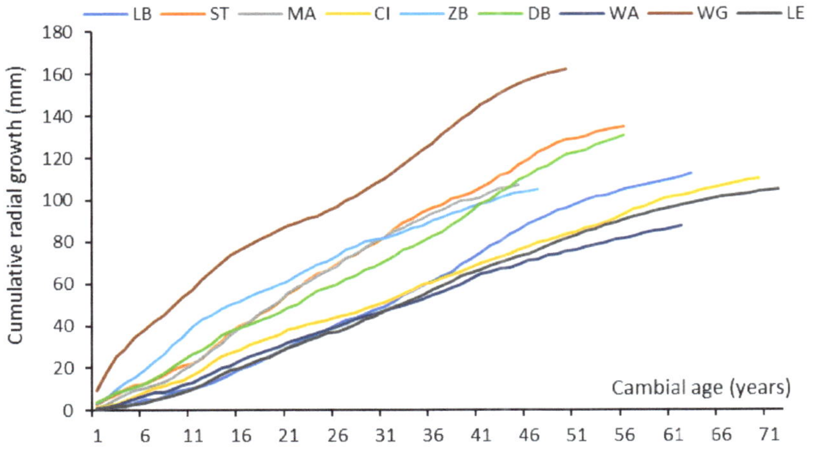

Figure 2. Cumulative radial growth of the hawthorn in Poland.

Table 3. Convergence of local chronologies of hawthorn as measured with *t* and *GL* (%) values.

t/GL	LB	ST	MA	CI	ZB	DB	WA	WG	LE
LB	-	4.49	3.35	9.44	3.30	4.22	3.05	2.93	3.19
ST	71	-	9.07	5.31	7.32	5.13	4.34	2.57	4.82
MA	71	86	-	4.45	3.06	10.99	5.28	3.63	7.57
CI	82	66	66	-	5.14	4.33	2.16	3.60	3.41
ZB	74	80	75	74	-	2.64	4.15	5.20	3.82
DB	66	76	96	66	72	-	3.17	3.34	4.34
WA	67	80	82	73	76	80	-	9.88	9.49
WG	72	74	75	69	74	69	78	-	7.50
LE	71	86	80	67	76	71	82	80	-

3.2. Correlation and Response Function Analysis

In the correlation and response function analysis, only negative values are noted as statistically significant for air temperature (Figure 3). In the winter period (December of the previous year, pXII, and January, I), there are single negative correlations. From May to August, however, negative values were noted at all study sites. Higher temperature does not favor the formation of wide rings. The average determination coefficient for all populations equals 28% (from 14% for LB to 36% for LE), and this is the lowest value among the analyzed relationships for the three weather elements.

Positive values of correlation coefficients prevail for precipitation. In the summer of the year preceding growth, there are positive (especially in August) and negative correlation values. From May to July, however, positive correlation values are observed for each site. Positive values indicate that tree-ring width increases with precipitation. On average, the r^2 coefficient equals 45% (from 34 to 53%), and this is the highest value among the analyzed weather elements.

Finding a clear relationship pattern in the analysis of the relationship between insolation and growth is challenging. Consistent correlations occur only for three months: negative correlation values in February, positive values in April and negative values in June. These, however, occur simultaneously only in 3 out of 9 chronologies. In the remaining months, the values are either positive or negative, and the relationships are site specific. On average, the r^2 value equals 35%, ranging from 18 to 58%. The highest determination coefficient (58%) was observed for the CI plot. At this site, there are only three positive values, for July and December of the preceding year (pVII, pXII), and for April of the current year (IV) (Figure 3).

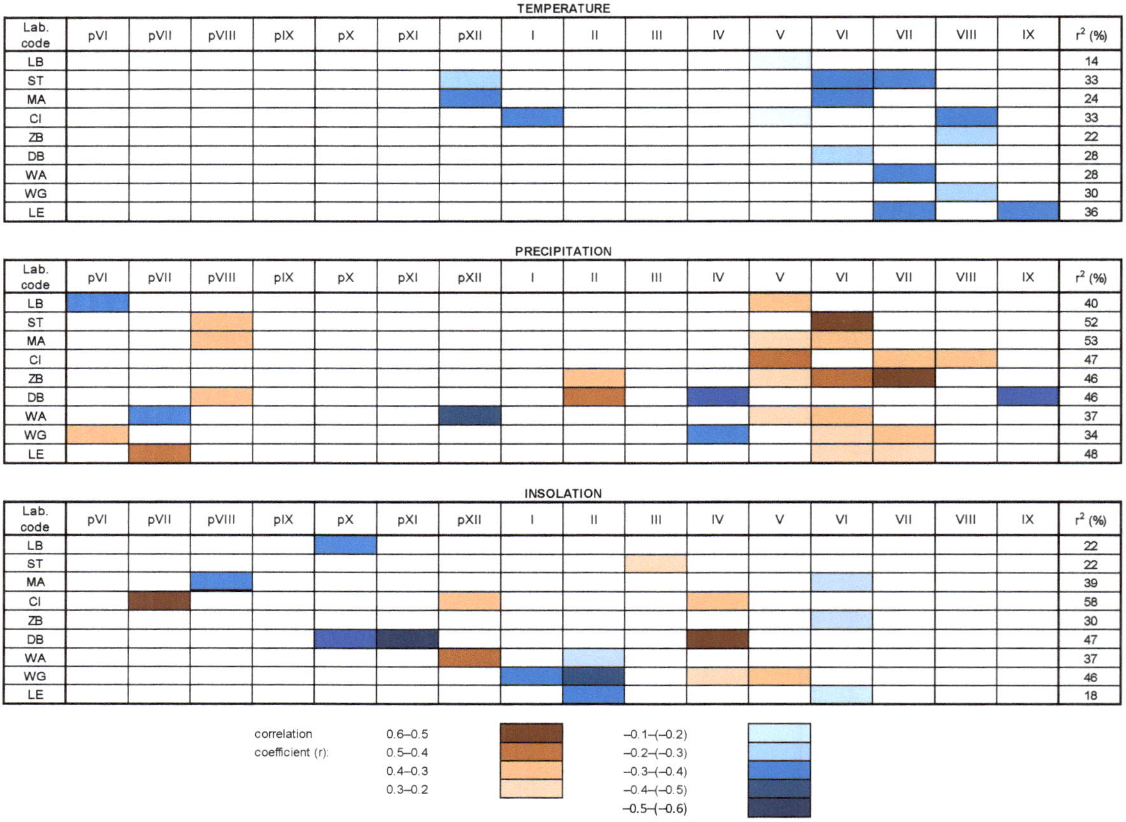

Figure 3. Results of correlation analyses (r) for the hawthorn chronologies through the period 1988–2020 (33 years) for temperature, precipitation and insolation. Only statistically significant values ($p \leq 0.05$) are shown; p, previous year; r^2, multiple regression coefficient of determination.

3.3. Regional Pointer Years

The analysis indicated 10 years, during which in at least 6 local chronologies, over 90% of trees had a lower growth compared to the preceding year (negative pointer years)—1992, 1994, 1998, 2001, 2003, 2008, 2010, 2012, 2015 and 2019—and 4 positive years, characterized by a positive growth trend compared to the preceding year—1999, 2002, 2009 and 2013 (Figures 4 and 5). The analysis of weather conditions in the study area during the designated pointer years enabled us to link the occurrence of negative years predominantly with the occurrence of drought during the summer period. The shortage of rainfall in the spring–summer period (from May to August) was the reason for decreased tree-ring width. Most frequently, negative pointer years are also years with a low annual rainfall sum (considerably lower than average). The average annual air temperature in negative pointer years was most frequently close to average or above average, the temperature of winter and early spring was variable, and summer temperatures were higher than average. The insolation during negative pointer years has a lower significance, as for various years the annual sum of insolation is higher or lower than average. It is frequent, however, for high insolation values to occur in the spring and summer months. The year 1992 may serve as an example of a negative regional pointer year for the hawthorn population in Poland. In this year, negative growth trends were noted in the following chronologies: LB, ST, MA, CI, ZB, DB and LE (7 out of 9 chronologies). The average annual air temperature for this year was higher than average at all the studied weather stations, the winter was warm, and the summer months were very warm or even hot (e.g., August). The annual rainfall sum was lower than average in all regions (a dry year), and severe rainfall shortages were noted for the period from May or June to the end of summer, especially in August. The annual insolation sum varies depending on the weather station (higher than average for some stations, lower than average for others). Elevated insolation is noted for the summer months, however.

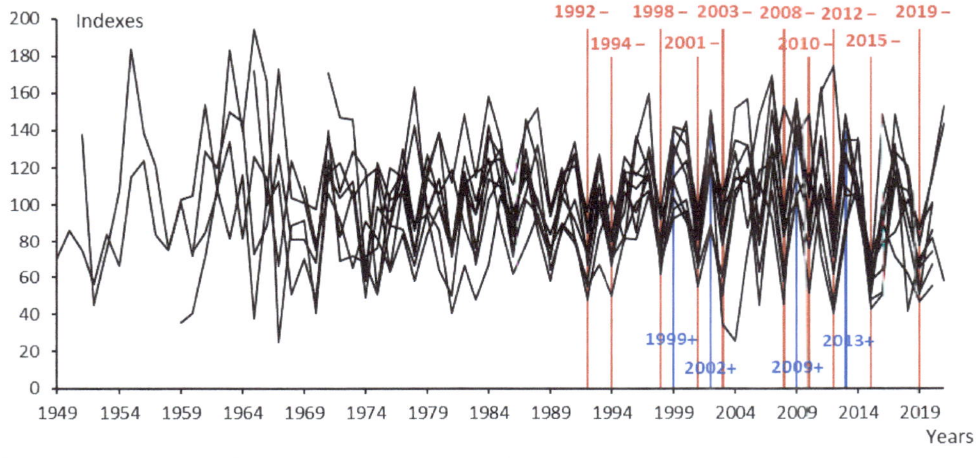

Figure 4. Indexed local hawthorn chronologies from Poland and regional pointer years (blue bars and + denote positive years; red bars and—denote negative years).

Air temperature and rainfall in the summer months are the most significant weather elements in the positive pointer years. The average annual temperature during these years is close to the multi-year average or slightly higher, the winter months are cold or warm, and from May to August, the air temperature is often close to average or slightly higher, but with no extreme values. Positive pointer years are years with higher than average annual rainfall sum (humid years). In the summer months, precipitation totals are also higher than average, although single months with rainfall shortages also occur. The insolation during these years (both annual and monthly values) is variable. The year 2013 may serve as an example of a positive regional pointer year in the hawthorn populations in Poland as it occurs in 7 out of 9 chronologies (LB, CI, ZB, DB, WA, WG and LE). The year 2013 was slightly warmer than the multi-year average, the winter was long and cold (negative average temperatures were noted as late as March), and the temperature in the summer months was average or slightly higher than average. It was a humid year (annual rainfall sums are considerably higher than average), but importantly, high rainfall sums are noted also for the summer months, although slight rainfall shortages occurred for single months at several stations. The annual number of hours of sunshine is higher than average or close to average, and in the vegetation season, there are both months with very low and very high insolation values, although high values are noted more frequently.

3.4. Dendrochronological Regions

Based on 9 indexed local hawthorn chronologies (for a common period spanning 33 years from 1988 to 2020), we made an attempt to distinguish dendrochronological regions using cluster analysis. The clustering procedure was based on the unweighted pair-group method using arithmetic averages (UPGMA), and the similarity function was computed using the Pearson correlation method. All computations were performed using Statistica (version 13.3). Based on the analysis of the linkage distance relative to Pearson's 1-r linkage steps, the breakpoint was determined at the height of about 3.5 linkage distance values (Figure 6).

Years /Plots	LB	ST	MA	CI	ZB	DB	WA	WG	LE
1988									-
1989	-	-	-		-				-
1990	+	+	+						+
1991									
1992	-	-	-	-	-	-			-
1993	+	+					+		+
1994	-	-	-	-		-	-		
1995	+			+	+		+		+
1996	-								
1997	+								
1998	-	-	-	-		-			-
1999			+	+	+		+	+	+
2000									
2001	-	-	-		-	-	-	-	-
2002		+	+	+		+		+	+
2003	-	-	-	-	-	-	-	-	-
2004						+	+		
2005	-								
2006						-	-		-
2007						+			+
2008	-		-	-		-		-	-
2009	+	+		+	+	+		+	
2010		-	-		-	-	-	-	-
2011		+		-			+		+
2012	-	-	-	-		-		-	-
2013	+			+	+	+	+	+	+
2014									
2015	-	-	-	-	-	-	-	-	-
2016		+		+	+				
2017						+			+
2018									
2019	-		-	-			-	-	-
2020	+								

Figure 5. Summary of pointer years for the local chronologies of the hawthorn in Poland during the period 1988–2020 (33 years). Blue—positive pointer years, orange—negative pointer years, bold years—regional pointer years.

Using a 0.35 linkage distance on the indexed chronology clustering dendrogram, we identified three clusters/regions: I—the western part of the study area; II—the CI study plot; III—the eastern part of the study area (Figure 7). Region I includes the sites: LB, MA, LE, WA, ST and ZB. These populations are located in the northern, central and southern parts of the country, at various elevations (from 45 to 501 m a.s.l.). Their age varies (from 45–47 to 72 years). Also, their average annual growth varies (from 1.42 to 2.41 mm/year). All the sites clustered together in cluster I are located in the western part of the country characterized by a milder climate than the eastern part of Poland. Cluster II comprises only one site, CI. The separation of this site from region I is likely caused by habitat or genetic conditions. Region III includes two sites: WG and DB. These tree populations are growing at various elevations (115 and 451 m a.s.l.), are of similar age (50 and 56 years), and are characterized by rather high growth rates (2.34 and 3.25 mm/year). They are also the easternmost study sites.

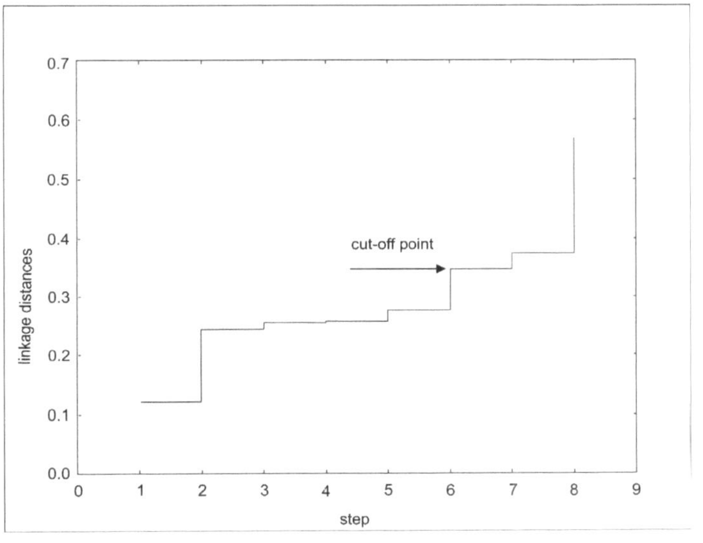

Figure 6. Linkage distances (the one minus Pearson's correlation distance) for the local hawthorn chronologies from Poland, through the period 1988–2020.

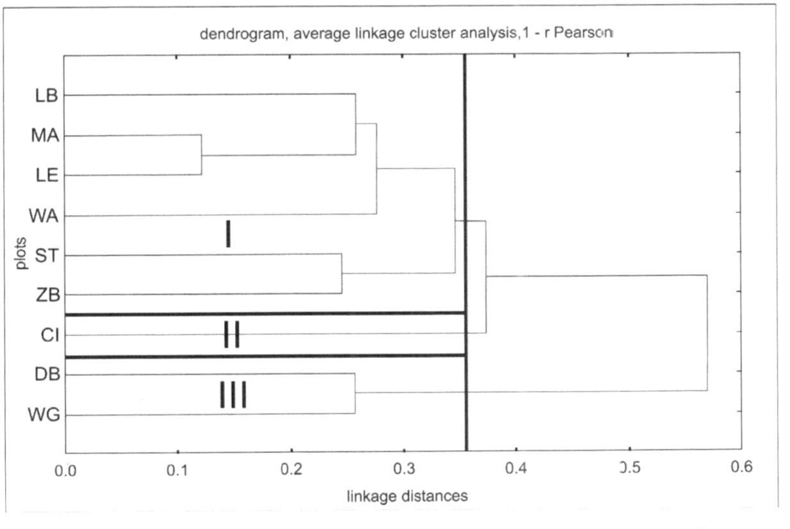

Figure 7. Dendrogram of indexed hawthorn chronologies (average linkage cluster analysis, the one-minus Pearson correlation coefficient distance, period covered: 1988–2020). Dendrochronological regions denoted with Roman numerals (I, III) and area (II).

30. Taggar, G.K.; Arora, R. Insect Biotypes and Host Plant Resistance. In *Breeding Insect Resistant Crops for Sustainable Agriculture*; Arora, R., Sandhu, S., Eds.; Springer: Singapore, 2017. [CrossRef]
31. *LBD_Measure*, version 1.0; Laboratorium Datowań Bezwzględnych: Kraków, Poland, 2020.
32. Holmes, R.J. Computer-assisted quality control in tree-ring dating and measurement. *Tree-Ring Bull.* **1983**, *43*, 69–78.
33. Holmes, R.J. *Dendrochronology Program Library*; Users Manual; University of Arizona: Tucson, AZ, USA, 1994; Available online: https://www.ltrr.arizona.edu/software.html (accessed on 9 November 2020).
34. Grissino-Mayer, H.D. Evaluating Crossdating accuracy: A manual and tutorial for the compuret program COFECHA. *Tree-Ring Res.* **2001**, *57*, 205–221.
35. Cook, E.R.; Holmes, R.L. Guide for computer program ARSTAN. In *The International Tree-Ring Data Bank Program Library, version 2.0*; User's, Manual; Grissino-Mayer, H.D., Holmes, R.L., Fritts, H.C., Eds.; Laboratory of Tree-Ring Research: Tucson, AZ, USA, 1996; pp. 75–87.
36. Walanus, A. Instrukcja Obsługi Programu TCS. In *Program TCS do Obliczania lat Wskaźnikowych [TCS Program User Manual. TCS Program for Calculating Pointer Years]*; TCS: Kraków, Poland, 2002.
37. Wigley, T.M.L.; Briffa, K.R.; Jones, P.D. On the average value of correlated time series, with applications in dendroclimatology and hydrometeorology. *J. Clim. Appl. Meteorol.* **1984**, *23*, 201–213. [CrossRef]
38. Jetschke, G.; van der Maaten, E.; van der Maaten-Theunissen, M. Towards the extremes: A critical analysis of pointer year detection methods. *Dendrochronologia* **2019**, *53*, 55–62. [CrossRef]
39. Kaennel, M.; Schweingruber, F.H. *Multilingual Glossary of Dendrochronology*; WSL FNP; Haupt: Bern, Switzerland, 1990; pp. 1–467. ISBN 3-258-05259-X.
40. Fritts, H.C. *Tree Rings and Climate*; Academic Press: New York, NY, USA, 1976; pp. 1–582.
41. Schweingruber, F.H. *Tree Rings: Basics and Applications of Dendrochronology*; Kluwer Academic Publishers: Dordrecht, The Netherlands, 1989; pp. 1–276. ISBN 978-0-7923-0559-0.
42. Mills, C.M. High Morlaggan: Dendrochronology of Hawthorn & Rowan. Report for Morlaggan Rural Settlement Group. 2011. pp. 1–19. Available online: https://highmorlaggan.files.wordpress.com/2016/05/high-morlaggan-dendrochronology-report-v1_coralie-mills.pdf (accessed on 27 September 2023).
43. Decuyper, M.; Cornelissen, P.; Sass-Klaassen, U. Establishment and growth of hawthorn in floodplains in the Netherlands. *Dendrochronologia* **2014**, *32*, 173–180. [CrossRef]
44. Nazari, N.; Bahmani, M.; Kahyani, S.; Humar, M. Effect of site conditions on the properties of hawthorn (Crataegus azarolus L.) wood. *J. For. Sci.* **2021**, *67*, 113–124. [CrossRef]
45. Good, J.E.G.; Bryant, R.; Carlill, P. Distribution, longevity and survival of upland hawthorn (*Crataegus monogyna*) scrub in North Wales in relation to sheep grazing. *J. Appl. Ecol.* **1990**, *27*, 272–283. [CrossRef]
46. Williams, P.A.; Buxton, R.P. Hawthorn (*Crataegus monogyna*) Populations in Mid-Canterbury. *N. Zeal. J. Ecol.* **1986**, *9*, 11–17.
47. Grissino-Mayer, H.D. An updated list of species used in tree-ring research. *Tree-Ring Bull.* **1993**, *53*, 17–43.
48. Schweingruber, F.H. *Microscopic Wood Anatomy*; Swiss Federal Institute for Forest, Snow and Landscape Research: Birmensdorf, Switzerland, 1990; pp. 1–226.
49. Available online: www.monumentaltrees.com/en/fra/mayenne/saintmarssurlafutaie/2810_church (accessed on 27 September 2023).
50. Barniak, J.; Jureczko, A. Impact of air pollution on forest stands in the vicinity of Wodzisław Śląski and Rybnik, Poland. *Geol. Geophys. Environ.* **2020**, *45*, 283. [CrossRef]
51. Cornelissen, P.; Decuyper, M.; Sýkora, K.; Bokdam, J.; Berendse, F. Forest development in a restored floodplain: Effects of grazing, inundation and vegetation. *Eur. J. Environ. Sci.* **2019**, *9*, 21–32. [CrossRef]

MDPI AG
Grosspeteranlage 5
4052 Basel
Switzerland
Tel.: +41 61 683 77 34

Forests Editorial Office
E-mail: forests@mdpi.com
www.mdpi.com/journal/forests

MDPI

www.ingramcontent.com/pod-product-compliance
Lightning Source LLC
Chambersburg PA
CBHW041401210926
51850CB00013B/3145
* 9 7 8 3 7 2 5 8 2 7 8 5 5 *